What Others Are Saying About

Astrology and the Archetypal Power of Numbers

How can you take a tour of the counting numbers, each in turn, and yet have the sense that you have been exploring the very nonlinear reaches of a fractal galaxy? In his new book, *Astrology and the Archetypal Power of Numbers, Part One*, Landwehr has masterfully woven a diverse collection of mythological teachings, modern social commentary, and personal life lessons into a coherent perspective of human spirit and consciousness.

Landwehr's tremendous breadth of knowledge and experience with esoterica, philosophy, Jungian psychology, eastern, western and native American religion/mythology, and modern science and scientific culture endows this work with a palpably subjective objectivity. Or is it objective subjectivity?

Materials science has a branch of research called 'Nondestructive Evaluation' whose goal is to analyze a material without breaking it apart. It is a very interdisciplinary 'holy-grail' study that science has come to only in the last half century. Landwehr writes about human existence from a place where nondestructive evaluation is not a holy-grail, but a fundamental tenet; his work is a priori (and unapologetically) subjective, and this beautiful book would not work otherwise.

Yes we begin at 0, and yes, we progress linearly through 9, but this as a framework for a very nonlinear, enlightening and sometimes breathtaking journey through the landscape of humanness. **– Dr. Mark Arnold, Associate Professor of Mathematics, University of Arkansas - Fayetteville**

Astrology and the Archetypal Power of Numbers, Part One is to be read as an epic poem that moves into unknown realms below, above and beyond our usual daily lives. This impressive work of scholarship is not an easy read, but then whoever said that deep knowledge is undemanding?

As source for meditation and reflection, this work can help facilitate personal and compassionate reactions within one's daily life. The book offers a new language for looking and living everyday in a deeper and more meaningful spiritual context within the cosmos.

This is a handbook to be taken along when going in search, research and into the mysteries of the soul's journey. This tome is not for light bedside reading, but it is a volume to carry with you whenever you work with your own dream worlds. The work offers a practical application of classical Jungian archetypal theory along with numerous references and sources from major Jungian scholars including von Franz, Neumann, Eliade and more.

Like a long slow thermal mineral bath, this is the type of experience each reader will want to come back to, again and again.. – **Dr. Jonathan Paul de Vierville, Professor of Humanities, History and Interdisciplinary Studies at St. Philips College, San Antonio, TX**

If you pick up Joe Landwehr's, *Astrology and the Archetypal Power of Numbers*, expecting to skim through New Age fluff, you may soon find yourself at the bottom of an ocean floor of wisdom you hadn't bargained for. But if you can stay with it, before you reach the surface for your first breath, you will have realized that the book's author is one of the few who has the intelligence and commitment to rescue the invaluable kerygmas of the Great Wisdom Traditions that modernist and post-modernist perspectives have all but have erased from contemporary minds. He legitimizes the core ideas in astrology and numbers as archetypes by interfacing them with his profound knowledge of several schools of psychology, Eastern religions, mythology, awareness of environmental issues and courageous self-exploration.

Landwehr not only displays a wide range of speculation, he is a superb discursive writer. This book is for the intellectual and the lay person whom technocratic knowledge and/or relativism does not resonate and who desires a fuller awareness. – **Dr. Rockey Robbins, Associate Professor of Counseling Psychology at University of Oklahoma – Norman**

For each one of us, our progress through life, through all the challenges we face in a lifetime, might be seen as an enactment of fate. We explore our embodied state until eventually we pass through the gateway of our death and back to spirit. Along the way, the choices we make, the actions we take, in efforts to modify our fate, eventually crystallize. The progression is sometimes a cyclical process, sometimes linear, sometimes non-linear. The hope is that out of the chaos of all our efforts to become something, we discover that we already are everything. That we are creative beings with the potential to live in harmony with all that constitutes the manifest world. That spirit and matter are in the end inseparable.

In this his latest book, Joe is demonstrating how numbers can be seen as catalysts for consciousness on the soul's journey. Individually or collectively, we make choices as we move through the various number 'Realms', and those choices have consequences. Joe's belief is that by becoming more conscious of this, often unconscious, choosing, we have the potential to influence outcomes both for ourselves and perhaps even for our species' survival. Into a background canvas of fascinating metaphysical and mythological associations, Joe weaves both threads of self and threads of society to show just how interconnected these are. His empathy for human striving and his concern for the planet that is our home make his writing vital and urgent. - **Sara Firman, Editor & student at The Astropoetic School**

Astrology and the Archetypal Power of Numbers

Part I: A Contemporary Reformulation of Pythagorean Number Theory

Joe Landwehr

Ancient Tower Press

Mountain View, Missouri

www.ancient-tower-press.com

FIRST EDITION
First Printing, 2010

Cover Design by Anne Marie Forrester
Interior Design by Joe Landwehr
Copy-Editing by Sara Firman

Publisher's Cataloging-in-Publication
(Provided by Quality Books, Inc.)

Landwehr, Joe.
    Astrology and the archetypal power of numbers. Part
one, A contemporary reformulation of Pythagorean number
theory / Joe Landwehr.
        p. cm. ~ (The astropoetic series ; v. 2)
    Includes bibliographical references and index.
    LCCN 2010913875
    ISBN-13: 978-0-9747626-2-3
    ISBN-10: 0-9747626-2-8

    1. Astrology. 2. Pythagorean theorem.   I. Title.
II. Series: Astropoetic series ; v. 2.

    BF1711.L36 2011   133.5
                          QBI10-600194

Ancient Tower Press
230 West First Street #119
Mountain View, Missouri 65548

www.ancient-tower-press.com

About the Author

J oe Landwehr is an astrologer of 40 years experience, seeking an eclectic integration of astrology, spiritual psychology and ancient wisdom teachings. Trained as a scientist in undergraduate school at Worcester Polytechnic Institute, Joe became increasingly disenchanted with the limitations of science, changing his major and turning to a study of philosophy, psychology, and world literature in his junior year.

After an initiatory reading in 1971 by an astrologer in Ashland, Oregon known only as Sunny Blue Boy, Joe began his own study of astrology in earnest. Moving to southern California in 1973, Joe became the resident astrologer at the Kundalini Research Institute, taught kundalini yoga and meditation in the surrounding colleges, and obtained a masters degree in Marriage, Family and Child Counseling. In graduate school, Joe encountered the same limitations of science and mathematics, in a different form as applied to the study of human psychology – and began to formulate the ideas that eventually led to the writing of this book.

Upon leaving the ashram in 1977, Joe moved to Florida, where he continued to study, write and teach. In 1978, he was introduced to the teachings of Swami Muktananda, and began to supplement his knowledge of yogic philosophy with the theory and practice of Siddha yoga – evolving a perspective that was shared in his previous book, *Tracking the Soul With an Astrology of Consciousness*.

Over the next 15 years, Joe developed an international mail order astrology practice, combining a Rudhyarian approach to astrology with a foundation in spiritual psychology.

In 1993, Joe started an intimately personal correspondence course called The Eye of the Centaur, and for the next 8 years, worked intensively with a small number of dedicated students, willing to let the crucible of their own lives be their classroom. In 2001, Joe took a sabbatical from his course to research and write his first book. In 2005, Joe resumed teaching his correspondence course, now transformed into The Astropoetic School of Soul Discovery. Since then, the course has undergone a complete revision to incorporate new principles of astropoetics not taught elsewhere.

In 2004, Joe published *The Seven Gates of Soul: Reclaiming the Poetry of Everyday Life* – a 6,000 year history of ideas about the soul, drawn from religion, philosophy, science, psychology and an intuitive form of astrology called astropoetics. *Tracking the Soul* was the sequel to that book, and the first in The Astropoetic Series, to which this current volume, to be published in two parts, is the second installment.

Additional information about astropoetics, The Astropoetic School, and the author's current workshop schedule can be found at www.astropoetics.com.

Acknowledgements

As a precocious child, perhaps in some ways far too serious for my age, I spent many hours in the library pondering the mysteries of numbers. The myriad ways in which they combined, mutated into other numbers, split apart again in new combinations, and revealed repeating patterns of mystery and intrigue all seemed like magic to me. Indeed, it was. I wanted to know this magic, and innocently believed that I would learn it in school. Alas, I soon found out, it was not the magic of numbers that was taught in school, but rather their utilitarian use as drones in the arsenal of science. Nonetheless, I was fortunate in my youth to have several teachers of math, who were as inspired by numbers as I wanted to be. Their names have long since been forgotten, but I thank them now.

I thank my calculus teacher at Worcester Polytechnic Institute, whose name I have also forgotten – the mad, but brilliant man who looked like Einstein on a bad hair day – who walked into class one day, crumpled a piece of paper and asked us to describe this impossible shape mathematically for our final exam. Of course, no one could, but the very idea of it warped my mind just enough to be able slip into an alternate reality where numbers were the portal.

I thank my statistics teacher in graduate school, who shall also remain nameless, and whose linear teaching style propelled me out of the claustrophobia of graduate school long enough to get my non-linear bearings. I eventually finished my degree, but not before accepting the fact that I was just not going to learn what I most needed to know in my formal education.

I thank all the well-intentioned minions of academia I studied numbers with for the unwitting part they have played in my liberation from any need to think inside the box.

I thank Pythagoras for taking numbers from the very start outside the box in which they are neatly packaged now; Iamblichus for collating Pythagoras' teachings about number in *The Theology of Arithmetic*; and Robin Waterfield for translating this work into English, and thus making it accessible to me.

I thank Anne Marie Forrester for another eye-catching cover design, based on the image from a classic photo taken looking out of Jam Up Cave, a half mile down the road from where I live.

Last but not least, I especially thank my editor, faithful companion and confidante Sara Firman, for believing in me, mirroring the power of my own ideas back to me, and helping me fine-tune my message.

This book is dedicated to Odin, my Number
One
and Sulis, my Number Two

Table of Figures

Table of Contents

CONTENTS

CONTENTS

CONTENTS

Civilized human beings have found in moments of disaster, in the thunder of avalanches and voice of storms at sea, that when all the many subtle, complex and impressive devices they have contrived for their security and intelligence against unpredictable forces of nature have all broken down, there is organic in the natural world around them another system to which their long rejected instincts compel them to turn for guidance. The great trouble is that this natural system is encoded in symbols on which the world has long since turned its back... and in the turning, lost the key to the cypher.

from *A Far-Off Place* by Laurens van der Post, New York: Harcourt Brace, 1974

Figure 1: The Rune Sowelu
Wholeness, the Sun, the ultimate goal of spiritual evolution

Preface

This book is the second in a series of books outlining a language of spiritual psychology that I call astropoetics. Astropoetics is a hybrid word meant to describe a poetic approach to astrology, applied to the human quest for meaning, purpose, and a deeper sense of connection to all of life and to the larger universe of which it is part.

A Word to Non-Astrologers and Skeptics

Not everything that can be counted counts,
and not everything that counts can be counted[1].

Although astrology has fallen into disfavor and has been seriously maligned – usually by those who know little or nothing about it – it has also been taken seriously by some of the greatest minds our culture has ever produced, including Hippocrates: the father of modern medicine; Nicholas Copernicus: the father of modern astronomy; and Carl Jung: the father of modern psychology. Through the 17th century, it was taught at every major university in the civilized world – not as the irrational superstition it is considered to be today – but as essential to an understanding of the natural order of things. Despite the institutionalized cultural bias against it, and centuries of vociferous rejection as a worthy discipline by religion, science, and the media, some form of astrology has existed in virtually every culture on the planet since humans first looked upward in wonder to the night-time sky, and it continues to flourish today in the popular imagination.

Belgian archeologist and historian Franz Cumont once called astrology "the most persistent hallucination which has ever haunted the human brain," intending to put to rest what he considered to be our neurotic fascination with it. "How could this absurd doctrine arise, develop, spread, and force itself on superior intellects for century after century?" he wondered. Indeed, this is a question worth asking, along with the corollary question (mine, not his): "Is it really all that absurd to think that we might actually be a part of the cosmos at which we – astrologers, scientists, 'superior intellects' and ordinary human beings from every culture and walk of life – continue to marvel?"

The very premise on which astrology is based – namely that there is a discernible relationship between the evolving cosmic order (the macrocosm) and the journey of an individual soul (the microcosm) – has been disavowed by science. And yet, the absence of this premise in contemporary discourse contributes greatly to the alienation and disenchantment of postmodern humans, who are often cast adrift in a cold and lifeless universe to which they

◇◇◇

do not feel as though they belong. Without a language with which to understand our personal connection to the cosmos – such as astrology – we lack the profound sense of being an integral part of the natural order of things that gave meaning and purpose to the lives of our indigenous ancestors, who were not too sophisticated or jaded to look up into the nighttime sky and marvel at the patterns they saw coalescing and dispersing there. In discussing why young people today (those born since 1980) have little interest in the scientific worldview – with which astrology is generally sharply contrasted – economist Jeremy Rifkin (608) clearly spells out the problem:

> [T]he scientific method [is] an approach to learning that has been nearly deified in the centuries following the European Enlightenment. Children are introduced to the scientific method in middle school and informed that it is the only accurate process by which to gather knowledge and learn about the real world around us. The scientific observer is never a participant in the reality he or she observes, but only a voyeur. As for the world he or she observes, it is a cold, uncaring place, devoid of awe, compassion, or sense of purpose. Even life itself is made lifeless to better dissect its component parts. We are left with a purely material world, which is quantifiable but without quality. The scientific method is at odds with virtually everything we know about our own nature and the nature of the world. It denies the relational aspect of reality, prohibits participation, and makes no room for empathic imagination. Students in effect are asked to become aliens in the world.

Although it is rarely considered as a viable alternative to science, astrology addresses these issues. It provides a more open-ended way to gather information about the world around us and about ourselves – not in an objective voyeuristic way, but as active participants in our own lives, and in the larger evolutionary processes unfolding within and around us. It facilitates an understanding of self that is filled with purpose. It propels us on a quest of meaning, wonder, and self-discovery. It connects us to others, the entire web of life, and the cosmic order in which we play our part. It does not dissect the world into fragmented pieces, but rather shows how all the pieces fit together into an integrated, cohesive whole. It unabashedly discusses quality. And it triggers the empathic imagination through its willingness to explore the hard facts, not just literally, but as metaphors and symbolic portals to a multi-dimensional truth larger than can be comprehended by the rational mind. Astrology – in its essence, if not always in its practice – provides the foundation for a psychology of soul that offers everything that is missing in science.

The first modern psychologist of soul Carl Jung once said (*Archetypes* 344, footnote 168):

> *I do not hesitate to take the synchronistic phenomena that underlie astrology seriously. Just as there is an eminently psychological reason for the existence of alchemy,*

◇◇◇

so too in the case of astrology. Nowadays, it is no longer interesting to know how these two fields are aberrations; we should rather investigate the psychological foundations on which they rest.

Astropoetics

In this series of books, I have taken Jung's advice, and sought to carefully develop a working integration of astrology and psychology – especially Jungian psychology, with an eye toward illuminating the soul's experience as it struggles with both the human predicament and its spiritual longing to be an integral part of something larger than itself. I call this integration of astrology and spiritual psychology "astropoetics."

As an astrologer of nearly 40 years experience, I believe the soul can be observed with uncanny clarity in relation to its movement through various cycles, which wax and wane in predictable rhythm, and are mirrored in the movement of planets and other heavenly bodies through the sky. Unlike some astrologers, I also believe that the mystery of soul is best approached through a language that is poetic in its use of words. A poetic language conducts its quest for the truth obliquely – through simile and metaphor, image and symbol, suggestion and allusion, rather than direct, dogmatic statement of fact. When astrology is approached poetically – as a right-brain contemplation of imagery and symbolism, rather than as an interpretative system based on authoritative prepackaged definitions – it becomes a potent language of soul that allows for the possibility of deep and penetrating self-discovery. It is this possibility – of knowing oneself with the insight of a poet, speaking a language both pragmatic and steeped in compelling mystery – that pulses at the heart of the astropoetic journey.

Astropoetics is the study of the relationship between cycles and patterns in the cosmos and life experience on earth, understood poetically through an intuitive, intimately subjective, and image-oriented perspective. Astropoetics draws its inspiration, its rationale, and its imagery not just from astrology, but from a broad range of sources, including but not limited to number, astronomy, mythology, psychology, the wisdom teachings from all traditions, poetry, literature, music, art, film, other dimensions of contemporary culture, and especially from the poetic nuances of everyday experience, both individual and collective. *The Astropoetic Series* takes as its ambitious mission, the integration of these diverse sources of linguistic food for the soul, within an astrological context.

Because this book builds on concepts introduced in my previous books, it seems useful to provide a brief summary of the discussion so far, for those who have not yet read them. Some of these concepts will be merely useful, and others essential, to understanding the arguments put forth here. While I would recommend reading these books in the order they were written, this preface should suffice to bring the reader who is not inclined to do that, up to

◇◇

speed. This preface is also oriented toward non-astrologers reading this book, whose primary interest is not astrology, but Pythagorean number theory. There are reasons why I link these two subjects, which should hopefully become more apparent as you read on.

The Seven Gates of Soul

Prior to *The Astropoetic Series*, I wrote a book called *The Seven Gates of Soul: Reclaiming the Poetry of Everyday Life*, which outlines a 6,000-year history of ideas about the soul, culled from Eastern, Western and indigenous religions; and European philosophy from around 600 BCE through the 20th century. In distillation of this perennial discussion, I arrived at my own definition of soul as the inhabitation of Spirit (Conscious Intelligence, both immanent and transcendent) within a mortal body. It is this embodied soul – the human being struggling to make sense of everyday life, as well as longing to glimpse a spiritual context in which everyday life evolves a larger, more deeply meaningful sense of purpose – for which *Seven Gates*, and the books that follow are being written. It is this embodied soul for which a language is necessary – one that is capable of describing its journey in intimately personal terms.

In *Seven Gates*, I discuss the ways in which science and psychology, born in the late 19th century as a science, have disavowed the soul, and left postmodern humans largely alienated in a world to which they do not belong, and without a language of soul in which to discuss their experiences. I show how an astropoetic approach to astrology – the cornerstone cast aside by the architects of Western culture – could once again provide such a language.

The seven gates in the title of this book are a reference to the mythological descent of the Sumerian goddess Innana into the Underworld, a journey that required her to pass through seven gates, shedding some outer garment and some aspect of her identity in order to stand naked before the unveiled truth. In the book, Innana's seven gates are meant to be a metaphor for various illusions about the soul that must be shed before a true language of soul can exist. In summary, the seven gates are:

1. **The religious notion of immortality**, which can be seen with an appropriate level of detachment, as a lopsided identification of soul with Spirit. I define soul instead as an alchemical fusion of Spirit and matter (the body), which is subject to the vulnerability that necessarily comes with mortality, but which also presents a unique opportunity to bring Spirit more deeply into the embodied world through our conscious participation in it. A language of soul must encompass the experience of the body and the embodied life, and at the same time, provide a window through which we can glimpse Spirit at work within ourselves and within the world.

◇◇

2. All the **conditioned patterns of judgment** that religion and psychology have projected onto the soul's process. The soul's job is not to be good in a moral sense, but to attune itself from moment to moment to the ever-shifting balance between light and dark, male and female, hot and cold – and every other polarity – that characterizes the embodied life. A language of soul must affirm both the innate human desire for pleasure, and the learning opportunity presented by the pain that comes with our inevitable mistakes.

3. The requirement that our language of soul speak exclusively to **the rational mind**. The soul's truth is often paradoxical, irrational, and rooted in sensory, emotional, and imaginal experiences. By itself, the rational mind is a poor tool with which to approach the soul's complexity, and any language of soul must speak not just to the rational mind, but also to the senses, the emotions, and the imagination – that is to say, must be poetic in nature and tone.

4. Science's demand for **objectivity**. The soul's experience is highly subjective, and does not necessarily conform to consensual notions of reality shared by society or the culture at large. Contrary to science's insistence, objective truth is not the only truth worth pursuing, nor necessarily the most relevant. Science's emphasis on empirical observation can be useful to the soul, but only if conducted within the subjective context of a deeply personal and often highly idiosyncratic life process. A language of soul must be adaptable enough to reflect the exquisite individuality of each soul using it, as well as those dimensions of the human experience that are fairly universal.

5. The scientific preoccupation with **causality**. Contrary to popular misconception, nothing astrological (or symbolic, for that matter) causes anything to happen. Instead, the information required by the soul to proceed wisely in its journey is reflected not just in the movement of planets, but also through the relationships we have with others, the movies we watch, the books we read, our dreams, chance encounters, conversations overheard by chance, the flight pattern of birds, tea leaves in a cup, the synchronicities we experience daily, and through countless other channels that broadcast simultaneously, as a reflection of the order of the cosmos and the soul's participation in it. The soul exists within a nexus of relationships, and any true language of soul must reflect this nexus and facilitate its exploration.

6. Our **linear notions of space and time**. The resonant space in which the soul lives is not measured by distance, height, breadth or depth, but rather by affinity, contrast, and aversion. Time is not chronological, but cyclical and bound by related memories – often distant in time – rather than a progression of suc-

◇◇

cessive events. A language of soul must assess these dimensions of time and space qualitatively, rather than through the quantifiable parameters measured by science.

7. **The notion that human life can be anything but a spiritual experience.** A true language of soul must recognize that everything that happens is an opportunity for the soul to learn, grow, and deepen its relationship to itself, to the Greater Whole of which it is part, and to the Spirit that encompasses all, out of which everything arises, and into which everything returns.

Lastly, in *Seven Gates*, I demonstrated how astropoetics could potentially fulfill every one of these requirements. I will not repeat those arguments here, but instead refer the interested reader to the book. Meanwhile, passing through these seven gates of misconception – perpetrated by religion, science, and psychology – equips us to begin to construct a language that allows us to breathe more freely, and discover soul on its own terms, apart from the conditioning imposed upon us by our culture.

Tracking the Soul With An Astrology of Consciousness

The key to this discovery lies in the exercise of consciousness. The first book in *The Astropoetic Series* addresses the issue of consciousness, which I define as a set of perceptual filters that condition our sense of who we are, what we are doing with our lives, and what our soul's journey is about. Our answers to these fundamental questions – and our understanding of ourselves – shift as consciousness changes, and consciousness changes gradually, sometimes abruptly, through a lifetime of experience. The information that we receive through our senses, emotions, and the images that arise within and without on a daily basis is designed to guide us along a continuum from utter separation, isolation, scarcity, vulnerability and disempowerment at one end to a supreme sense of unity and connectedness with all that is, abundance, invulnerability and empowerment at the other. How and where we move along this continuum is a matter of consciousness.

The lessons and issues that arise for each of us along this continuum will in some sense be reflected by our birthchart – our personal map of the heavens at the time and place of our birth. But how well we interpret and respond to the birthchart, and how much we learn from it will be entirely up to us – and again, largely a matter of the consciousness that we bring to it. In this regard, a useful language of soul is one that can reflect not just the map of the terrain we will traverse in our soul's journey, but also how conscious we are in any given moment, how open we are to the information available to us, and which set of filters we are using.

◇◇◇

In the first book of *The Astropoetic Series*, my intent was to address the question of consciousness, and show how astrology could be integrated with an understanding of the *chakra* system from yogic philosophy, in order to improve our language of soul. Although consciousness can be understood in a number of ways, my preference for approaching it from the perspective of yogic philosophy arises from the decade of my life that I devoted to its study, under the tutelage of Yogi Bhajan and Swami Muktananda – a decade in which I was also learning and beginning to practice astrology. As I demonstrate in some detail in *Tracking the Soul*, astrology and the *chakra* system supplement each other in ways that create a synergistic effect that is greater than the sum of the parts.

More specifically, astrology provides an unparalleled capacity to personalize the spiritual process, and to time it. Each chart is a unique signature of possibilities for spiritual growth belonging to a particular soul, and encoded within each chart is a timetable for the outworking of those possibilities. But astrology lacks an explicit understanding of the way in which consciousness alters the meaning of its symbolism, nor does it inherently include a discussion of consciousness as a framework for spiritual evolution. Individual astrologers may bring a sense of this to their work, but it is not intrinsically a part of the astrological language.

Meanwhile, the *chakra* system – developed over thousands of years of yogic practice as a universal framework – is lacking in individual nuance or any sense of timing, but it does provide a way of assessing consciousness. Thus combining astrology's exquisite sense of individuality and timing with yogic philosophy's sophisticated model of consciousness provides a more complete system – that I call the astro-*chakra* system – than either discipline alone.

Observing life from within the framework of the astro-*chakra* system allows the hungry soul to perceive the poetry of its everyday life with more nuanced clarity. Since my understanding of numbers in this book is predicated on an understanding of the *chakras* in Part One, and the astro-*chakra* system in Part Two, it will be helpful to the reader of this book, who has not read the others, to have at least a rudimentary knowledge of the *chakras*. Again, I would recommend reading *Tracking the Soul* before you read this book, but short of that, the following synopsis should help you comprehend the sections of this book, where I refer to the *chakras*.

In the yogic system, there are seven major *chakras*, which can be understood as descriptive of various sets of perceptual filters, worldviews and psychological frameworks that govern the movement of a given soul through life. Yogic philosophy tends to consider the *chakras* hierarchically, as an evolutionary ladder upward toward ever-higher levels of consciousness. Within the astro-chakra system, I have found it more useful to consider the *chakras* in a circular arrangement – suggesting their equality as co-existent states of consciousness that interpenetrate and often interact with each other. Within this context, the seven *chakras* can be briefly described as follows:

PREFACE

◇◇◇

Red

1. The first *chakra* is concerned primarily with matters of **survival** – both on the physical plane, and on every other level on which survival appears to be an issue. As with any *chakra*, this is a matter of subjective perception rather than objective reality. When the first *chakra* is emphasized, we see the world through a set of filters that allows us to secure our survival when it appears to be threatened, and to establish a safe and secure perimeter to our existence in which it is possible to construct a life of stability and continuity

orange

2. The second *chakra* is concerned primarily with the creation of **a pleasurable and fulfilling existence**. The second *chakra* is often associated with sexuality, but also encompasses our natural predispositions, innate desires, and natural rhythms. When the second *chakra* is emphasized, we see the world through a set of filters designed to maximize pleasure, minimize pain, and organize our lives in ways that are enjoyable, emotionally rewarding, rich and abundant, in every possible sense of that word.

Yellow

3. The third *chakra* is concerned with the cultivation of **a functional ego** and an area of personal competence, talent and/or expertise. At this level of consciousness, we are concerned with learning, acquiring life skills, establishing our place within the world, and making a creative contribution to it. When the third *chakra* is emphasized, we see the world through a set of filters that illuminate both the opportunities for advancement within the world and the challenges and obstacles to our advancement.

green

4. In the fourth *chakra*, we enter the world of **relationships**, including on a deeper level, our relationship to ourselves and to the greater world in which we live. In this *chakra*, we learn how to negotiate our relationships, seek a deeper connection to what James Hillman has called the "soul's code" or calling, and cultivate a nascent awareness of being on a spiritual path. When the fourth *chakra* is emphasized, we see the world through a set of filters that allows us to see and understand how we are connected to everyone and everything around us.

blue

5. In the fifth *chakra*, we are challenged to harvest and utilize whatever **wisdom** we have managed to earn by paying attention to our life experiences and learning from them. Here, we are required to walk our talk and to share what we have learned. When the fifth *chakra* is emphasized, we see the world through a set of filters in which we recognize everyone we meet and everything that happens to us as an opportunity to learn and to grow in consciousness.

indigo

6. In the sixth *chakra*, we learn to see the relative nature of our most cherished **belief systems**, and begin to evolve toward maximum flexibility in both our un-

derstanding of reality and our capacity to respond to it. In the sixth *chakra*, we begin to shed what American Buddhist nun Pema Chodron calls the "story line of our lives" (35) along with our ego-bound identities to embrace a larger, more compassionate identification with the *anima mundi*, or soul of the world. When the sixth *chakra* is emphasized, we see reality through a set of filters in which the illusion that we are separate from the world, from each other, and from Spirit, begins to fall away.

7. In the seventh *chakra*, we identify completely with **Spirit** – understood here as That Which Contains All That Is. This is, of course, an elusive state of being, even for those who have attained a classic version of enlightenment, nirvana, or transcendence. It is here that we face Buddha's conundrum: whether to dissolve back into Spirit or to take the vow of the *bodhisattva* and work for the enlightenment of all sentient beings. When the seventh *chakra* is emphasized, we see the world through a set of perceptual filters in which this conundrum resolves itself beyond the reach of any words we might use to articulate it.

violet

Each *chakra* can be experienced as a set of opportunities for seeing and relating to the world in a particular way – which has both advantages and disadvantages – or as a set of issues encountered within that particular perceptual framework. Often we will have both experiences simultaneously. If, for example, I am relating to the world through my fourth *chakra*, I will see opportunities for meaningful relationship with others everywhere around me. I will also be acutely aware of the ways in which relationships are difficult for me – my inability to love myself and others, or trust them, or share myself with them, and so on.

Both our issues in relationship, and our predispositions toward relationship can be seen in our astrological birthcharts through what I refer to in *Tracking the Soul* as its fourth *chakra* signature. Each *chakra* has its own signature, which may or may not be emphasized in a given chart at birth, but which may also be triggered at various times in one's life. Whether or not we choose to view life astrologically, the *chakra* system can be useful in assessing the ways in which we choose to live our lives within the larger world around us.

Astrology and the Archetypal Power of Numbers

The *chakras* are also useful in understanding how the larger world around us is constructed. Taken as a whole, the *chakra* system suggests that there are many ways to understand the world around us, and that how we perceive it will largely depend on the consciousness that we bring to it. Yet the world also has its own reality, apart from our perceptions. This is the objective reality measured by science in the quantitative language of numbers.

PREFACE

◇◇

Mathematicians and scientists use numbers quantitatively in order to measure various physical parameters of the world, without any reference to the interface between consciousness and reality. But there is another sense of number that is also worth considering for its inclusion of this important missing piece. This sense of number – originally taught by Pythagoras – was also known to Jung, who said (*Synchronicity* 41-42):

> I must confess that I incline to the view that numbers were as much found as invented, and that in consequence, they possess a relative autonomy analogous to that of the archetypes. They would then have, in common with the latter, the quality of being pre-existent to consciousness, and hence, on occasion, of conditioning it rather than being conditioned by it. . . . Accordingly, it would seem that natural numbers have an archetypal character. If so, then not only would certain numbers have a relation to and an effect on certain archetypes, but the reverse would also be true. The first case is equivalent to number magic, but the second is equivalent to inquiring whether numbers, in conjunction with the combination of archetypes found in astrology, would show a tendency to behave in a special way.

It is in pursuit of this inquiry that this book is being written, as the second in *The Astro-poetic Series*. Since, as Jung has indicated, the study of number is – at least from an archetypal point of view – also a study of consciousness, the inquiry about numbers in *Astrology and the Archetypal Power of Numbers* will build upon the foundation for a language of soul in the astro-charka system as outlined in *Tracking the Soul*.

In the same way that each *chakra* represents a particular worldview by a soul conscious at that level, each number represents a particular dimension of the world that becomes accessible to a soul who enters the Realm[2] it governs. Entry to these Realms requires a soul to integrate various *chakras* in order to arrive at a level of consciousness through which they can be perceived. As we work through the various issues associated with our *chakras* as reflected in our birthchart – the map of the heavens at the time and place of our birth – we rise to the level of consciousness necessary to make a difference in the world within one or more of these Realms. How exactly this works will become more evident in a general way in Part One, and a more specific way in Part Two (a separate volume), as we consider the astrology of each Realm.

In this book, however, we will go beyond a consideration of individual psychology to consider the ways in which number is "pre-existent to consciousness," which is to say, encoded in our collective psychology as human beings, and in the construction of the universe of which we are conscious. The more deeply we go into this great mystery, the more apparent it will become that not only does the world shape us into becoming who we think we are, but that as we participate consciously within the world, we help shape the *anima mundi*, or soul of the

◇◇

world. It is my contention that numbers are the key to this two-way interaction between soul and world, and that the birthchart – understood as a map of the Number Realms – is the key to approaching this fundamental relationship more consciously.

As the poet Muriel Rukeyser once said, "The universe is made of stories, not of atoms." It is my contention in this book – and apparently also Jung's – that these stories (yours, mine, and ours) are laid out numerically in a way that only a poet can fully appreciate. Let us then don the poet's cap as we follow Pythagoras into the archetypal story of numbers that may hopefully one day soon, once again be not just for counting.

Endnotes

1. This quote is often attributed to Einstein. According to Wikipedia, "A number of recent books claim that Einstein had a sign with these words in his office in Princeton, but until a reliable historical source can be found to support this, skepticism is warranted. The earliest source on Google Books that mentions the quote in association with Einstein and Princeton is Charles A. Garfield's 1986 book *Peak Performers: The New Heroes of American Business*, in which he wrote on p. 156: 'Albert Einstein liked to underscore the micro/macro partnership with a remark from Sir George Pickering that he chalked on the blackboard in his office at the Institute for Advanced Studies at Princeton'." Whether or not this quote can be attributed to Einstein, a moment's reflection should convince anyone but the most dogmatic scientist that it is true.

2. In this book, I will capitalize the word "Realm" when it refers to the archetypal dimension of number, and to emphasize the idea that these dimensions were considered by the Pythagoreans to be aspects of Spirit, or the "One," as they referred to it.

There is no linear evolution; there is only a circumambulation of the self.

from *Memories, Dreams and Reflections* by Carl Jung, New York: Vintage Books, 1965.

Astrology and the Archetypal Power of Numbers

Part 1: A Contemporary Reformulation of Pythagorean Number Theory

Figure 2: The Rune Perth

Initiation, uncertain meaning, a secret matter,
the rune related to terma teachings of all kinds

Introduction

In the 8th century CE, Guru Padmasmabhava is said by followers of the Nyingmapa sect of Tibetan Buddhism to have planted secret teachings all over Tibet. Padmasambhava intended these teachings – called *terma* – to be discovered by future incarnations of his 8th century disciples. *Terma* were written in a concentrated code called *dakini* language, through which a few seed syllables might contain within them volumes of information. Since then a 1,200-plus-year lineage of *tertons* – translators of *dakini* language – have spawned an entire literary tradition with its own taxonomy of teachings, and a small firestorm of controversy over which teachings are true and which are false.

In Namkhai Norbu's book *The Crystal and the Way of Light*, he tells a story about his uncle, a contemporary *terton*, who one day announces that he has received the inspiration to search for new *terma* (70). Inviting the whole community to join him on his expedition, he makes his way to a sheer mountain face and announces that the sacred text is just above them. After a moment's meditation, he takes a knife and throws it with all his might into the rock face. "There," he proclaims, "the *terma* is behind the knife." Several of the younger, stronger men of the community build a makeshift ladder, climb to the knife and retrieve it. As the knife comes out of the rock face, it dislodges a piece of rock, behind which is a smooth, round, luminescent ball about the size of a grapefruit. The *terton* instructs his helpers not to touch it, but using the knife and a blanket, they manage to lower it safely to the ground. The *terton* then wraps the ball in a white scarf and tells the wide-eyed crowd that the *terma* he sought is inside the ball.

Before and since, other *terma* have been found in mirrors, statues, skulls, earrings, onyx bones, and supposedly even undergarments worn by Padmasambhava. They could, in fact, be anywhere. As incredible as such a tradition might seem to the postmodern Westerner, the whole notion of *terma* is a profound metaphor for the ominpresence of an immanent God or Spirit everywhere throughout this manifest creation – speaking to us from every rock, bird and burning bush.

Most of the world's esoteric spiritual traditions identify with this presence as the Holy Grail of the soul's quest for self-knowledge. Spirit's numinous self-revelation is a gift freely given, they say, but not one easily received. An awareness of the presence of Spirit is elusive to all but those with eyes to see, and more importantly, with the intent necessary to pierce the veils of material obscurity behind which Spirit is hidden. Spirit is alive within every mirror, statue, bone and, yes, even every undergarment. It is alive and in addition potentially conscious of Itself within each one of us. But before we can realize what It is and who we truly are, we must recognize that beneath the dust and disguise, everything and everyone is

◇◇

a luminescent sphere with a *terma* teaching inside. We must remember our cosmic function as *tertons* – as revealers of the presence of Spirit.

Astrology as a Dakini Language of Number

To decode the *terma* teachings that abound everywhere around us, knowledge of a *dakini* language can be helpful. In my previous books, I have suggested that astrology is such a language because it holds as its most sacred premise the idea that everything in this embodied world is a reflection of a divine order unfolding. This order is discernable through an application of astrology within the context of a human life, an historical sequence of events, the movement of weather patterns through a given geographical clime, or countless other frameworks of revelation. Within the metaphor I am exploring here, each symbol in the astrological lexicon is a syllable of this *dakini* language – a potent seed, the cracking of which opens a floodgate of hidden knowledge.

Astrologers differ widely in their understanding of how this works, as well as what it is that they are actually seeing in a birthchart when they speak the astrological language. However, the potential exists in any given encounter with astrological symbolism for a *terma* explosion – a sudden opening of the intuitive mind to volumes of information, the unexpected revelation of meaningful patterns unfolding simultaneously on multiple levels in cyclical time. It is this discovery – or the possibility of this discovery – that unites all astrologers, regardless of their perspective or preferred school of thought – and makes each one a *terton* (intuitive revealer) of the astrological language.

In Part Two of this two-volume set – the second installment in a series outlining a more intimately personal, imagistic, and intuitive approach to astrology that I call astropoetics – I will elaborate this metaphor by suggesting that astrology's function as a *dakini* language is rooted in its use of numbers. Today, numbers are understood only in the scientific sense as measures of quantity but, at one time, they were conceived very differently as potent archetypal gateways through which the creative forces at work within the embodied world could be glimpsed, accessed, and harnessed with the power of focused intention. In Part One – which you are reading now, I will explore this archetypal sense of number. Leaving astrology's use of number aside for the moment, it is this more ancient knowledge of numbers to which we will turn our attention in this book.

The Source of Esoteric Knowledge About Number

According to Richard Heath, author of *Sacred Number and the Origins of Civilization*, this knowledge was known in prehistoric times, probably before the Egyptian calendar combining solar and lunar cycles arose somewhere in the 5th millennium BCE (Grun 3). Heath traces this knowledge back to Atlantis, the actual existence of which he acknowledges is denied by science. Atlantis, he says nonetheless, was the source of a culture based on number that rivaled science in its sophisticated understanding of the order of the cosmos. Writes Heath (4):

> ... the fall of Atlantis represents closure on an important period of human evolution that went from a Stone Age culture in which humans learned to count, often using marks on bones to track Moon time, to an Advanced Neolithic culture that was megalithic, in which the view that the world was a numerical creation led to a culture that measured the world and was capable of building according to the same principles.

These principles were known to Pythagoras, who extracted them from a 40-year study of Egyptian, Babylonian, Chaldean, Phoenician, Brahmanic, Hebrew, Greek, Zoroastrian, and Hyperborean (pre-Celtic British) mystery teachings, and passed them on to contemporary Western civilization. Pythagoras, whose name refers to Pythian Apollo, was supposedly conceived after his mother paid a visit to Delphi. He is rumored to be the actual son of the Greek god whose specialties were prophecy, mathematics, music, and medicine. In turn, Pythagoras fathered a lineage of wisdom teachers called the Pythagorean Succession, said to have survived 1,000 years after the Christian Emperor Justinian closed the last of the pagan schools in Europe (in 529 CE), at which point it went underground (Opsopaus, Part V:1). Traces of the teaching were passed on through Plato and other members of the lineage to several generations of Neoplatonists, the Hermetic magical tradition, and the esoteric schools of Christianity, Judaism, and Islam (Guthrie 13).

Pythagorian teachings spanned an appropriately Apollonian range from hygiene, diet, and the early Greek medicine of the four elements, to number theory to geometry to music to astronomy to metaphysics. Though he wrote nothing down, various students preserved bits and pieces of his oral teaching and passed them on in written form. To confuse matters, as seems to be the case in the dissemination of teachings from many great teachers, there is also a rich, profligate tradition of pseudopigrapha – false teachings forged in Pythagoras' name – probably during a revival of Neopythagoreanism between 150 BCE and 100 CE (Huffman).

Pythagoras' knowledge of numbers was further developed by Nicomachus of Gerasa and other disciples, whose teachings in turn were written down by Neoplatonist teacher Iamblichus of Chalcis in a book called *The Theology of Arithmetic*, translated into English by Robin Waterfield – a work that is generally believed to be an accurate reflection of Pythagoras' teachings[1]. Waterfield's text is my primary source of Pythagorean number theory. Other

◇◇

commentaries on Nichomachus' work with numbers were made by Asclepius of Tralles and Philoponus in the 6th century BCE, but translations of these are not available (Huffman).

A second strain of this knowledge – though not entirely separate from that espoused by Pythagoras, who studied and absorbed it – arose from the Kabalistic tradition of esoteric Judaism, said to have been handed down to Moses on Mount Sinai along with the Ten Commandments (Hall CXIII). The sacred power of number was known in other parts of the world as well, and forms the basis for various ceremonial rituals, religious practices, and calendrical observances - especially among indigenous peoples. The ancient Maya, for example, had an entire series of gods associated with various numbers (Sharer 539), while the Teutonic tribes of northern Europe used their knowledge of number to enhance the power of runic spells (Gundarsson 192).

Although much of the original knowledge about number has been lost, or diluted through intellectual processing of information best experienced energetically, numbers remain a potent set of *dakini* syllables for those with the capacity to open to them on a level beyond the grasp of the rational mind. According to Aetius (1st or 2nd century BCE), Pythagorus "assumed as first principles the numbers and the symmetries existing among them, which he calls harmonies, and the elements compounded of both, that are called geometrical" (qtd in Waterfield 9). Harmonies can be understood as relationships between the numbers, while geometry is the projection of numbers into three-dimensional space. Number, harmony and geometry "are the three primary means that Plato proposed that the Divine Artificer . . . used to proportion the world soul" or *anima mundi* (Plato, Timaeus 36, qtd. in Waterfield 9). Numbers, in other words, are the source code through which the world is infused with the presence of Spirit.

Vestiges of this knowledge are encoded within music (as first revealed by Pythagoras), the construction of language, sacred geometry, and the architecture of medieval Europe, Islamic mosques and Hopi kivas. A more overt form of this knowledge exists today as numerology. As is true with most popularized expressions of occult teachings, what we are left with in our modern understanding of numerology is but the echo of an original teaching that reverberated at a deeper level. Perhaps this original teaching can be found in the ancient Greek practice of arithmology – an application of Pythagorean ideas to number, arithmetic and geometry distinct from, but related to, the essence of mathematics as a cosmic art. Arithmology can be linked to mathematics in the same way that astrology is linked to astronomy, or alchemy to chemistry. Arithmology first considers basic mathematical principles related to each number, and then speculates about the allegorical implications of these principles.

Within a world dominated by science, such speculations – indeed all metaphysical speculations about the more cosmic implications of natural order – have become taboo. As Waterfield reminds us, however (26):

◇◇◇

It should be noted that while the Pythagorean attempt to give meaning to peculiar properties of number is unfashionably mystical, such peculiar properties have not been explained or explained away by modern mathematicians. They exist, and one either ignores them and gets on with doing mathematics, or gives them significance, which is what arithmologists do.

In this book, I explore the peculiar speculations of arithmology alongside other distinctly non-mathematical sources in order to recover the true nature of numbers as a source of creative potency. I approach the essence of number obliquely, so that we can gain a more visceral sense of it, through our peripheral vision and our non-rational faculties. I believe the original teachings are meant to facilitate an experience of alignment to the forces to which they allude, and it is this alignment that I want to evoke through my exploration of number.

Once numbers are revealed experientially to be the *dakini* syllables that they are, in Part Two we will then explore their relationship to astrology, and show how they expose a deeper layer of meaning within the birthchart. This is an ambitious undertaking, and I do not claim as deep an understanding of the secret knowledge of numbers as those who have gone before me. Let's just say that poking around the ruins of this ancient knowledge is a calling I feel drawn to pursue, and like the Fool of the Tarot – associated with the number Zero in most decks – I am ready to step off the cliff into the Abyss of unknowing to see what might be on the other side of the chasm into which I leap.

My Personal Quest For a Deeper Understanding of Number

I first became aware of this chasm in my sophomore year of college. I was studying chemistry, hoping one day to become a research chemist. Who knows, thought I in my foolish youth – perhaps I would be the one to discover a cure for cancer. On the road to this lofty goal, lay a rather pedestrian titration experiment in organic chemistry. I was charged with the task of adding a certain reagent, measured drop by drop, to a flask of sodium permanganate in order to neutralize its acidity, and then determine the concentration of the original solution. It is possible to know when you have reached the point of neutrality because what begins as a dark purple liquid turns clear.

Within the scientific context in which I was conducting this experiment, what was important was the quantification made possible by the numbers. Knowing the volume of reagent necessary to neutralize the acid, I could determine mathematically the molar density (units of reagent per volume of solution) of the original material. But what I experienced instead that day in the laboratory was a moment of personal epiphany. I became enthralled with the magical transformation I witnessed as the purple solution began changing – first to

◇◇

burgundy, then magenta, then blood red, then orange, then pale yellow, then clear. Since this was only one of a series of experiments I had a limited amount of time to complete, my lab instructor was insistent that I move on with the procedure. But I was so excited by this apparent miracle, that I had to repeat it several times. By the time I had satisfied myself that what I was seeing was real, I had unwittingly stumbled into a realization that there was much more to numbers than could be explained by their quantitative properties. For me, this experiment was a *terma* teaching that changed my life.

In the days that followed, I wrestled with self-doubt. I had thought I wanted to be a chemist, but how could I resign myself to a pursuit so dry that it failed to appreciate the miraculous beauty of the transformations it measured? How could I relate such magic to the mundane chemical formulas that described them, and how was it that something with the power to move me the way this chemical reaction did could be reduced to a mere number? There simply had to be more to the story than I was being told. The more I thought about it, the more the academic journey toward a degree in chemistry seemed a Faustian bargain I could not make. I changed my major to English, and for the remainder of my college career, I studied literature, philosophy, and psychology – finding much more gold at the end of the day for all my efforts than I was permitted to find in the laboratory.

Eventually, I continued my education, working toward a Masters degree in Marriage, Family and Child Counseling. Again, however, I came up against a major stumbling block involving numbers – statistics. I was just beginning to appreciate the complexity of the human psyche – illuminated by some rather clever and imaginative theories, and given poignancy by the fledging work I had begun to do as a therapist-in-training – when I was asked by my teachers to reduce this intriguing complexity to a pedestrian set of numbers. Meanwhile, many of the more fascinating ideas I encountered – particularly those put forth by Jung, Assaglioli, the humanistic/transpersonal school and the radical brief therapy movement being pioneered by students of Milton Erickson – seemed to point toward the psychological equivalent of that sophomore titration experiment.

It seemed to me that those who had the most interesting ideas about the possibilities of human transformation were not concerned with the quantitative dimensions of the human psyche. Instead, they focused on its irrational power to evoke and wrestle with archetypal forces that changed their shape as consciousness was brought to bear upon them. As far as I could tell, this power of the psyche could not be measured by science, but rather had to be felt, sensed by intuitive faculties, and spoken of obliquely through story, metaphor, and analogy. A true transformation of the psyche's wounds could not be brought about by manipulation or coercion. It had to be evoked through a compassionate dance of intentional courtship and openness to the mutability of color, tone, and luminescence in the tender soul at the bottom of the flask of circumstance in the laboratory of life.

⬥⬥⬥

Yet here I was, being taught to measure these wounds, the external factors that may or may not have caused them, and the effectiveness of various strong-arm tactics of intervention in terms of number. Unlike chemistry or physics, psychology could not be made to conform to immutable laws that would always ensure the same result each time an experiment was performed, despite the collective efforts of generations of psychologists who believed it should be able to. As a concession to the softness of the science beneath psychology, its truths were measured instead by statistics – number used, not as a statement of irrefutable fact, but as a measure of our ability to know or not know something definitive about human nature.

From Control to Conscious Alignment With Natural Law

Why would it be important to measure human nature according to various sets of probability? Because, as I learned early on in my study of psychology, the goal – at least from a clinical perspective – was control: control of the uncomfortable symptoms of the wounded human condition; control of unruly and potentially disruptive emotions; and control of undesirable behaviors. In the darker, more chilling visions of the scientific psychological establishment – as sketched for us, for example, by Aldous Huxley in *Brave New World* and more recently by the Wachowski brothers in the movie *The Matrix* and its sequels – control extended to the human soul and the course of its journey from cradle to grave.

As pointed out in my earlier book *The Seven Gates of Soul*, the history of the early development of psychology was an attempt to wrestle the unruly human soul into submission to scientific standards of measurement, predictability, and ultimately control (Landwehr 239-260). Wilhelm Wundt, widely considered the father of modern psychology, was one of the first to insist that psychology was as much a science as physics and chemistry. In defiance of common sense, he believed ultimately there was no difference between inner experience and the outer phenomena by which inner experience could be inferred (Kim). Such a radical posture was necessary because outer phenomena such as behavior and physiological response to stimuli were measurable and thus within the reach of science, while inner experience was not. Despite the fact that this assertion was increasingly untenable to successive generations of therapists dealing with the actual experiences of real human beings, Wundt's psychological scientism set the tone for all future investigations of the human psyche. Although "psyche" means "soul," the soul had been effectively eliminated from the discipline of psychology centuries before I began my study of it.

What I was offered in graduate school instead of the discussion of soul I craved was a reductionist insistence on describing the endless variety of tastes, smells, and sounds on life's table as a set of numbers. I couldn't do it. I dropped out, and fled to northern California where I indulged myself in distinctly non-mathematical pastimes. Eventually, I returned to

◇◇◇

finish what I had started. But I never did get a satisfactory answer to the obvious question: "What could numbers possibly tell me about the human soul?" Numbers had their own undeniable attraction in the mysterious ways in which they combine with each other and shape-shift into other numbers. At the same time, I could not help but feel that something vital was missing in our understanding of them, especially as they were employed by science.

Numbers, as I had been taught to use them, were solely for the purpose of counting, measuring, comparing, and contrasting, according to quantifiable units of measurement. Because I had witnessed the transformations behind the numbers – the magic that numbers were employed to measure – it seemed to me that numbers were more fundamentally an invitation to understand how each individual piece of the whole fabric of this apparent reality was related to every other piece, and how each piece revealed the presence of the whole. How was it that all these wondrous colors could emerge from the same substance? How is it that a soul caught in the debilitating snare of some psychological trauma could also be loving, creative, intelligent, and at times free from trauma? How do all these various and diverse, sometimes contradictory states of being, live in juxtaposition to each other within the same individual? If numbers were going to be used to investigate the nature of reality, both around us and within us, then they ought to somehow speak to its inherent interconnectedness and multi-dimensionality.

Numbers As the Gateway to an Understanding of Multidimensionality

In the Pythagorean approach, number is seen not as a counting mechanism, but as a living force. In the foreword to Robin Waterfield's translation of *The Theology of Arithmetic*, Keith Critchlow finds an expression of this more holistic approach to number in the famous arrangement of ten dots in a triangle. Pythagoras used this pattern – called the Decad or Tetraktys – to illustrate the unity of the first ten numbers. Critchlow says "that ideal number is not necessarily subject to a sequential or causal progression from one through ten, but is rather a unity with ten essential and potential qualities, simultaneously present...." (10) The intriguing implication behind these words is that everything science uses number to measure, participates in this Unity, and encompasses these same universal qualities.

Within this worldview, the observer cannot be separate from what is being observed in the way that science insists she must be. The most intriguing implication of the Pythagorean understanding of number is that everything I am capable of observing in some way reflects back to me who I am, because both it and I are embodiments of this same Unity, elaborated and illuminated in different ways by the same numbers. Or as put by the Jewish Neoplatonist Philo of Alexandria, "Both in the world and in man the decad is all" (qtd in Waterfield 21).

◇◇

As I observed and measured the changing colors in my sophomore titration experiment, something parallel to that transformation occurred in me. As a therapist, every time I was able to facilitate growth in a client, it was because on some level of my own being, I resonated with the changes in them that wanted to take place. If numbers are a Unity, within which simultaneous qualities intermingle and perpetually shift in relation to each other, then subject and object must both share in that dance.

Science remains blind to this possibility whenever science insists on objectivity - requiring its practitioners to pretend that they are not part of the world they observe, and that what they observe is not influenced by their observation. Science's own experiments have proven that this is an untenable pretension, but most scientists keep on pretending anyway. By contrast, in the original sense in which numbers were understood - by Pythagoras, by Moses, and by the indigenous peoples who employed them in their ceremonies and rituals - they were instead a key to more conscious participation in the world. Participation by number was more conscious because each number was a *dakini* syllable within which natural law was written, and to become conscious of the true meaning of numbers required the seeker of truth to align herself with natural law. Unlike science, which insists on maintaining intellectual distance, engaging the *terma* teaching of number is not a spectator sport.

If we imagine this *terma* teaching to be a thunderstorm, the difference in these two approaches becomes rather tangible. A scientist witnesses and documents this thunderstorm from the comfort and safety of a room protected from the weather. One who is interested in knowing the original power of number, however, can only do so by entering the storm and getting wet. Such an act would make no objective sense, but be filled with subjective meaning, in the same way that communing with the gods is different than worshipping them.

As a student of both chemistry and psychology, I was brought to the altar by the high priests of education and compelled to worship. All the while, what my soul craved was communion. I wanted to tear the altar down and embrace what it was built to honor. Numbers were offered me in college as the ritual tools of worship. Never was it hinted that they could also become the ritual tools of communion, if I was willing to cross the subject-object barrier and allow them to permeate me. It is to reclaim this missing piece of my education that this book is being written.

Part One (this volume) explores individual numbers from Zero to Nine[2]. Each chapter begins where possible with references to the original teachings of Pythagoras, and to other sources that hint at the true creative power of number. My exploration then extends into a free form dialogue designed to elicit an experience of each number in its archetypal essence.

Part Two (a separate volume) will show how these experiences reverberate through the birthchart, and are embedded as implicit dakini syllables at the heart of the astrological lan-

◇◇◇

guage. Part Two includes a series of case studies – of both significant moments in history and of individuals who have altered history – showing how these principles can be used to enhance our understanding of the soul's journey as it is revealed in the birthchart, and to facilitate the soul's participation in our collective evolution.

The Journey Begins

Imagine, as we begin this adventure, that I have just led you to a sheer rock face that appears to be indistinguishable from the other rock faces that surround it. Suddenly I announce that the *terma* teaching we seek is above us. I close my eyes, and rest for a moment in what seems like shimmering stillness about to explode. I reach into a sheath attached to my belt, pull out a knife upon which are inscribed a few ancestral symbols, then with all my might, I throw the knife high above me into the rock. Together we fashion a ladder, and from the ground, I encourage you to climb. Upon removing the knife, you notice something shining from within the opening that was created by the knife. I tell you to carefully dig out the opening, and you tell me there is a luminescent ball there. I pass a red silk scarf up to you, and ask you to remove the ball without touching it. You bring it back down to the ground, and hand it to me. As we gaze upon it, it is as though we are standing before a doorway in the rock face before us. Using the luminescent ball as our source of light, we enter.

Much to our surprise, immediately we begin to fall. The luminescent ball appears to float above us, receding as we plunge into utter darkness.

Endnotes

1. Although Iamblichus was essential in preserving what we now know about the Pythagorean perspective on the power of number, his work was often a compilation of sources to whom he did not always give credit. According to Robin Waterfield, "Whole sections are taken from *The Theology of Arithmetic* of the famous and influential mathematician and philosopher Nicomachus of Gerasa, and from the *On the Decad* of Iamblichus' teacher, Anatolius, Bishop of Laodicea. These two sources, which occupy the majority of our treatise, are linked by text whose origin is at best conjectural, but some of which could very well be lecture notes – perhaps even from lectures delivered by Iamblichus" (23). Elsewhere, Waterfield calls this seminal work but the "tip of an iceberg" (30), suggesting that taken as a whole, this compilation of ideas is far more provocative than definitive of a vast tradition that unfortunately has mostly been lost.

2. I will capitalize numbers when I refer to them in the archetypal sense to distinguish them from their application as quantitative units, used for counting and measuring.

Chapter Zero: Into the Abyss

In the beginning, there was nothing. Or so say many of the world's great cosmological traditions. According to a Greek poem, rumored to be the composition of Orpheus, it is said that before the dawn of Creation, a great black-winged bird called Nyx hovered over a vast darkness without form and void. Unmated, this bird laid an egg out of which golden-winged Eros flew. From the two halves of its shell, Ouranos (Heaven) and Gaia (Earth) were born, and the manifest world came into existence.

In the Teutonic tradition, there was a bottomless well, called Nvergelmir that existed in a region of frost and mist. Out of this well several ice-rivers flowed into Ginnunga-gap, a great Abyss to the north, where they met warmer air and melted. In this melting, life was quickened into the form of a being called Ymir, who began the lineage of a race of frost giants, ancestors to the Viking clans.

Polynesian mythology speaks of Po, a great Void or Chaos, without light, heat, or sound, without form or motion, out of which the world as we know it gradually took shape. The Shinto world begins with a reed – The Eternal Ruling Lord – that arises from a state of nothingness, depicted as an endless ocean of oil. Many of the Pueblo peoples believed that their ancestors emerged from a great hole, or *sipapu*, and began their lives in quest of the Earth's navel, or the center of the world. The Bible tells us that before Creation, "the earth was a formless Void" (Genesis 1:2). Around the planet, stories like this of the great Void or Abyss at the beginning of Creation are common.

The timeless mystery at the heart of these stories can be symbolized by the number Zero. It is difficult to speak of this mystery, because nothingness is preverbal, existing before language, and before mouths and minds capable of speaking or understanding language. All words – even the word "nothing" – point to something. Thus the mythological traditions mentioned above begin with an image – a vast darkness, a bottomless well, an endless ocean of oil, a great hole – meant to represent that nameless, shapeless Void out of which creation arises. These images, myths and stories take us to the edge of the Abyss, where knowing arises not in the mind, but as a cellular memory of the place from where we came – the origin of all things, which is No-thing. As it says in the Kabala (Matt 67):

> The depth of primordial being is called Boundless. Because of its concealment from all creatures above and below, it is also called Nothingness. If one asks, "What is it?" the answer is, "Nothing," meaning: No one can understand anything about it – except the belief that it exists. Its existence cannot be grasped by anyone other than it. Therefore its name is "I am becoming."

◇◇

A similar passage in the *Rig Veda*, India's oldest scripture, tells us (10.129, qtd in Oxtoby *Eastern* 24):

> *Darkness was hidden by darkness in the beginning; with no distinguishing sign, all this was water. The life force that was covered with emptiness, that one arises through the power of heat.*
>
> *Desire came upon that one in the beginning; that was the first seed of mind. Poets seeking in their heart with wisdom found the bond of existence in non-existence.*
>
> *Their cord was extended across. Was there below? Was there above? There were seed-placers; there were powers. There was impulse beneath; there was going forth above.*
>
> *Who really knows? Who will here proclaim it? Whence was it produced? Whence is this creation? The gods came afterwards, with the creation of this universe. Who knows whence it has arisen?*
>
> *Whence this creation has arisen – perhaps it formed itself, or perhaps it did not – the one who looks down on it, in the highest heaven, only he knows – or perhaps he does not know.*

The Myth of Orpheus

Though Pythagoras does not speak of the number Zero, he was heavily influenced by Orphism, an early Greek mystery religion arising from the experience and teachings of the legendary musician Orpheus. The story of Orpheus - particularly of his descent into the Underworld in search of his wife, Eurydice - speaks directly to the experience of the number Zero, and so it is with this story that we begin our quest, as we feel ourselves falling into the Abyss at the gateway of the *terma* teaching of number.

As the story goes, Orpheus' new bride Eurydice is bitten by a snake. She dies and descends into the Underworld, and Orpheus follows, pleading his case for her release on his lyre. Ovid describes his persuasiveness in *Metamorphoses* (qtd in Morford & Lenardon 299):

> *As he made his plea and sang his words to the tune of his lyre, the bloodless spirits wept; Tantalus stopped reaching for the receding waters, the wheel of Ixion stopped in wonder, the vultures ceased tearing at the liver of Titys, the Danaid descendants of Belus left their urns empty, and you, O Sisyphus, sat on your stone. Then for the first time, the story has it, the cheeks of the Eumenides were moist with tears as they were overcome by his song, and the king who rules these lower regions and his regal wife could not endure his pleas or their refusal.*

Hades and Persephone agree to let Eurydice return with Orpheus to the world of the living, provided he lead her out of the Underworld without looking at her. He fails, and Eurydice falls again into the Abyss – this time without any further hope of rescue. As painful as Orpheus' first loss was, this second loss was far worse, since he knew it was irreversible. It was also Orpheus' true initiation into the archetypal meaning of the number Zero, for it is only when we have lost what is most precious to us, with no possibility of retrieving it, that we understand what "nothing" truly means.

Before this critical juncture, we are riding a wave of continuity that allows us to pretend to be who we think we are. Until the wave is broken, we cannot begin the journey back to the Source, because we are too enamored of the endless process of becoming. Within this embodied world in which the soul finds its home, becoming this or that is always a seductive possibility, but to return to the place out of which becoming arises – the origin of every something in No-thing – we must stop attempting to become and simply be.

> At the second death of his wife, Orpheus was stunned . . . The ferryman kept Orpheus back as he begged in vain, wishing to cross over once again; yet he remained seated on the bank for seven days, unkempt and without food, the gift of Ceres; anxiety, deep grief, and tears were his nourishment as he bewailed the cruelty of the gods of Erebus. He then withdrew to the mountains. . . (Ovid, qtd in Morford & Lenardon 301)

As sons and daughters of the culture of positive thinking, we want to believe that Orpheus can begin again. Surely, in time, he can put this grief behind him, find someone new and resume his life. Alas it was not to be. Three years after his wife's death, Orpheus aroused such fury in the women whose attentions he refused – followers of Dionysus – that they stoned him to death and tore him limb from limb. Eventually his severed head, still singing, washed ashore on the island of Lesbos, where it was retrieved by Apollo. Apollo enshrined Orpheus' head as an oracle, and subsequently repressed its prophecies. Once having entered the Realm of Zero, Orpheus' story had no happy ending.

In death, however, the legend of Orpheus grew. Having survived a journey to the Underworld, it was believed that Orpheus knew something about death and the afterlife that no other mortal did. A vast body of literature ascribed to him – erroneously or not – became the basis for a mystery cult, popular throughout Greece and the Mediterranean region in the 6th century BCE, just as Pythagoras was gathering the threads of his own teaching. From the Orphic tradition, Pythagoras took, among other pieces, a deep appreciation for the power of music, and a belief in metempsychosis, an early Western conceptualization of reincarnation. Orpheus' escape from the realm of the dead through the power of music had fired the imagination of countless followers throughout the ancient world, Pythagoras among them.

◇◇

Into the Depth of the Abyss

Before we go on with our exploration of the becoming process – which in its essence is about perpetual rebirth – it behooves us to linger a bit with the experience of utter No-thingness that precedes it. In *The Seven Gates of Soul*, I discuss this place of devastating loss as an opening to soul. We move into this place when we lose something dear to us, especially if it involves the death of a loved one or the eminence of our own death. A sense of death – metaphorical, if not literal – is also evoked when we lose a job, end a relationship, experience a health crisis, or let go of a youthful dream. The Realm of Zero does not have to be a morose place, or a place of abject despair, although this is often our initial reaction to it. It does necessitate the recognition that life as we knew it no longer exists – or can exist. There is no turning back once we fall into the Abyss. There is only working through – being with our feelings, letting go, waiting for the spark of renewal to illuminate the darkness with no guarantee that it will, somehow enduring and gradually coming out the other side – or not.

We humans are resilient creatures. We can adapt to almost anything, if we have to, and prevail, all the stronger for our ordeals. But in the Realm of Zero, we must leave the past behind and begin again. Viktor Frankl, renowned psychiatrist in the decades following World War II, tells of such a realization upon his entry into a Nazi concentration camp (20-21):

> I tried to take one of the old prisoners into my confidence. Approaching him furtively, I pointed to the roll of paper in the inner pocket of my coat and said, "Look, this is the manuscript of a scientific book. I know what you will say; that I should be grateful to escape with my life, that that should be all I can expect of fate. But I cannot help myself. I must keep this manuscript at all costs; it contains my life's work. Do you understand that?"
>
> Yes, he was beginning to understand. A grin spread slowly over his face, first piteous, then more amused, mocking, insulting, until he bellowed one word at me in answer to my question, a word that was ever present in the vocabulary of the camp inmates: "Shit!" At that moment I saw the plain truth and did what marked the culminating point of the first phase of my psychological reaction: I struck out my whole former life.

In my previous book, *Tracking the Soul*, I speak of the dark night of the soul, and quote my former teacher Swami Muktananda from his autobiography, as he describes the devastating effect of his dark night (Landwehr 72):

> I was in a very strange state. I was seized by restlessness. My whole body ached and every pore felt as if it were pierced by needles. I don't know why this suddenly happened. Where had my rapture, my ecstasy, gone? My pride and my elation had been

◇◇◇

taken away, and I was suddenly the same poor, miserable wretch that I had been be-
fore meeting Nityananda . . . Swami Muktananda felt like a ruler gazing out as if in
a dream at the ruins of his once-beautiful and beloved city, now destroyed by fate . . .

I have had several such moments in my life – as have we all. Often these moments can be linked astrologically to difficult patterns in the birthchart, which are periodically triggered by the movement of planets in the current sky. After the abrupt ending of a 13-year relationship, for example, I discovered, as part of the same astrological cycle, the death of my father and the burning down of a tipi that was my home – events that pushed me over the precipice of the Abyss for a while, until I could regroup and continue on with a life that was utterly different than the one I left behind. I will talk more about the astrology related to the Realm of Zero in Part Two. Meanwhile, wherever death or radical change intrudes upon the continuity of everyday life – whether through the actual death of a loved one, getting fired from a job, divorcing a spouse, being diagnosed with cancer, going bankrupt, undergoing a dark night of the soul, losing a home, or falling into the dark hole of depression – the number Zero reverberates at the root of our experience.

Our Shared History in the Abyss

As devastating as these individual moments can be, humanity as a whole has faced similar collective openings many more times than history can count. These openings have brought countless individuals to their knees at the edge of the Abyss, weeping and wailing for all that had fallen into it. Since the beginning of recorded time, natural and man-made disasters have killed many millions and changed everything for those left behind. Here are just a few of the more memorable:

From 1346 to 1352 – the Bubonic Plague, or Black Death, killed 25,000,000 people, including one-third the population of Europe (Scaruffi). The worldwide influenza pandemic of 1918 left over 25,000,000 people dead (Scaruffi).

12,000,000 people died in Hitler's concentration camps; 13,000,000 in the cultural purges of Stalinist Russia (1934-1939); and 49,000,000 in Maoist China from 1958 to 1969. (Scaruffi). A government-sponsored genocide in Sudan has killed upwards of 400,000 people and forced another 2,500,000 to become refugees (Bauer).

World War II claimed 62,000,000 lives altogether; the Mongolian conquest of China in the 13th century killed nearly as many (60,000,000); the Ameri-

◇◇

can phase of the Vietnam War killed over 2,000,000 people (Wikipedia). As of the end of 2009, the war in Iraq has killed over 4,400 Americans (Anti-War) and an estimated 100,000 Iraqis (Iraq Body Count).

In 1931, a flood of the Yellow River in China killed up to an estimated 4,000,000 people (Wikipedia). More recently, a tsunami in Southeast Asia (2004) killed over 230,000 people (Scaruffi); an earthquake in Haiti (2010) killed about the same number (Wikipedia); and Hurricanes Katrina and Rita (2005) destroyed over 40,000 homes and the lives of their inhabitants (Institute of Southern Studies).

Up to 200,000 people (and probably still counting) died in the aftermath of the world's worst nuclear accident in Chernobyl in 1986; over 18,000 died in the Bhopal disaster in India in 1984 (Wikipedia).

As horrific as these figures are, this is one place where numbers do not tell the whole truth. For figures alone – considered quantitatively – cannot begin to hint at the depth of human suffering to be found at the bottom of the Abyss. More telling are firsthand accounts that individualize these numbers. A weekly diary of the bloodbath in Iraq (from early 2007), for example, edges us a bit closer to the actual experience of Zero (Iraqi Body Count):

Monday, 19 February: 127 are reported killed, 13 of them members of a Sunni family shot dead near Falluja, as they are returning from a funeral. In Ramadi, 11 people are killed when a suicide bomber explodes his car at a police station; 47 are killed or found dead in Baghdad, 6 in a Mahmudiya market, while 9 are killed in two attacks in Tal Afar. Among Monday's victims are 5 children.

Tuesday, 20 February: 64 are killed; a suicide bomber kills 7 mourners during a funeral, 11 are killed by car bombs in Baghdad, and 9 lose their lives after a chlorine bomb attack in Taji. In Diyala, 3 lorry drivers are killed and their lorries set on fire. Another 25 bodies are found in Baghdad. US and Iraqi forces bombard Moqtada al-Sadr's offices and confiscate his documents.

Wednesday, 21 February: Another 80 civilians die, 16 of them at a bombing in a market in Najaf. In Balad Ruz, gunmen kill 17 men during a raid, while another chlorine bomb kills 6 people in Baghdad.

Thursday, 22 February: There are 56 reported civilian deaths, including the death of a child in Kut. US air strikes kill up to 14 civilians in Ramadi, in the process of killing 12 suspected insurgents. 37 bodies are found in Bagh-dad, Hilla, Kirkuk, Mosul and Khalis.

Friday, 23 February: Only 25 are killed, 2 of them young Iraqi boys killed by US fire during clashes with insurgents. Just 5 bodies are found in the streets of Baghdad.

Saturday, 24 February: More than 120 people are killed. In the 7th major attack this year, 56 lose their lives when a truck bomb explodes outside a mosque in Habaniya. Among the dead are 5 children. In Kut, a policeman and his 12-year-old son are shot dead outside their house, 8 policemen are shot dead near Baghdad International Airport, 30 people are killed in vari-ous incidents in Baghdad, and 23 bodies are found in Baghdad and Mosul. On Saturday, another 12 unidentified bodies are buried in Kut.

Sunday 25 February: Around 115 civilians die. A suicide bomber kills 41 people, most of them young female students, at the entrance of the Eco-nomics College of Mustansiriya University. There are reports of 10 people killed by mortars in Abu Dsheer in Baghdad, while 44 bodies are found in Mosul, Baghdad, Enjana and Sulaimaniya.

Imagine for a moment that you are the mother of one of the young female students who has just been killed. A month before, your son was stabbed in the back while praying in a mosque; two months before that, your husband was blown to pieces by a car bomb while driving his lorry home from work. You will never forget his mangled body, his charred face unrecognizable. If not for his wedding ring, which had your name inscribed on it, you would not have known it was him. Both your parents were killed last year in an American air strike, which falsely targeted a local shoe factory as the hideout of insurgents. Your home has been broken into several times – whether by American soldiers or Iraqi freedom fighters, you don't know. Who knows what they were looking for? Because you know you might be next, you have trouble sleeping. You often wake up sobbing uncontrollably. You have developed a nervous tic on the left side of your face, and any sudden noise makes your heart beat rapidly. You try not to think what tomorrow might bring, and in any case, your grief at the death of your daughter, in the wake of this ongoing litany of deaths in your family and violations of your safety, is all-consuming in this moment. You are living in the Realm of Zero.

Psychological Strategies for Coping With the Abyss

To feel the emotional reality of the Realm of Zero – outrage, inconsolable grief, horror, panic, or unrelenting anxiety – can be devastating. The Abyss requires a response from the

◇◇

very depth of our being, but immobilization and denial are often our first stunned reaction to the overwhelming and incomprehensible, followed rapidly by what is now called post-traumatic stress disorder (PTSD). The thought, "this can't possibly be happening," is common. Immersion in the Realm of Zero for any length of time can result in a profound desire to escape into a less challenged existence that simply no longer exists. In the Realm of Zrro, we can't go backwards, but neither can we immediately see a clear path forward.

Denial is also often the official response of governments, and those ostensibly in charge. What is done cannot be undone, and to take responsibility for it would require major retribution; committed, sustained compassion; and/or a major rearrangement of political priorities – a response that few governments seem willing to make, especially where there is money to be lost by changing the status quo. Typical of this attitude, is Brigadeer General Mark Kimmitt's advice to Iraqis who saw unrelenting TV images of innocent civilians killed by coalition troops. He simply shrugged and suggested that they "Change the channel" (NYT April 12, 2004 qtd by Iraq Body Count). For those living in the Realm of Zero, changing the channel is not possible, even long after the ongoing crisis they face has left the evening news.

Similar to denial is the coping strategy of numbness. The body responds to pain through the production of endorphins, which have a morphine-like effect in numbing sensitivity (Cleland). Psychologically, it is natural to shut down and become emotionally immune to further suffering. The alternative can be unsustainable vulnerability, or in extreme circumstances, even a rapid descent into insanity. Viktor Frankl describes the experience of an inmate who had entered what he calls the "second stage of his psychological reactions" to the Realm of Zero (32-33):

> He stood unmoved while a twelve-year-old boy was carried in who had been forced to stand at attention for hours in the snow or to work outside with bare feet because there were no shoes for him in the camp. His toes had become frostbitten, and the doctor on duty picked off the black gangrenous stumps with tweezers, one by one. Disgust, horror and pity are emotions that our spectator could not really feel any more. The sufferers, the dying and the dead, became such commonplace sights to him after a few weeks of camp life that they could not move him any more.

In *Tracking the Soul* I write of the necessity for protection at the level of the first *chakra*, when survival – physical or psychological – is threatened. The same mechanism is at work here, in the early stages of response to the Realm of Zero. Numbing is essentially a second *chakra* (pleasure center) strategy for coping with first *chakra* pain by artificially introducing a counterbalancing force of mitigation through pleasure. Mitigation by pleasure cannot adequately address a threat to survival, but it can provide momentary respite from pain. When the suffering in the Realm of Zero becomes unbearable, and temporary measures like numbing no longer work, addictions of all kinds often begin. Suicide is perhaps the most extreme coping strategy in the Realm of Zero – an utterly illogical attempt to end pain by destroying the very life one is attempting to protect.

Unfortunately, these are short-term strategies at best. Denial, psychological numbness and addiction all ultimately serve to deepen the hellish nature of the experience. Even suicide must be seen as a short-term solution within the context of a spiritual worldview, for the soul that could not face what it experienced in the Realm of Zero will return in some fashion to face it again. Despite the impossible circumstances in which one suddenly finds oneself, life in the Realm of Zero requires a response. Those embroiled in war must strive for peace; those surviving genocide must reach out to one another and find a way to heal; those picking up the pieces in the wake of a merciless pandemic must find new reason to celebrate life and build a stronger immunity to protect it.

There is often no readily apparent pathway to resolution in the Realm of Zero, but there is also no true exit – either personal or collective. We are in this together, and like it or not – we are in this for the long haul. Even a personal death by natural causes is no ultimate escape, since future generations will have to deal with whatever we leave behind. The Realm of Zero – often thrusting us into the stark face of death – requires us to decide how we are going to live while breath still flows through our body, and how we are going to leave the world for our children and our children's children. Our answers to these questions – often posed in the most unbearable circumstances imaginable – make all the difference in every other Realm we might subsequently enter.

In 1983, a devastating tornado destroyed 600 homes in the town where I lived. While I joined dozens of other volunteers with the Red Cross and helped cut 100-year old tree limbs that had fallen onto these houses with my chainsaw, scalpers went door-to-door, charging exorbitant fees for repair work they knew the insurance companies would finance. Still other shady opportunists looted local stores that had been damaged. What is it that makes one person respond to such devastation from a place of greed, or parasitic opportunism, and another from a place of compassion and caring? It doesn't appear to be a matter of circumstance, nor even of upbringing or conditioning, but something more fundamental than that. I believe it is a matter of consciousness.

The Abyss as a Catalyst to a Shift in Consciousness

In *Tracking the Soul*, I described consciousness in terms of the seven *chakras* postulated by Hindu yogic tradition, associating a specific psychology with each. In this book, I extend that metaphor by suggesting that each number exists as the catalyst for a pressing collective quantum leap in consciousness that serves as an overarching context for any personal

◇◇◇

changes we make. Numbers are – in their essence as *dakini* syllables – archetypal forces of nature, impersonal in their function, yet profoundly personal in the experience of their effects. Where a given number predominates in an individual birthchart – in ways I will explain in Part Two – there will be psychic pressure toward healing, transmuting or otherwise resolving personal issues impacted within the pattern emphasizing the number. As the individual then responds to this pressure, there will be reverberations capable of influencing the collective atmosphere in which we are all seeking to evolve. From an astrological perspective, many of the patterns emphasizing a given number will be generational patterns, pressuring various subpopulations toward growth.

In Part Two of this two-volume set, I will show how core issues related to specific *chakras* are triggered within a larger context informed by number. For now, I just wish to suggest that the number Zero – and each subsequent number – will imply pressure toward the transmutation of consciousness that is global in nature, regardless of the particular astrological factors or life circumstances involved. More specifically, the number Zero will tend to induce a powerful first *chakra* awakening on a societal scale, and bring to the surface any latent personal issues that are harbored there.

Normally, first *chakra* issues – related to survival – are addressed by gradually allowing a deeper penetration of consciousness by Spirit. This in turn, allows us to approach our mortality – and the deaths of those we love – more consciously. Within the context of yogic philosophy, the various levels of penetration are called *koshas*. The *koshas* are discussed extensively in *Tracking the Soul*, but to recapitulate briefly here, they are: 1) *annamaya kosha*, or the physical body; 2) *pranamaya kosha*, or the energy body; 3) *manomaya kosha*, or the sensory-emotional, psychological realm; 4) *vijnanamaya kosha*, or the symbolic realm accessed through the intuitive mind; and 5) *anandamaya kosha*, or a state of complete identification with Spirit. As we move through successive *koshas* at the level of the first *chakra*, our understanding of survival changes – from a preoccupation with the death of the body (at *annamaya kosha*) to an experience of the continuity of life, even as the body falls away (at *anandamaya kosha*).

In dealing with a garden-variety first *chakra* issue, one can make progress toward healing and resolution by shifting awareness to a deeper *kosha*. If, for example, you are faced with the impending death of your father, you may find yourself focusing primarily on the physical level – making sure he is comfortable and taking his medications, helping him pay outstanding bills and tie up loose ends, checking to see that his Will and other legal documents are in order, etc. Aside from the necessity for doing this, such a focus might also work as an unconscious strategy to avoid deeper feelings that could be overwhelming in their intensity. Over time, if your father's illness is extended, you may find your energy flagging while outer demands refuse to let up and perhaps even intensify. This combination of pressures, internal and external, will tend to bring up whatever unresolved emotional issues still exist between you and your father, and toward the end you may find yourself speaking more candidly about them. As you and your father work toward resolution, you may begin to feel yourself bathed in an increasing aura of peace, acceptance and serenity despite the outer crisis that brings you together, so that eventually when he does die, you are able to arrive at a sense of psychological completion.

When such a first *chakra* opportunity is experienced within the context of a full-blown immersion in the Realm of Zero, however, the shift required of you is more demanding. Imagine for a moment that you are the woman in Iraq we mentioned earlier. While your father is dying, you are also in the middle of a war that has killed many of your friends and family, and that threatens your own safety. In this intensified scenario you do not have the time or psychic space to move through a natural progression in your response to crisis. Your life has been blown wide open and you are now staring out at the disintegrating world around you through a gaping hole where your first *chakra* once was, and your sense of security has been shattered. Obviously, such an intensified scenario calls for a more drastic strategy.

You could, as discussed earlier, move into denial and attempt to address your first *chakra* dilemma by escaping from it. In terms of the astro-*chakra* system presented in *Tracking the Soul*, this would essentially mean shutting down the first *chakra* (through denial) and taking refuge in the second *chakra*. One might speculate, for example, that this was one factor in fueling a sudden a shift from rice production to opium cultivation in southern Iraq at the height of US involvement there (Cockburn). Or you could jump to the third *chakra*, become hypervigilant, buy a gun, and defend your space to your last dying breath. Many Iraqis – affected by the death of those they care about, and supported by a religious fanaticism that sees martyrdom as a higher calling – choose to join the insurgency and defend what is left of their homeland. This is a justifiable response, but it does not lead to a resolution of the intense first *chakra* dilemma posed by the reverberation of the number Zero. It only intensifies and perpetuates it.

The Power of An Open Heart

A shift in consciousness – capable of moving yourself and those around you out of the Realm of Zero into the Realm of One – requires a more dramatic and a more collective leap. You, and a significant number of those around you, must find the inner strength to move from the first *chakra* to the fourth *chakra*. In the fourth *chakra*, the desire to escape or seek revenge is replaced by an opening of the compassionate heart. This is a seemingly impossible feat in the midst of the Realm of Zero, but it is not beyond the reach of those for whom spiritual evolution is primary. Within the Realm of Zero, opening the heart is actually a matter of survival.

CHAPTER ZERO

◇◇

As Viktor Frankl describes this possibility (104-105):

We who lived in concentration camps can remember the men who walked through the huts comforting others, giving away their last piece of bread. They may have been few in number, but they offer sufficient proof that everything can be taken from a man but one thing: the last of the human freedoms – to chose one's attitude in any given set of circumstances . . . .

Any man can, even under . . . (the worst of) circumstances, decide what shall become of him – mentally and spiritually. He may retain his human dignity even in a concentration camp. Dostoevski said once, "There is only one thing that I dread: not to be worthy of my sufferings." These words frequently came to my mind after I became acquainted with those martyrs whose behavior in camp, whose suffering and death, bore witness to the fact that they were worthy of their sufferings; the way they bore their suffering was a genuine inner achievement. It is this spiritual freedom – which cannot be taken away – that makes life meaningful and purposeful.

The inner achievement of which Frankl speaks is the filtering of suffering through the heart. First, we must open to it, allow it to move us - to tears, to trembling, to despair, to heartbreak, to devastation - to an intimate acquaintance with what Buddhists call *bodhichitta*, or what American Buddhist nun Pema Chodron calls the "soft spot of vulnerability." Though we live in a macho, reactionary culture that often views any hint of softness in the face of suffering as weakness, it actually takes great strength to allow ourselves to descend into this place - which stripped of any spiritual connotation or cultural baggage is nothing more or less than an embrace of our human fragility. Despite all our posturing to the contrary, we are inherently vulnerable creatures, separated in any given moment from our death by the thinnest possible protective skin, never more than one or two steps away from the Abyss. In the end, to acknowledge this truth is an act of much greater courage than fighting it. Such an acknowledgment is not surrender to victimhood or abrogation of responsibility. It is an intentional commitment to be worthy of our suffering by meeting it with an open heart.

Nor is living with an open heart merely a matter of passive receptivity, for an open heart quickly fills to overflowing. One cannot truly entertain the suffering in the world for any length of time without either shutting down or responding to it. An open heart demands that we respond, as humanely as possible. Shutting down is often easier, but shutting down greatly diminishes us.

Like Frankl's inmate in the second stage of his psychological reactions to the Realm of Zero, many of us have become numb to suffering, simply because it is all around us, as pervasive as the air we breathe. Most of us simply can't carry the weight of the world on top of our own personal burdens, so we choose not to see it, or if we do see it, we look the other way. The suffering of the world parades before us on the nightly news as a litany of rapid-fire invitations to the Realm of Zero, sandwiched between commercials promising the good life, and it fails to register. We talk about it sometimes, just to vent, but rarely do we let it in. When we trip over it personally, through some tragedy in our own lives, we take a pill so that we can continue on with business as usual. We avoid the Abyss, wherever possible, because it slows us down, makes us outwardly unproductive, and dissolves the fragile boundaries of our illusion of control. We also avoid it, because information processed through the heart requires heartfelt action – compassion, forgiveness, altruistic caring, selfless generosity, self-acceptance, nonjudgmental toleration of others, and so on. Heartfelt action takes time, often entails what seems to be a detour on our journey toward accomplishment or demands a postponement of more pressing personal agendas. Such responses are not the stuff of which worldly conquests are made. They are, however, the soil in which the soul roots itself.

Turning the Other Cheek

Like Orpheus, until we have survived a journey to the Underworld and back, with our heart still open, there is nothing within us durable enough to withstand death, nothing vibrant enough to sing its way to a becoming that will outlast the body. Many are swallowed whole by the Realm of Zero, their bones spit upon the trash pile of human waste. History records these casualties of Zero as numbers, and posterity often fails to remember them as anything but numbers. Few pass the test of being reduced to nothing, and still emerge with anything meaningful to pass on. Those that do are those who, against all odds, manage to let the Realm of Zero pass through them like the wind – feeling the rattle in every nook and cranny, but ultimately offering no resistance. This is the true meaning of Christ's admonition to turn the other cheek. One has no other cheek worth turning, after passing through the Realm of Zero, unless one has felt the sting in the first cheek as a personal invitation to a deepening of love – for self and for those at the giving end of the sting.

Turning the other cheek does not mean rolling over and playing dead in the face of atrocity, nor allowing those who commit atrocities to walk away without paying for their crimes. It does mean that the motivation for responding to atrocity must come from a deeper, more liberating place than denial or the desire for revenge. Christ's admonition to turn the other cheek, love thy enemy, and forgive thy crucifiers was meant to replace the old paradigm of "an eye for an eye" – which as the bumper sticker quoting Gandhi reads, is a sure recipe for making the whole world blind. Few since the time of Christ – including many Christians – have taken these admonitions to heart. But it is possible, and it has been done. Indeed, if the world is ever to emerge from the nightmare Realm of Zero into which it has blindly plunged too many times for history to record, it must become a way of being for us all.

CHAPTER ZERO

<><><><><><><><><><><><><><><><><><><><><><><><><><><><><><><><><><><><><><><><><><><><>

The Power of Forgiveness

Nowhere was this possibility seen more clearly in recent times than during the dismantling of apartheid in South Africa, and the restorative justice practiced by the Mandela government that followed. Most people know about Nelson Mandela's incredible personal capacity to forgive his jailers after 27 years of imprisonment, and rightfully consider him to be a great man because of it. But the reconciliation and healing sought by the South African people would not have been possible were Mandela not merely the poster celebrity for an attitude that was apparently widespread within that particular dark corner of the Realm of Zero. As Bishop Desmond Tutu observes:

> Mr. Mandela has not been the only person committed to forgiveness and reconciliation. Less well-known people (in my theology no one is "ordinary," for each one of us is created in the image of God) are the real heroes and heroines of our struggle.

> There was a Mrs. Savage who was injured in a hand-grenade attack by one of the liberation movements. She was so badly injured that her children bathed her, clothed her, and fed her. She could not go through a security checkpoint at the airport because she still had shrapnel in her and all sorts of alarms would have been set off. She told us [at the Truth and Reconciliation Commission] that she would like to meet the perpetrator – she, a white woman, and he almost certainly, a black perpetrator – in the spirit of forgiveness. She would like to forgive him and then extraordinarily she added, "And I hope he forgives me." Now that is almost mind-boggling.

It is mind-boggling because we are used to processing this kind of information with the mind, not the heart. The mind is an organ of separation and reflexive duality. It sees injustice as a matter of us vs. them and demands a form of retribution that merely shuffles roles between the victims and the perpetrators, without ever addressing the dichotomy that separates one from the other. We will talk more about this dichotomy when we move into the Realm of Two. Meanwhile, in the Realm of Zero, the mind is only capable of moving suffering from one side of the line to the other, while the real challenge and the only real rite of passage from Zero to One is to erase the line entirely. This is only possible when we process our experience through the heart, which unlike the mind is an organ of synthesis and all-inclusive unity. As Bishop Tutu further explains, the South African heart was moved toward a collective experience of healing:

> . . . because we believe in Ubuntu – the . . . idea that we are all caught up in a delicate network of interdependence. We say, "A person is a person through other persons." I need you in order to be me and you need me in order to be you.

◇◇◇

The greatest good is communal corporate harmony, and resentment, anger, revenge are corrosive of this harmony.... Retribution leads to a cycle of reprisal, leading to counter-reprisal in an inexorable movement, as in Rwanda, Northern Ireland, and in the former Yugoslavia. The only thing that can break that cycle, making possible a new beginning, is forgiveness. Without forgiveness there is no future.

We have been appalled at the depths of depravity revealed by the testimonies before the Truth and Reconciliation Commission. Yes, we human beings have a remarkable capacity for evil – we have refined ways of being mean and nasty to one another. There have been genocides, holocausts, slavery, racism, wars, oppression and injustice.

But that, mercifully, is not the whole story about us. We were exhilarated as we heard people who had suffered grievously, who by rights should have been baying for the blood of their tormentors, utter words of forgiveness, reveal an extraordinary willing-ness to work for reconciliation, demonstrating magnanimity and nobility of spirit.

Yes, wonderfully, exhilaratingly, we have this extraordinary capacity for good. Funda-mentally, we are good; we are made for love, for compassion, for caring, for sharing, for peace and reconciliation, for transcendence, for the beautiful, for the true and the good.

The affirmation of this *heart chakra* truth is not a sweet, but ultimately naïve sentiment. It is – at the level of the first *chakra* in the Realm of Zero – our only real hope of survival as a civilization and a species. I cannot ever hope to alleviate my suffering by passing it on to you, yet this has largely been the history of our world, ever since Cain killed Abel and then answered Yahweh's question, "Where is your brother?" by responding, "I do not know. Am I my brother's keeper? (Genesis 4:9). The key to the lock on the door to the Realm of Zero is to realize that "yes, you are your brother's keeper," extending the concept of "brother" to include every living thing, and then realizing that everything in this embodied world is alive. Such a notion is a bit of a stretch for the mind, but the open heart knows it to be true. Until and unless we individually and collectively find the capacity of heart to respond to the suf-fering in the world as the "keepers" of each other – Palestinians and Israelis; Christians and Moslems; blacks and whites; haves and have-nots; men and women; humans, wolves and microbes – we will continue to circulate through the hell hole of Zero with no real hope of redemption or salvation.

The Power of Non-Attachment

Another useful strategy for coping in the Realm of Zero – complementary to the open heart – lies in the Buddhist ideal of non-attachment. Outside of a spiritual context, non-

◇◇

attachment is often associated with withdrawal from the world, or a posture of aloof indifference. Taken to its extreme, this attitude can be summarized by the omnipresent Valley girl expression, "Whatever." This apparently generational mindset is the antithesis of the caring involvement and passionate commitment to alleviate the suffering of the world that naturally arises from an open heart. It is the sacrilegious cousin of the spiritual ideal of non-attachment. Non-attachment in the Buddhist sense of the word is not cavalier detachment. It is a profound understanding about the true nature of this embodied world that allows us to exercise our compassion in a sustainable way.

Non-attachment was summed up beautifully by fellow astrologer Brad Kochunas in a sermon delivered at his Unitarian church:

> The idea of detachment in spirituality should not be construed as withdrawing from the world but rather recognizing that it all goes away and being able to radically accept this. In so doing, we can love the world fiercely in all its ephemeral radiance.

A week before my mother died in the hospital, I watched appalled as she was forced to eat institutionalized food with all the appeal of regurgitated cardboard. Acting from sheer compassion, my sister and I smuggled in some real food for her, including a fresh Florida orange. I will never forget the absolute delight on my mother's face as she devoured that orange, the juice dribbling shamelessly down her chin. I was enjoying her as much as she was enjoying it, both of us aware that this moment of "ephemeral radiance" was just about all we had left. It was so very precious, precisely because it was – as Brad put it – about to go away.

The power of non-attachment in these moments comes as we center ourselves within them. We do not look back. We do not look ahead. We are simply here now with what is. The capacity to do this is hailed by spiritual traditions around the world as the key to communion with the divine. It is also a lost art. As filled with the opportunity for spiritual awakening as the moment is, very rarely are we content to dwell there. Why? Because we are attached to our becoming, and thoroughly invested in the illusion that it will continue forever. If today is not going so well, there is always tomorrow. If we don't get it in this life, there is always the next. If looking forward depresses us too much, we look backwards to memories of happier days. The mind – symbolized astrologically by quicksilver Mercury – is a slippery substance, and all but the most extraordinary of present moments are poor containers for it.

In the Realm of Zero, however, there is no past – as Frankl pointed out when he tried to share the importance of his manuscript with a fellow inmate – and there is no future, as anyone who has ever been close to death can attest. There is only now, which often seems interminable and intolerable. How can we possibly cope with such an experience other than through denial, numbness or suicidal martyrdom – anything to end it all, and be done with it? The Buddhist answer to this dilemma is non-attachment – accepting on a very deep level

◇◇

of our being the fundamental truth that it all goes away, or as George Harrison once put it, that "all things must pass."

Sunyata

The simple awareness that all things pass is not hard to come by. A moment's reflection on your life - any life - should be enough to convince even the most attached among us that this is true. Acceptance of the impermanence of this life, though much harder, is the real key to spiritual triumph in the Realm of Zero that we are seeking. When life is going well, it is hard to accept its impermanence because we want our enjoyment to last forever, and we mistakenly believe that our pleasure depends upon the people, circumstances and/or experiences that seem to be responsible for it. When life is going badly - and those people, circumstances and/or experiences go away - we suffer, because we mistakenly believe that our enjoyment and our pleasure depend upon that which is no more.

Such beliefs are not rational. Intellectually we may know that life goes on, and we will somehow make it through to better times again. But emotionally, we attach ourselves to the impermanent, believing against all evidence to the contrary that what we cherish most in life will last as long as we need it to. It generally will not. As a consequence, we will suffer when our attachments are ripped away from us by the natural flow of life. This is even more intensely true to the extent that we attempt to maintain our attachments in the Realm of Zero.

In *Tracking the Soul*, I consider beliefs as the province of the sixth *chakra*. In the sixth *chakra*, our task is to question our beliefs, and discard those that are no longer useful. In the Realm of Zero, it is our false belief in permanence - our wishful thinking about it - that we must leave behind. Or put another way, we must learn to see this apparently solid world as the "dust in the wind" that it ultimately is. The study of quantum mechanics tells us that what appears to be solid is actually a universe of infinitesimal particles, separated by lots of empty space. The particles themselves are capable of transmuting at any given moment into energy, which is even more transitory in its movement through space. What is this scientific conclusion, if not a materialistic understanding that everything we consider to be something is at core, really nothing at all?"

The same understanding is articulated in a more spiritual context by Mahayana Buddhists, when they speak of *sunyata*, or as it is roughly translated into English, "emptiness." Actually, the word *sunyata* was originally derived from the root verb, "sui," meaning "to swell," thus referring simultaneously to the experience of that which is swollen, or pregnant with unlimited possibility, and that which is empty. Its meaning is a paradox, or a point of entry into that place of deeply resonant truth where the mind cannot go. *Sunyata* is the Nothing of the Kabala, which is also known as "I am becoming," its apparent opposite.

CHAPTER ZERO

<><><><><><><><><><><><><><><><><><><><><><><><><><><><><><><><><><><><><><><><><><><><><><><>

The paradox of *sunyata* is described in a key passage of *The Heart Sutra of the Perfection of Wisdom*, a pithy, but potent text embraced by the Hwa Yen Buddhists of China (Conze):

> *Form is emptiness and the very emptiness is form; emptiness does not differ from form, form does not differ from emptiness. Whatever is emptiness, that is form, the same is true of feelings, perceptions, impulses, and consciousness.*

Form (or the various containers for the life we live – roles, relationships, bodies, civilizations, and so forth) is emptiness, according to the Hwa Yen Buddhists. Form exists within time and space, and is subject to perpetual change. Form exists for a limited duration of time, which is virtually no time at all when measured against eternity. Likewise form occupies a limited range of space, which essentially disappears as it is swallowed up by the infinite universe that contains all form. At the same time, form can only be perceived by a consciousness that itself is constantly changing. Our experience of any form is often far more dependent upon our shifting perceptions than anything substantial within the form itself. Ultimately then, all form is void or empty. The original emptiness with which this Creation began, continues to pervade every nook and cranny of this seemingly tangible universe.

Yet, so thoroughly are we attached to form, that it is virtually impossible for us to experience this original emptiness, while within the human body, which is itself only a passing form, or through the human mind, which is conditioned to perceive in terms of form. Every aspect of our lives, every experience, every encounter we have within the context of this earth-plane existence necessarily takes place within some form. All the people that we care about, all our possessions, every aspect of every circumstance – social, professional, religious, ethnic, political, financial, or otherwise – all constitute external forms to which we all feel some degree of attachment. Earth, sky, ocean, minerals, plants, and animals, all the various objects and phenomena created by man, atoms, molecules, planets, stars, galaxies, black holes, pulsars and quasars, indeed the very universe in which we and everything else we know, imagine to exist, or cannot begin to imagine – all of these things are, on one level or another, forms. More subtle, yet no less vigorous in their hold upon our consciousness are the inner forms. Feelings, sensory experiences, dreams, memories, fantasies, meditative states, anticipations and desires, thoughts, attitudes, beliefs, self-images and images of other people, places and things, all constitute forms for the duration in which they exist.

According to the Hwa Yen Buddhists, all of these things are empty – not that they do not exist, not as the Hindu Vedantists would argue, that they are illusions – but that regardless of the form these various phenomena take, the reality is *sunyata*, emptiness, No-thingness, the Void out of which all things spring and to which all things ultimately return.

Not only is form emptiness, say the Hwa Yen Buddhists, but emptiness is form. The essence of *sunyata* is revealed, for those with eyes to see, through the evolution of forms. It is

◇◇

only because all things are devoid of any kind of continuous or immutable identity of their own, that the movement of Spirit through form becomes possible. The essential emptiness of form is ultimately what allows the soul to undertake its evolutionary journey using form as its vehicle. It is only as form after form after form is shed, that the true nature of reality, or *sunyata* reveals itself.

Seeing Life in the Realm of Zero as Sunyata

Nowhere is the shedding of form after form after form more intensely realized than in the Realm of Zero. To conceptualize it this way - as a loved one lies dying in your arms, bombs bursting all around you, your house in flames, and everything you have hoped and dreamed suddenly cast into the Abyss - can seem cold, or perhaps too abstract to be real. These are not forms that are dying, but flesh and blood people we have loved. How is the concept of form different than say, the "collateral damage" of military jargon? As such a lofty concept is entertained by the mind, there is no difference. Intellectually, the concept of *sunyata* can easily be but yet another subtle strategy for distancing ourselves from unbearable pain.

When the heart is also engaged - as it must be in order to truly experience *sunyata* - the falling away of form reveals and releases the essence of whatever it is that is passing, and we take that gift inside, where it will be preserved through the life that we carry on. That precious moment - watching my mother eat an orange - is now a part of me, though the form, the temporary embodiment of Spirit that was my mother, is no more. Right now - in this paragraph that you are reading - I am passing it on to you, knowing that when I am gone, some of you, maybe just one of you, will carry it forward. Forms come and go, but nothing essential is ever lost. Knowing that - not in our minds, but in our hearts and bones - is what allows us to pass through the Realm of Zero unscathed.

Had Orpheus been able to practice the art of *sunyata*, he would not have looked back at his beloved. He would have understood that what he wanted to look back upon was never there in the first place. His beloved Eurydice had been every bit as real as he was. But the Eurydice that Orpheus wanted back was an empty form into which he had projected his hopes and dreams. The Eurydice that followed him out of the Underworld was likewise only an empty form, waiting to be filled again with living Spirit. This is not the Eurydice to which Orpheus was attached and when his attachment turned him around, there was only emptiness to greet him.

If form is emptiness and emptiness is form, you might wonder, what is the point of living? Why the charade of attachment to form we call life, when at death, the illusion only falls away? Why invest in anything, knowing that emptiness lie at the heart of it? How is the response of a compassionate heart compatible with the knowledge that the object of our

◇◇◇

compassion is but an empty form, soon to be molted in the wake of our compassion? In theory, we can celebrate the height of spiritual attainment in the Buddhist concept of *sunyata*. In practice, the actual experience of it can either liberate us or plunge us into the dark pit of despair. Sometimes the line between the two can be rather thin, and walking that line a choice we must renew at each step.

The Way of the Fool

One answer to this dilemma can be found in the teachings of Yaqui shaman Don Juan about controlled folly (Castenada *A Separate Reality* 106-107). By following the path with heart, it is possible to move through the world, fully aware of its inherent emptiness, as though such movement mattered. This may seem like a paradoxical or foolish way to live, and indeed it is to the rational Western mind. To one who is willing to entertain the radical attitude of controlled folly, the dance of paradox becomes the pathway of meaningful action within the Realm of Zero. We do this not to change the world, but to empower ourselves as agents of change, and to help alleviate suffering within the world of form. Empty though it is, this world of form still cries out for our enlightened compassion. Before we can respond in a meaningful way, we must get clear about what we are doing and why we are doing it.

In *The Seven Gates of Soul*, I write of the futility of attempting to change the world (Landwehr 174):

> *Though individual souls have undoubtedly grown to live better lives because . . . great beings showed the way, the world per se is every bit the same battleground between light and dark today that it was 2,000 years ago, or indeed since creation set all the polarities that define our world in motion. In fact, we could reasonably argue that given our increased technological capacity to destroy ourselves, the entire battle has intensified. The faces of the protagonists change endlessly, we occasionally shift arenas, the challenges evolve, and from time to time the forces of both light and darkness will appear to wax and wane, but as long as the embodied world exists, the battle will rage on, beyond the reach of cause and effect. It's not that the efforts of these brave souls were in vain, but the world is hardwired for reflex action, and there is nothing that any of us can do to change that.*

Be that as it may, within the Realm of Zero, we are called to respond to the suffering in the world. Our own personal melodramas matter very little in an Abyss where suffering is the everyday norm, and to be meaningful at all, our actions must rise above the perpetual siren song of our own self-preoccupation. Our choice in the Realm of Zero is either to respond from a place of compassion exercised with non-attachment – doing what we can to alleviate the suffering of others without looking back – or to allow ourselves to be swallowed up by the insanity of a world governed by denial and meaningless self-interest. Following the latter path, we contribute blindly to its perpetuation. Following the path of controlled folly, we roll up our sleeves and ask how we can help.

Frankl describes the existential dilemma we face in the Realm of Zero, where emptiness prevails, but life demands our response nonetheless (121):

> Woe to him who saw no more sense in this life, no aim, no purpose, and therefore no point in carrying on. He was soon lost. The typical reply with which a man rejected all encouraging arguments was, "I have nothing to expect from life any more." What sort of answer can one give to that?
>
> What was really needed was a fundamental change in our attitude toward life. We had to learn from ourselves and, furthermore, we had to teach the despairing men, that it did not really matter what we expected from life, but rather what life expected from us.

In the grand scheme of things, our contribution will matter less to the world than it will to us personally as souls, seeking to maximize our experience in this world as an opportunity for spiritual growth. If we are brave – or foolhardy – enough to make this choice, then perhaps extraordinary things can happen. History teaches us that for every person who rises to the occasion, there will be ten that take the easy way out. The odds are against us. The only way we can expect to face those odds and put our positive foot forward into the Abyss is by exercising controlled folly. Knowing full well the impossible nature of our mission, we can sign up any way.

From this perspective, the only way through the Abyss is the way of the Fool. In most traditional renditions of the Fool of the Tarot, he is depicted as a hapless fellow, stepping blithely into the Abyss, a meager bundle of essential possessions on a stick over his back, the sun blazing overhead, a dog nipping at his heels. This picture suggests inattention and oblivion to impending catastrophe, but that is only how we see the Fool as we stand outside his experience. In actuality, the Fool has made a conscious choice to embrace the experience of emptiness. To him, whether he lives or dies, it is all the same. In death as in life, the Fool makes a mockery of our attachment to form. While we laugh at his antics, we secretly fear the freedom he has attained. The only power to be found in the relentless psychological undertow of the Realm of Zero is the power of one who has nothing left to lose. Despite the universal desire not to want to suffer fools, the Fool is the archetype of this power.

Knowing that all is empty, we have nothing to lose in exercising our compassion toward that which is shedding form. Given the prevailing paradigm that "he who dies with the most

◇◇

toys, wins," such shameless selflessness is seen as foolish. Seen through the eyes of the soul, the Fool is wiser than we give him credit for. He is choosing to fill his empty life with living heart rather than dead stuff or soul-deadening self-aggrandizement.

In the West, we have relegated the Fool to the status of entertainment, but in ancient cultures the Fool served a much more transformative function. The *heyokas* of the Dakota tribe in the US, for example, and the *keshore* of Africa held positions of great respect. Their mockery was a constant call to attention, a constant reminder of attachment as well as an invitation to let go. The effects of their antics upon those to whom they turned their attention was profoundly therapeutic because there was reverence in their cultures for the power of the Abyss.

In the British Christmas-tide dramas, the Fool would willingly undergo death each year, so that his people might live. He would descend into the Underworld, taking with him the accumulated evils, burdens, and difficulties of the year just ending, and precipitate a cultural state of *sunyata* in which the new year could be free from the limiting residues of the old. Although this is a Christian festival, in our fear of the Abyss, we have long since divorced Christ from his essential function as purgative Fool. Instead, we speak euphemistically of Christ's role as savior, and thank him for dying for our sins. Ironically, it was Christ who showed us that we already live in the Abyss (where moth and rust doth corrupt), and that it is only our willingness to follow the path with heart through the valley of the shadow of death that marks our way through and onto a more solid path of becoming.

Paradoxically, the source of our becoming – and of Creation itself – is the very Abyss through which we seek safe passage. The descent into the Underworld described in ancient myths is a return to the source of all fertility, a reconnection with the omnipotent ground of our being. Until we are empty, we cannot be filled. Until we have accepted the essential emptiness of all form, anything we create will carry about it a sense of desperation in denial of the Void. The Realm of Zero exists to teach us humility in the face of awesome powers beyond the reach of our striving, so that as we do enter the embodied world in order to become who we are meant to be, we carry within us an instinctual measure of what is truly important and what is not.

It is only as we learn to live with death as our advisor (Castenada *A Separate Reality* 182-183), that we are empowered to choose a life of meaning and purpose. It is only when we become empty that we are empowered to move through this world as co-creators of it. It is only as we pass through the Realm of Zero consciously that our every act of controlled folly will have its roots sunk deeply into the primal No-thingness out of which this world first sprang and into which it ultimately returns.

Chapter One: Into Being

The concept of the One is widely understood throughout the world's mystical literature as a numerical depiction of God or Spirit. More particularly, the One is understood as an immanent god, dwelling at the very heart of the manifest world. In Pythagorean arithmology, this idea is grounded in allegorical speculation about a unique mathematical property of the number One – namely the fact that multiplying any number by 1 does not change it, i.e. $3 \times 1 = 3$; $4 \times 1 = 4$, etc[1]. Pythagoreans called the One, Monad, "because of its stability, since it preserves the specific identity of any number with which it is conjoined" (Waterfield 35). At the same time, every number contains 1 as a factor. Thus the number One brings stability to the manifest world through its omnipresence, yet allows everything to be what it is, without interference.

Having just emerged from the Realm of Zero, where all is chaos, we can appreciate how the stability of One represents an evolutionary improvement. Keep in mind that successive numbers are not necessarily a progression forward, so much as the revelation of concurrent layers of truth. In any case, the chief or sole deity in any religious pantheon – the embodiment of the One – is usually depicted as a Creator god who brings stability to the chaos of the Abyss.

The Creator is evoked as a metaphysical explanation for the creation of a world governed by natural law, in which human beings are often, if not always, the crowning jewel. The Creator god serves as an intermediary between the Realm of Zero – of death, annihilation and non-being – and the process of becoming that also emanates from No-thing and then blossoms in the Realm of One. It is the stability of the One that tames the Abyss and allows the process of becoming to proceed.

The first questions we must ask as we seek an understanding of the number One are: "Who is this Creator god?" "How and why does He/She/It arise out of nothing?" and more fundamentally "Does this Creator god – or any god for that matter – actually exist?"

The answer to this last question is largely a matter of belief. Those who are religious by temperament or choice would argue that the existence of their chosen God is self-evident, though when pressed would have trouble providing conclusive evidence to a non-believer. The non-believer would likely be inclined to treat all gods (and goddesses) as projections of the human mind. Since this question has been debated – at times rather hotly – ever since human beings were evolved enough to think and speak their thoughts, it would be presumptuous of me to propose a definitive resolution here. Instead, I would like to suggest a third approach to this question that will prove helpful to us in talking about this Creator god – and the number One – in a meaningful way.

◇◇

God as Projection

Let's suppose at the outset that both believers and non-believers are correct. The Creator god exists as a projection of the human mind. This does not preclude the possibility that such a god also exists in its own right, beyond the reach of the human mind. It does recognize that, since we can only know such a god through the faculties that are available to us, we must leave the question of this god's independent existence unanswered. It is part of the ultimately unknowable mystery that any god essentially is.

As philosopher-astrologer Richard Tarnas describes the necessity for this caveat (48):

> Within the established structure of the modern world view, no matter how subjectively convincing might be the psychological evidence for a transcendent spiritual dimension, an archetypal realm, an anima mundi, a universal religious impulse, or the existence of God, the discoveries of psychology could reveal nothing with certainty about the actual construction of reality. . . . Human spirituality and religion were still, in effect, confined to the subjective universe. What existed beyond this could not be said.

On the other hand, the assertion that the Creator god may only exist as a subjective projection of the mind does not make it any less real. As discussed in *The Seven Gates of Soul*, the soul essentially creates the world in its own image. Soul responds primarily not to the factual, objective reality measured by science, nor by any psychology rooted in science, but to its own subjective interpretation of the raw data. Often that which cannot be explained by the soul through a mere observation of appearances is assigned to the realm of the spiritual. The largest existential questions the soul is capable of posing – "Who am I? What is this world into which I was born, and how did it come to be as it is? What is my place within this world?" – are those most likely to be answered through a subjective projection of belief in some greater Being, wiser, more enduring, less bound by skin, distance and time than we are.

According to Richard Heath, our gods are a projection that arises as a consequence of our lost knowledge of number (3):

> In the dark ages that followed the destruction of Atlantis . . . the numerical powers that the ancient culture understood as giving rise to all creation became symbolized as gods. The early historical period became filled with symbols, iconographies, and myths that were degenerate components of the numerical worldview. Numerical realities were characterized by association to corresponding objects and human characteristics. Scholars have identified storm gods, fertility gods, trickster gods, and so on, but not the system that lay behind them.

◇◇◇

This system contains answers to the soul's existential questions, for those who know how to read it. For those who don't – which includes nearly everyone alive today – the projection of our gods and goddesses, particularly our Creator gods, serves as a gateway through which we can evoke a memory of this ancient lost knowledge. If our deities serve the psychological function of bringing order and stability to the chaos of the Abyss, perhaps it is the numinous power of number encoded in their symbols, iconographies and myths that ultimately has this power to do so.

Science – which vociferously disavows this power – treats our projections of God or Spirit as mere superstition, awaiting the discovery of new facts capable of dispelling them. Sigmund Freud considered spiritual belief to be a form of psychopathology, a neurotic obsession that healthy individuals – and eventually society as a whole – would outgrow (4-12). Never mind that this belief in God or Spirit or some Greater Reality is prevalent throughout the world – a central feature of nearly every culture on the planet. Our human impulse to believe in some form of divinity has existed as far back as anthropologists have been able to trace the human experience, and continues to exist despite 120 years or so of psychoanalytical input, and about 300 years of scientific dominance. An impulse as durable as this cannot be explained away by science.

Whether the Creator god exists in any objective sense or not, it exists for each of us to the extent that we can imagine such a possibility and project it into the manifest world. Whether our subjective projections are real, the tendency to project – indeed the need to do so – is not only real, but a defining characteristic of our species. In terms of our discussion here, let's recognize this human impulse to be the animating principle within the Realm of One.

The Projection of God as a Call to Consciousness

So what then is this Creator god? We agreed to speculate that our gods are a projection of the human mind. But surely the images of the gods we conjure are not *just* a product of the mind, certainly not the rational mind. For all its merits, the rational mind, which depends upon logic and analytical faculties, is incapable of understanding something that by its nature transcends logic and analysis. To be imagined at all, any god must engage the whole being, or the soul – which as discussed in *The Seven Gates of Soul*, includes far more than the mind. Such a god must also speak to our senses, our feelings, our imagination, our deepest desires, and our intuitive sense of connection to a larger, more all-encompassing Whole.

To conjure and project any god at all, we must employ a creative power that is only available to us when all our faculties are engaged. If we call the sum total of these faculties "consciousness," then the Creator god is a projection of the highest, most enlightened, most integrated level of conscious creativity we are capable of accessing. Pythagoreans called this

◇◇

archetype, the One. They alternately referred to the One as "artificer" and "modeler," terms appropriate, not just to the One Itself, but to any act of creation, including the imaginative projection of a Creator god.

To truly imagine a Creator god – not just to intellectualize its existence or borrow it from some pre-existing religion, but to make it real for us and infuse it with the power of an archetype – we must become creative ourselves. Even to the non-believer, it must be apparent that the act of drawing forth something out of nothing – the nature of any projection – mimics the very act that we attribute to this god. Aside from whether we think our belief in such a god is a sign of spiritual evolution or neurosis, the act of evoking it is a demonstration of a creative consciousness at work. In this regard, what Jungian psychologist Erich Neumann says about archetypes in general (295) must be even more deeply true about the central archetype of God or Spirit:

> When instincts are centrally represented, i.e. when they appear as images, Jung calls them archetypes. Archetypes take the form of images only where consciousness is present; in other words, the plastic self-portrayal of instincts is a psychic process of a higher order. It presupposes an organ capable of perceiving these primordial images. This organ is consciousness, which on that account is associated with eye, light, and sun symbols, so that in mythological cosmogony the origin of consciousness and the coming of light are one and the same.

If the Creator God – the intermediary between death, annihilation and non-being and the possibility of becoming – arises through an act of consciousness, then perhaps our psychic purpose in evoking it is our need to become conscious. Consciousness is our vehicle for shining a light in the darkness, making sense of the incomprehensible, forging a plan for dealing with the intolerable, rising to the occasion demanded by our circumstances, transcending the limitations imposed upon us by our suffering, and becoming more than we previously imagined ourselves to be through responding creatively to the unimaginable Abyss in the Realm of Zero.

In a sense, conjuring the Creator god – or any god – is like priming the cosmic pump. If we create a god by projecting our image of it into the Abyss, then perhaps it will create an ordered world out of that Abyss for us. Could it be that we imagine this Creator god because we need this ordered world to exist, and have an archetypal instinct toward ordering it? A god that is capable of doing this is also capable of inspiring us to do it. Within the psychological definition of god as psychic projection that we are entertaining here, the two are one and the same. Who this god is – its attributes, its cosmic function, and its strategy for doing what we have created it to do – will depend upon how we perceive the Abyss into which we are attempting to shine a light.

◇◇

The Projection of God as the Embodiment of Our Strategy For Dealing With the Abyss

As we saw in the last chapter, the Abyss is often perceived as an intolerable and unsustainable state of debilitating chaos. One instinctual mechanism for dealing with the Abyss is to seek an escape from it. To the extent that a desire to escape the horror of the Abyss fuels our projection of a Creator god, then that god becomes the vehicle of our escape. Not only does this god create the world, but it – or its surrogate – saves us from the Abyss of suffering at the heart of this world.

As pointed out in my previous book *The Seven Gates of Soul* (Landwehr 39-55), escaping the intolerable No-thingness of the Underworld after death was one factor triggering the birth of some religions. These same religions – Christianity, Judaism, Islam, Zoroastrianism, Hinduism, Jainism, and to some extent, Buddhism – extended the concept of escape to the Creation, so spiritual redemption was conceived as a quest for an escape from the body and the embodied world. The story of Christ's crucifixion and resurrection, for example, is a model for the spiritual opportunity for escape available to everyman seeking a reunion with his god in heaven. Getting off the wheel of karma and transcending the cycle of rebirth is likewise the goal of most Eastern religions based on this model.

A second strategy presented in the last chapter for dealing with the psychological intensity of the Abyss is to be found in the concept of control. Control in this context means mastering and taming the forces of chaos, and bringing them into orderly obedience to human will. When this is the psychic need that drives the projection of a god, then that god naturally becomes the embodiment of a benevolent force of control. The Egyptian god/king, Osiris, for example, along with his consort, supreme goddess Isis, brought civilization to Egypt – abolishing cannibalism, introducing agriculture and marriage, and passing laws that brought the unruly forces of chaos into a manageable order. Prometheus – the maker of human beings in Greek mythology – served a similar function. Before Zeus punished him for stealing fire, Prometheus taught knowledge of astronomy, mathematics, language, agriculture, animal husbandry, shipbuilding and other practical skills to empower mortals to push back the chaos of the Abyss. Where the need is for control, the human imagination conjures deities like Osiris and Prometheus with the capacity to teach humans how to control their environment. Incidentally, one of the many names for the One proposed by the Pythagoreans was Prometheus, quite in keeping with its arithmological function as a provider of stability and order.

As discussed in the last chapter, escape and control are not the only possible responses to the chaos of the Abyss. When the strategy of choice for dealing with the Abyss is compas-

◇◇

sion, and/or awareness of the emptiness of form, then gods and religions arise to meet those psychic needs as well. For those with ears to hear Christ's deeper message, ample demonstrations of the transformative power of compassion are available. The Buddha's Eightfold Path – right view, right thought, right speech, right conduct, right livelihood, right effort, right mindfulness and right meditation – was a prescription for living in harmony with all other sentient beings with whom we share this planet. Lao Tzu posited a yin way of being based on yielding, rather than asserting one's will; opening to the truth of the heart, rather than reasoning one's way to the truth of the mind; and following the flow of nature, rather than the structure imposed by human invention.

While neither Christ, nor Buddha, nor Lao Tzu are strictly speaking, Creator gods, we nonetheless project our understanding of them through our psychic need for a more compassionate, more benevolent, more enlightened world. They in turn, serve as intermediaries between the Realm of Zero and the Realm of One, in which our becoming takes shape, and guide us in the creation of a world that is in harmony with our conscious perception of it.

As we move from the Realm of Zero to the Realm of One, we might postulate that the true force – not just behind our projection of a Creator god, but also behind the act of creation itself – is the evocation of consciousness. As Jungian analyst, Marie-Louise von Franz declares at the beginning of her seminal work on creation myths – echoing Neumann – "it is sometimes revealed very clearly to us that [creation myths] represent unconscious and preconscious processes which describe not the origin of our cosmos, but the origin of man's conscious awareness of the world" (8).

Types of Creation Myths

Though we may speak of the theoretical concept of the One, in practice, which One will depend upon the consciousness that we are projecting. Our gods are but the concentrated collective images of our perception of the world into which we have been born - either responsible for that which lies beyond our control, or a supreme resource for dealing more effectively with this same overwhelming reality, or both. Human needs in this regard will be perceived somewhat differently in each culture, and the gods we conjure consequently a reflection of those needs. By the same token, our creation myths will reflect each culture's understanding of the relationship between gods, human beings, and the world they co-inhabit. According to mythologists, Scott Leonard and Michael McClure (vii, comments in brackets are mine):

> While all creation myths tell us how we got here, by whose agency, and how human
> beings fit in the grand scheme of things, characteristic qualities of each culture inflect

◇◇◇

these stories. For example, Native American creation myths tend to dramatize the idea that everything is intelligent, that the human being is related to brothers bear and fox as well as to the rocks and trees, while Egyptian and classical Greek accounts only imply human creation in their portrayals of how the various gods gave cosmic order to primordial chaos [the Abyss]. These distinct cosmological approaches shape a culture's social organization, religious beliefs, art, customs, and way of relating to outsiders.

Leonard and McClure acknowledge four types of creation myth identified by religious historian Mircea Eliade, then present a more elaborate system based on Eliade and further developed by feminist anthropologist Marta Weigle. In their more elaborate scheme, they discuss the following possibilities (Leonard & McClure 32-43):

Accretion or Conjunction Myths: creation arises through the synthesis of primal elements, i.e. the confluence of fire and ice in Eddic mythology produces the primordial giant Ymir, out of which creation was subsequently fashioned by dismemberment.

Secretion Myths: creation is formed from divine emissions such as vomit, spit, urination, defecation, masturbation or web-spinning, i.e. the Egyptian god Aten (or Atem) masturbating or spitting to form the additional gods, Shu and Tefnut.

Sacrifice Myths: creation proceeds through the sacrifice of the Creator god; i.e. the Chinese Creator-giant Pan-Ku grows for 18,000 years, expanding the universe with him. When he finally dies, his skull becomes the dome of the sky, his flesh the soil, his bones the rocks and mountains, and his hair Earth's vegetation.

Division or Consummation Myths: creation emerges from a cosmic egg or through the union of earth and sky, i.e. a cosmic egg appears on the primeval sea (an image of the Abyss) in Hindu cosmology, then cracks open to reveal Atman, a Creator man-god. In Greek mythology, creation is formed through the union of sky god Ouranos and earth goddess Gaia.

Earth-Diver Myths: the Creator god or his agent dives to the bottom of the primeval sea and returns with a bit of sand or mud, out of which creation eventually grows, i.e. Turtle dives to the bottom of the sea in a Maidu (California tribe) creation myth to retrieve a few grains of sand, out of which Earth-Starter fashions a pebble, which grows to become the Earth.

CHAPTER ONE

◇◇

Emergence Myths: the first people, often including the Creator god, emerge from a cramped womb world into this one, i.e. after four tries, two Hopi brothers build a ladder, climb and dig their way out of the first world, where "people kept stepping on one another," into successive larger worlds until they reach the fourth one, ours, where they proceed to make the Sun, Moon and other elements of creation as we know it.

Dual Creator Myths: two Creator gods create the world through cooperation or competition, i.e. the African Basonge tribe tells of the rivalry between Kolombo mui fangi and Mwile, who alternately produce a series of ever more challenging aspects of creation through daring each other to top the last achievement.

Deus Faber Myths: creation is fashioned through the efforts of a solitary Creator god, who is a master craftsman, i.e. Yahweh of Genesis, who accomplishes the impossible in seven days, or the "Fourfold Unfolding" of the Mayan *Popul Vuh*, who tries clay, then wood, then corn before finding a suitable substance out of which to make humans.

Ex Nihilo Myths: the Creator god magically produces creation out of nothing, i.e. Taiko-mol (Solitude Walker) of the Yuki (northern California tribe), who emerges from the foam on the primeval sea, plants four crooks in the shape of a swastika, speaks a magical word which brings the Earth into being, then fashions the plants and animals from feathers in his headdress. As a final *coup de grace*, Taiko-mol fashions a house out of mahogany, then announces, "Tomorrow there will be laughter and singing in this house." The following morning, when the sun rose, the house was full of people.

Because the consciousness that conjured these Creator gods' stories vary from type to type, so too does the nature of the Realm of One that arises out of the Realm of Zero in allegiance to these myths. It is beyond the scope of this book to analyze the collective psychology associated with each myth, but I can suggest that each one is worthy of deep investigation. The sheer variety of possibilities suggests that the Realm of One is richly multi-dimensional, and beyond the capacity of any one myth or type of myth to fully encompass. Difficulties arise when cultures subscribing to differing Creator gods or Creation myths come into conflict with each other. This is especially true when a given culture tries to prove that theirs is the One True Myth, and then seeks to impose their conjured Creator god on others.

This, of course, is the sad history of the human race – played out on countless stages as holy wars between Christians and pagans, as the American policy of genocide against indig-

⟡⟡

enous peoples, and as contributing factors to current political stand-offs between Israelis and Palestinians, Sunnis, Shiites and Kurds in postwar Iraq, Pakistanis and Sikhs in the Punjab region of northern India, Tutsis and Hutus in Rwanda, and Tibetans and Chinese – to name just a few of many possible examples. Ultimately, any dogmatic assertion of the One And Only Truth, will pose a painful catalyst to perpetuation of the trauma of Zero in the name of One.

Looking Beyond the Projection of God to Its Source in the One

To mean anything worthwhile, the archetypal number One will have to stand for something more than a fiercely defended attachment to our projections of truth. Let us see if we can penetrate to a deeper level of understanding that transcends this religious in-fighting and out-fighting in the name of the One – which really belongs not to the Realm of One at all, but to the Realm of Zero we are leaving behind. If the One is potentially the source of consciousness, then we ought to find within it, not just salvation from the Abyss, but the keys to self-salvation through the exercise of consciousness.

According to the Kabala, it is only possible to approach the One through a process of elimination. As explained by esotericist Manly Palmer Hall, "That which remains – when every knowable thing has been removed – is Ain Soph, the eternal state of Being" (LXVII). This is similar to the Hindu concept of *neti neti* – not that, not that – a path toward knowledge of the One traveled by eliminating everything the One was not. If the One arises out of the Realm of Zero, then it is easy to see the appeal of such a path. For whatever God is, it is certainly not – we would like to believe – reflected in the wretched circumstances of the Abyss. Indeed, as Joseph Campbell points out, this was the case in the culture in which the concept of *neti neti* arose (*Oriental* 285):

> It can be supposed that in the Indus Valley . . . a mood of world- and life-negation overcame many of the native non-Aryan population in their period of collapse, when the Vedic warrior folk arrived, c. 1500 – 1200 BC. But whereas in neither Egypt nor Mesopotamia does anyone seem to have found a practical answer to the problem of escape from sorrow, in India yoga supplied the means. Instead of striving for mythic identity with any being or principle of the object world, the meditating world-deniers now began – perhaps already c. 1000 BC – the great (and I believe, uniquely) Indian adventure of the negative way: "not that, not that" (neti neti).

Though originally conceived as a path of escape from the Abyss, the concept of *neti neti* taken to its ultimate conclusion, led to an experience quite different. For at its core – after peeling away every possible layer of identity – is the realization that neither the subject that

43

◇◇

is negating, nor the object of her negation, actually exist. This is similar to the Buddhist concept of *sunyata*, explored in the last chapter, but with a twist. If the subject that is negating everything it is possible to negate doesn't exist, then who is it that remains to be aware of this? Stripped to the core, the practitioner of *neti neti* comes to the ultimate realization that something does in fact exist that cannot be eliminated. This something is the One.

When the One is realized through whatever means, then the practice of *neti neti* becomes transmuted into an awareness of *iti iti*, "it is here, it is here." Or as Campbell concludes (quoting the Tibetan Buddhist *Madhyamike Sastra*) in discussing the annihilation of both subject and object (*Oriental* 350-351):

> We have seen that two negatives make a positive and that when dualistic thought is wiped away and nirvana therewith realized, what appears to be the sorrow and impurity of the world (samsara) becomes the pure rapture of the void (nirvana):
>
> > The bound of nirvana is the bound of samsara.
> > Between the two, there is not the slightest difference.

What this means is that once we eliminate everything that is not the One, we find that the One is everywhere. Everything we previously thought was not the One, while we were busy eliminating, is now revealed to be yet another clever disguise behind which the One is hiding. Ain Soph, the Kabala's name for God, literally means "without (Ain) ceasing to exist (Soph)," suggesting that there is no place where God or Spirit is not the underlying reality beyond changing appearances (Kaplan 53).

If this is so, then what we have been calling "the sorrow and impurity of the world" is – from the perspective of the One – also "the pure rapture of the void." *Samsara* is *nirvana* and vice versa once we have eyes to see it that way. Zero is both the horror of the Abyss and the magnificent raw power of unmanifest Creation, pregnant with possibility. To fully digest this truth – which we cannot do with the mind – we must recognize that the same world that a moment ago appeared insanely tortured from our vantage point in the Realm of Zero is simultaneously the fertile womb of becoming, distorted only by our incomplete awareness of it.

Harnessing the Power of Consciousness to Illuminate the Realm of Zero

Ain Soph (or the creative power of the One in the Kabala) actually fills the void as the "unconditioned state of all things" (Hall CXVII), just as for the arithmologists, the number One participates in all numbers.

◇◇

> *In the process of creation the diffused life of Ain Soph retires from the circumference*
> *to the center of the circle and establishes a point, which is the first manifesting One*
> *– the primitive limitation of the all-pervading O. When the Divine Essence thus*
> *retires from the circular boundary to the center, It leaves behind the Abyss, or as the*
> *Qabbalists term it, the Great Privation. . . . About this radiance is darkness caused*
> *by the deprivation of the life which is drawn to the center to create the first point, or*
> *universal germ. The universal Ain Soph, therefore, no longer shines through space,*
> *but rather upon space from an established first point.*

Bring anything into relationship to That Which Cannot Be Eliminated, and we immediately understand it to be part of the Creation emanating from the One. We intuitively grasp this truth because what we are observing is now bathed by Light, and in that Light, the One and its emanation are obviously connected. But take that same reality out of the Light, and it now appears to be part of the Abyss or Great Privation. Light on – One; Light off – Zero.

Yet from the perspective of the One, there is never a time when anything we could possibly perceive is not bathed by Light. So how is it the switch gets turned off, and we find ourselves back in the Realm of Zero? If we evoke the One in order to become conscious, then perhaps it is we who flip the switch off, when we go unconscious.

We look to our gods and goddesses to shine a Light into the Abyss for us, and they do this job for which we have evoked them admirably – as long as we provide the eyes that are necessary to perceive the Light they shine. Without our eyes to see it, the Light is as dark as if not darker than any darkness. How dark it is, in any given moment, depends upon how much consciousness – individually and collectively – we are able, or unable, to muster.

Not only do we project these gods and goddesses because of our own need to become conscious, but we provide the vehicle through which they do the work we have conjured them to do. It is willingness to use the divine gift of consciousness – and only this – that makes the shining of Light meaningful. The One fills creation with Light; it is up to each of us to illuminate the darkness by using our gift of consciousness wisely. In this way, and only in this way, does the Abyss organize itself into a world stable enough to support the process of becoming.

It is interesting to recall here the arithmological notion that the One brings stability to all things – because multiplying any number by 1 will yield the same number. Multiply 0 x 1 and we still have 0. Infuse the Abyss with the Light of Spirit and you still have the Abyss. Pythagoreans understood this when they also called the One, "Chaos, which is Hesiod's first generator, because Chaos gives rise to everything else, as the monad does" (Waterfield 39). If Chaos (Zero) and the Monad (One) are one and the same, then how is it the Realm of One supports the process of becoming, while the Realm of Zero does not?

◇◇◇

Again the answer appears to be consciousness, but more specifically in this case, the focusing of consciousness with clear intention.   This aspect of the One – and by extension, of each of us co-creators – was referred to by Pythagorians by yet another name for the One: Intellect.   It appears that what the Pythagorians meant by intellect was "moral wisdom" (Waterfield 39), but we might take the liberty here to reinterpret that phrase as the wisdom to direct our divine gift of consciousness with clarity of intent – beginning with our choice of god or goddess.   For who we project into the Abyss – consciously or not – will be the guide to our passage through it, and thus set the standard by which we conduct ourselves.   We want to believe that our deities will do the work of stabilizing the Abyss for us, but in the end, as 1×0 still equals 0, the responsibility is ours.

Taking Responsibility for the Light

When, according the Kabala, "Ain Soph . . . no longer shines through space, but rather upon space from an established first point," it is – from the perspective of our role as projectors of deity – we who establish this point. We do this by selectively directing our awareness toward that which appears to draw it, and then interpreting what we see. That which we are able to see clearly as an emanation of Ain Soph becomes illuminated by Light, and appears to contribute to our becoming. That which we are not able to see clearly as an emanation of Ain Soph appears to be cast in darkness, and becomes associated with the Abyss. The One would not be the One if it did not encompass both the Light and the darkness. But it is our point of reference that determines where the line is between Zero and One – how much separation there is between the Abyss and the Light that fills the Abyss with the promise of becoming. If this is so, then where we establish our point of reference determines everything else that follows and casts the entire Creation in the light of a specific hue.

Most of us learn how this works through a process of *neti neti*. That is to say, we evolve through identification with a series of false focal points, realizing at each step of the way, that what we thought was so important or critical to our happiness or well-being, isn't. Perhaps this is really what the Tibetan Buddhists mean when they say, "the bound of *nirvana* is the bound of *samsara*. Between the two there is not the slightest bit of difference." Regardless of where we place the focal point of our consciousness – as the pivotal point around which Creation revolves – it produces only an illusion that is relative to our point of view. As that point of view shifts, in service to an evolving vision, it creates a new illusion.

We like to think that there is a sweet spot somewhere in this shifting of perspectives that will yield the final "aha" experience of something that is real in an objective sense, but in the end even the sweet spot is mobile and transitory. Eventually, even our bottom-line values will prove themselves relative, through circumstances that require us to relinquish them, modify them, or revert to their opposite. Though it is perhaps human nature to want there to be an ultimate, Absolute Truth against which everything else can be measured, what we find when we adopt such a reference point is that real life never quite fits the ideal. We can adjust the ideal, or settle for less, but there is no *nirvana* at the end of the trail. What is at the end of the trail, when we realize the ultimate futility of chasing after this absolute sweet spot, and have released all of our attachments to visions of it, is the power to choose anew in each moment.

Regardless of the form it takes in any given moment, this becomes the inviolable essence of the One – the one thing that is left after *neti neti* has run its course, and we have eliminated everything that the One is not. When even the subject "I" has been eliminated, the power to focus consciousness remains. This is implied in the understanding of death presented in the *Tibetan Book of the Dead* in which awareness is the key to navigation of the *bardo* state (Thurman 29). It is also implied in the Hindu concept of the days and nights of Brahman, in which creation dissolves into nothingness at night when Brahman inhales, and then reemerges as He exhales (Oxtoby *Eastern* 45). From a shamanic perspective, Don Juan describes this same possibility as the emanations of the Eagle, in which the focusing of awareness along a different emanation changes the nature of the reality that is perceived (Castenada *Fire From Within* 108). In each instance, that which is left after everything else has been eliminated chooses where to focus its awareness, and consequently, what kind of Creation will evolve within the Realm of One. On a more mundane level, each of us does this in each moment throughout our lives, and consequently creates the reality that we perceive.

In *Tracking the Soul* I discuss this power to choose the focal point around which our consciousness of the embodied world revolves as the attainment of an awakened sixth *chakra*. This is where our beliefs about the nature of the world become subject to review, and we become aware of the relative nature of those beliefs. It is in the sixth *chakra* that we recognize that we have been relating to the world through a subjective image, which is more amenable to creative manipulation than we have previously imagined. In this awareness is the opportunity to choose anew, though no guarantee that we will do so. Habits of consciousness run deeply, and have insidious ways of reinstating themselves, despite our determination to see life through different colored glasses. Even though we might speculate that mastery within the Realm of One is essentially a sixth *chakra* process, something more is required of us than just a flexible point of view – if we are to realize the One within the Realm of One as a pathway to liberation from the downward pull of the Abyss in the Realm of Zero.

Choosing Our God as a Pathway to Becoming

After realizing that any point of view we could possibly choose is relative, and ultimately an illusion, we must still choose. There is no liberation in the mere knowledge that *nirvana*

✧✧✧

is *samsara* or that *samsara* is *nirvana*, unless we agree to engage *samsara* in order to experience *nirvana* within it. We cannot cash in our chips before the game is played; we must play. This is the essence of the call to Don Juan's life of controlled folly, discussed in the preceding chapter. The only way out of the Abyss is through it, and any path will do, as long as we can feel its validity while we walk it.

How do we do this without simply getting lost in the game – without mistaking our pet illusions for reality? The only way I know is by first choosing what Don Juan calls a path of heart (Castenada *A Separate Reality* 122), and then by allowing the gods and goddesses we have projected to guide us on this path. We project them from a place of deeper wisdom within ourselves, so that we might have more stable access to this place. We ask them to shine a Light, and they in turn, ask us to use the Light they provide to direct our consciousness.

We stumble this way and that within our now illuminated Abyss, directing our awareness to one false reference point after another, until we eventually find ourselves back against the wall, wondering what went wrong. We got what we thought we wanted, but we're still living in chaos, still seeking direction. And all the while, our projected gods and goddesses remain silently and passively available to us. They are not only illuminators; they are guides and they will guide us if we ask.

We can – through an endless trial-and-error process of *neti neti* – eventually find a working focal point, around which chaos will settle into some semblance of natural order that resonates with who we are at the core of our being. Or we can follow as consciously as possible in the footsteps of the gods and goddesses with which we identify. The popular current quest, for example, for understanding, "What would Jesus do?" might be one such strategy for doing this, were it not mouthed sanctimoniously by those who believe in Jesus, and sacrilegiously by those who don't.

Belief in Jesus – or any other projected deity – should not be an excuse to rationalize one's behavior or proselytize a particular point of view. It should be a calling to identify so closely with your deity of choice that it serves as a pivotal frame of reference around which your life acquires the power to transmute the Abyss into an arena for becoming. These deities, projected by the soul for the purpose of evoking consciousness, can teach us everything we need to know to transmute the Abyss into an illuminated pathway, if we take them seriously enough to actually live their teaching.

To deliberately evoke a god or goddess is to choose a specific track of becoming. If I choose Jesus as my guide through the Realm of One, my experience will be distinctly different than it would be if I choose Hermes, or Shiva, or Ishtar. In each instance, creation will be cast in a different hue of Light, and I will become something different as I move toward that Light. In the end, it will not matter which god or goddess I chose, and at various points in my life I may chose differently. It will behoove me nonetheless to choose wisely at each juncture that a choice is required. It will also behoove me to understand that my choice entails a responsibility, and that I will be challenged accordingly.

In Part Two, we will see how the birthchart suggests a natural affinity for one or more of these deities, although at various times in the course of a normal life, other deities will demand attention and ascend into prominence. Regardless of how this plays out, it is the act of deliberately and intentionally choosing to evoke a particular deity - knowing that one ultimately becomes what one chooses - that constitutes the most conscious path possible through the Realm of One. Choosing a deity is different than simply projecting one, or unconsciously adopting one. It means intentionally identifying with a frame of reference, so that Ain Soph - or That Which Cannot Be Eliminated - can illuminate the Abyss, or the shadow of the valley of death, and then walk you through it, fearing no evil.

From the perspective of the *chakras*, elaborated in *Tracking the Soul* (and reviewed on p. xviii), this is a matter of "descending" from the sixth *chakra* into the fifth, where living one's truth becomes the focus. Having realized in the sixth *chakra*, the relativity of your choices, you can now align yourself with a "Higher Power" that transcends a merely relative point of view. This higher power will subsequently empower you to the extent that you are able to rise to the challenges it poses to you. Its truth will become your truth to the extent that you live it, knowing that it is a relative truth, but trusting and owning your resonance with it.

Following in the Footsteps of a God

The process begins with choosing a god or goddess to follow, but in practice it is the god or goddess who will choose you. Our lives within the Abyss take certain shapes that normally unbeknownst to us, mirror the stories of one or more deities. Our quest for escape, control, or more conscious and compassionate perspective within the Abyss, is at core, an unconscious yearning for contact with this deity. If we think in such terms, we may project such a deity or identify with some religion that does.

Chances are, we are already stumbling blindly in the footsteps of the very god or goddess who holds forth the greatest promise of guidance, whether we believe or not, whether we ever step foot inside of any house of worship or spiritual gathering, whether or not we consciously recognize our way through life as a path at all. This is so, because within the Realm of One, everything that is conscious - and here consciousness may be defined broadly enough to include rocks and trees, stars and earthworms - partakes in some way of the essence of the One. The more conscious we become of this, and the more deliberately our yearning for redemption from the Abyss becomes attached to an intentional quest, the more diligently our

CHAPTER ONE

✧✧✧

gods and goddesses will find us. It is our focused intention that makes this possible. When the illumination of Ain Soph is directed by clear intention, a path appears beneath our feet.

Following in the footsteps of any god – you will not find a well-mapped or well-lit path. Life does not usually announce itself as a path in the footsteps of a god until its traveler is moved to trace unfathomable experiences of mysterious numinosity backward to their Source. Even then, the connection may not be consciously made between a life path and a god, since there is not a standardized checklist against which one can measure it, and then say, "I see I am on this path, instead of that." It is more a gradual process of identification, as this clue and then that resonates emotionally and intuitively with a story about the god that on some level, appears to be your own. You must know the stories to make the connection, which may only crystallize years after the initial seeds have been planted, and the stories will reverberate differently at different times.

Nor is there necessarily a ready-found community following a given path, since the associations are deeply personal and will not be the same for everyone who resonates with the same god. Churches encourage a uniformity of worship, but each mystic who has ever cultivated a personal relationship with Jesus, or Buddha, or Aphrodite has had a unique relationship with their chosen deity. Until such a personal connection has been made, the gods one's culture has projected will tend to be adopted (or rejected) by default, and understood only on the intellectual level as an embodiment of certain ideas and principles. Once the deeper, more personal connection has been made, it will become apparent that these teachings are merely the feeble ritualized attempt to mimic an actual encounter with the god, which is an entirely different experience.

It will also become apparent that at various junctures in your life, you have already been seized by the god or goddess for purpose of initiation, instruction, or restructuring. It is this series of seizures and our awareness of their import in relation to the deity that has chosen you that then marks your path.

The German philosopher of religion Rudolf Otto describes this seizure (12-13, reversed italics are italicized in the original):

> We are dealing with something for which there is only one appropriate expression, *mysterium tremendum*. *The feeling of it may at times come sweeping like a gentle tide, pervading the mind with a tranquil mood of deepest worship* . . . *It may burst in sudden eruption up from the depths of the soul with spasms and convulsions, or lead to the strangest excitements, to intoxicated frenzy, to transport, and to ecstasy . . . It may become the hushed, trembling and speechless humility of the creature in the presence of that which is a* Mystery *inexpressible and above all creatures.*

This is not normally what we think of as God in the religious sense. The god that has been worshipped in most churches, mosques and shrines has been safely sanitized for mass consumption. Yet silently reverberating at the heart of nearly every tradition is the story of a seizure that would have brought a chill to your spine or a trembling to your soul, were you there to witness it. The Bible, for example, is filled with stories like the parting of the Red Sea; the killing of the first-born in every house not marked by sacrificial lamb's blood; a burning bush that spoke to Moses, and many others like them. In Greek mythology, we have the hero Perseus doing battle with Medusa, a terrifying Gorgon with snakes for hair; Pentheus, ruler of Thebes, torn limb from limb by the Maenads after attempting to imprison Dionysus; and the hunter Actaeon being turned into a stag after having witnessed Artemis bathing in a river, and then being attacked and torn to pieces by his own hounds. Joseph Campbell tells the story of Hindu mystic, Ramakrishna, who witnessed a beautiful woman ascend from the Ganges, give birth to an infant, nurse it, then turn into an ugly hag, devour her own child, and disappear back into the river (*Hero With a Thousand Faces* 115).

Such stories hint at the awesome power of these gods we have projected. Of course, there are also stories of beatific visions, rapturous moments of transport into paradise, and sublime fragrances arising from unexpected sources. For most us, yearning for a way out of the Abyss, it is the fearsome stories that shake us out of our self-limiting trances. Like the familiars of magicians, our gods and goddesses are endowed with capacities beyond our own, or at least beyond what we perceive our own to be. We enter relationship with them, so that they can teach us what we have forgotten, and so that as we access the deepest, most potent resources of creative consciousness available to us, we become like them. The relationship begins with a demonstration of power – *mysterium tremendum* – and a sudden inexplicable urge to follow. In their footsteps, we learn to navigate the Abyss as conscious travelers.

The Story of My Choice

In *The Seven Gates of Soul*, I speak of a personal identification with the Teutonic god Odin (Landwehr 369-370). This identification arises in part from a strong and pivotal square (angular relationshp of 90°) in my birthchart between Mercury and Mars, which together fuse at various times to become the fury of the wind, the cunning of the wolf and the visionary perspective of the raven, or a propensity for turning conventional wisdom on its head in imitation of Odin hanging upside down on the World Tree Yggdrasil in order to "channel" the runic language. The fact that this identification is meaningful to me will mean that often the world – which is neutral in its essence as an embodiment of the One – will appear in what we might call an Odinic Light, and require of me what we might call an Odinic response. It also means that my pathway from the terror of the Realm of Zero to the possibility of becoming which is realized in the Realm of One will entail taking various steps of growth in consciousness along what we might call an Odinic path.

In *The Well of Remembrance*, Ralph Metzner – who also admits to being "seized" by this god – describes Odin as "the truth-seeking wanderer, the vision quester, or questioner, who wandered through many worlds seeking knowledge and wisdom" (9). Upon at least one occasion, Odin paid for his knowledge when he exchanged an eye for the right to drink from the Well of Remembrance. As a consequence, Odin has been associated with the concept of shamanic self-sacrifice for admission to this knowledge. Through another shamanic ordeal – nine days of hanging upside down on the world tree, Yggdrasil, Odin "became a master of the power of speech (*Sprachgewalt*). . . . Speech is connected to breath, and breath manifests in the outer world as the movement of air, the winds. . . . Odin, like Hermes, was a classical example of a wind god, moving swiftly and mysteriously, at some times tempestuous and dangerous, at others benevolent and inspiring with his wisdom and enthusiasm" (115).

My own journey – documented at some length in *Tracking the Soul* (Landwehr 230-240) – began with an initiation on a pile of ashes in my grandfather's back yard. I had those ashes stuffed into my mouth by the neighborhood children, and stopped breathing. I was rushed to the hospital, contracted pneumonia, and learned to mistrust the world and shun winning. I call this an initiation, because in retrospect, the process of healing my lungs and the distorted belief system I developed in the wake of this incident, I have learned most of what I now feel compelled to share in these books that I am writing.

To restore my capacity to breathe, in the wake of the ashpile incident, I learned to play the saxophone, a wind instrument wielded by a strong lineage of Odinic musicians – Charlie Parker, John Coltrane, Sonny Rollins, etc. – each with their own tale of wandering in search of a wisdom to tell beyond words. Later, I joined a yogic *ashram* where *pranayama* or breath control and chanting were the basis for spiritual practice. Throughout my life, I have been drawn to mountains and canyons, where the wind is known to live. On one such trip, nearly 20 years ago now, in the harsh mountain deserts of New Mexico, seeking insight about the further healing of my lungs, I had what I would consider an Odinic experience (Landwehr *Hiding From the Wind* 222-223)

> I climbed to the top of a fully exposed red rock perch, and sat. I did not know what to expect, why I had come, or what I was doing here. I didn't have to know. The wind knew many things I had not yet found the courage to face. The wind blew straight to that hole inside of me where I hid, and whispered the truth about my fears. I glanced to the bottom of the sheer 200-foot drop that hung 2 feet to my left, and I felt fear. No question about it. Fear of death, of being killed by something that I could not see, of being swallowed alive by the unknown, and spat out into the dust, the semi-

digestible cud of some sacrificial appetizer. . . . Then came the tears, and along with a strange sensation of heat in my stomach, anger. I screamed into the faceless howl of my adversary/teacher. "Fake rage," the wind taunted. I screamed louder. I grew hoarse. The wind was right. What after all was I trying to prove? And to whom?

I moved to a more sheltered place behind the rocks I sat upon, but still open to the sky. After awhile, the wind fell quiet. I took off my clothes and allowed the warmth of the sun to bathe my tense and tired body. I drifted off into reverie, the faint flapping of wind in my ear. I awoke to a blast of sand chipping around the corner of my sanctuary. The wind had found me again.

I hurriedly dressed, and went back down the mountain to rethink my strategy. I returned to base camp to find my tent half down, and grew really angry. "You fucker," I screamed. "I've had just about enough of you for one day." In mock retreat, the wind died down long enough for me to secure my tent with rocks on each stake. I sat in my truck to eat some lunch, glad to be out of the wind completely, but feeling angry and trapped. My momentary respite was no triumph.

Then it dawned on me. I set out again, this time for a humbler battle station, the ridge on which I had done my vision quest ten months ago, a home of sorts, a place entirely etched in the photo gallery of my heart. I sat on a rock facing into the bowl of the box canyon, the red rock temple to my left. I took out my saxophone and played. I played the wind at its own game. When the wind rose, so did I, trilling shrilly in the upper register, squawking in my anger, rumbling through the bass notes of my fear, waxing melodic when the wind took a breath, taking off on solo flights when the wind left me alone, playing occasional duets with the birds that dared to peep in the teeth of danger. For the moment, caught in the fantasy of my own windsong, the wind ceased to be threatening. For the moment, breath for breath, the wind and I soared as one. On the way back to camp, the wind blew my hat off, but it was only the backslap of a foe who had already lost the war.

This may seem an odd way to relate to a god, at least within the context of traditional religious conditioning. But following in the footsteps of a god is different than worshipping one, in much the same way that an apprentice is different than a customer. The customer supplicates, pays the required fee, and eventually gets what she wants. The apprentice invites the god to challenge her, rises to the occasion, often against great inner doubt, then gradually evolves to meet the god as an equal. On such a path, the apprentice does not always get what she wants, but rather what she needs. The next step on the path rarely makes sense to the rational mind, nor does taking it often feel like a sensible thing to do, but there is always learning to be found that will bypass the mind and register directly in the belly and the bones.

CHAPTER ONE

<><><><><><><><><><><><><><><><><><><><><><><><><><><><><><><><><><><><><><><><><><><><><><><><><><><><>

In the midst of writing this chapter, my lungs started bothering me again. Over the course of my 57 years since the ashpile incident, I have learned to take this recurring health problem as a signal that Odin is knocking on my door, letting me know it is time for another lesson. Aside from the usual visits to my health practitioner of choice, I'd been asking the question, "What would Odin do?"

The answer came, with very little effort on my part, in the form of an invitation to travel with a friend of mine back to the southwest. Our destination: Havasu Falls in the Grand Canyon and then Chaco Canyon, home of the ruins of ancient Anasazi culture. I had no doubt that an Odinic adventure awaited me.

All I could really do in response to this invitation was to follow in the footsteps of my chosen god, and remain awake for opportunities to co-create this magnificent world in which I live. To the extent that each of us does this, the Abyss reveals itself to be but the veil between No-thing and becoming. I would be hopelessly naïve if I didn't recognize history as an endless procession of sagas in the footsteps of false gods, followed to disastrous dead ends still very much within the Abyss. Still I take hope in knowing that history is also a *neti neti* trail of trial and error, on a timescale I cannot easily fathom, with something essential that cannot be denied as our ultimate destination, and the option to choose anew at each step of the way. As history marches on, I continue to follow my own chosen path of controlled folly from Zero to One as consciously as possible.

Endnotes

1. In this book, numbers will be designated in two ways. When speaking of them in the archetypal sense, they will be written out and capitalized - i.e. One, Two, Three, etc. When referring to arithmological calculations or functions - i.e. 3 x 1 = 3, the number 7 as a prime number - or to numerical figures - i.e. 25,000,000 people dead in the influenza epidemic of 1918 - they will be notated numerically.

Chapter Two: Into Soul Space

Within the Realm of Two, life is a field of resonant interplay, in which we are attracted to various people, places, things, ideas, and experiences. This resonant attraction generally occurs in one of three ways. By affinity, we are attracted to that which is like us in some way. By contrast, we are attracted to that which is unlike us in some way, so that we may come into greater balance with ourselves. Wherever we have issues to address, we are attracted – usually unconsciously – to that which holds the potential for triggering our core wounds, so that we may work on them, and learn whatever lessons are harbored there[1].

That resonance at the heart of the Realm of Two was recognized by the Pythagorean arithmologists, when they called the Dyad, "Erato, for having attracted through love, the advance of the monad as form, it generates the rest of the results, starting with the triad and tetrad" (Waterfield 46). Resonance is made possible within the Realm of Two through the relationship that now exists between Monad and Dyad.

The number Two was "opposed and contrary to the monad beyond all other numerical terms (as matter is contrary to God, or body to incorporeality) . . ." (Waterfield 41). According to this principle of opposition, Pythagoreans considered the Monad representative of the male principle, while the Dyad was identified with the female principle (Opsopaus). In a similar way, the One became associated with the Sun, while the Dyad was considered to be of the Moon (Waterfield 47)[2]. The Realm of Two was predicated on the existence of polar opposites – the field of resonance in which all relationships take place. The concept of relationship itself only emerges in the Realm of Two, as male and female, Sun and Moon, light and darkness, and every other pair of opposites, now interact with each other.

Within the Realm of Two, the Pythagoreans identified a number of basic polarities, enumerated in a Table of Opposites, preserved by Aristotle in his *Metaphysics* (qtd in Guthrie 23):

Limited	Unlimited
Odd	Even
One	Plurality
Right	Left
Male	Female
At rest	Moving
Straight	Crooked
Light	Dark
Good	Bad
Square	Oblong

◇◇

Within the Realm of One, these contrasts do not exist, because in the One all opposites are reconciled. The One is both odd and even, male and female, light and dark fused into an integrated Whole. Within the Realm of Two, however, the Monad is understood to possess certain properties in contrast with the Dyad. These differences constitute a resonant field in which becoming is furthered through contrast. Resonance by affinity generally takes place within the Realm of One, where the gods we project are created in our likeness. In the Realm of Two, that to which we are attracted is essentially the Other – everything we are not, or more accurately, believe we are not.

Implied in this understanding of Other is a separation from the One, which paradoxically makes an awareness of the One possible. In the Realm of One, the One is All That Is, and we are as indistinguishable and inseparable from one another as drops of water within the ocean. It is only within the Realm of Two that we become aware of our individuality, and eventually – as we exercise this new faculty of consciousness – of our relationship to the One. Within the Realm of One, the imaginative projection of our gods and goddesses was seen as an act of consciousness. The actual projection of any deity, however, is only possible from within the Realm of Two. The One needs the Dyad in order to perceive itself; and we need the Dyad in order to perceive the One.

Consciousness as a Play of Opposites

Consciousness evolves within the Realm of Two as the interplay of opposites. It is only through an encounter with darkness that I can know there is such a thing as light. It is only as male interacts with female that gender identification becomes possible[3]. It is only as I notice something "moving" that I know what "at rest" means; only as I encounter something "crooked" that the concept of "straight" is relevant; only as I become aware of the difference between one side of the Aristotelian Table of Opposites and the other that I become aware at all. Within the Ream of Two, consciousness arises through contrast and comparison.

This simple model of consciousness offers two possibilities for navigating the Realm of Two. If we treat each member of each polarity as a necessary and useful complement to the other, we have the possibility for creative synthesis, love, mutual respect, compassionate tolerance of differences, and a synergistic interaction in which the Whole becomes more than the sum of its parts. If, on the other hand, we treat one member of each polarity as superior to the other – e.g. light is better than dark; male is superior to female; good is good and bad is bad – then we have a recipe for conflict, for hatred, disrespect and cruel intolerance, and for an antagonistic interaction in which the Whole becomes compromised and fragmented.

In the last chapter, That Which Remains After All Else Has Been Eliminated – or Ain Soph, the essence of the One – was observed to be the power of choice. There is, however, nothing to choose until we get to the Realm of Two, where our hard-won prerogative is immediately tested. For choosing equality between the opposites is a recipe for becoming, which can be furthered within the Realm of Two. Choosing inequality is a sure ticket back to the Realm of Zero.

Although Two is honored as necessary to the evolutionary process throughout the Pythagorean literature, there is also a sense of inequality implied within the system. In theory, Pythagoras honored all the numbers as necessary to the process of creative evolution. At the same time, however, he saw the overriding spiritual goal of this life to be reunion with the One, and the process that led through the Dyad away from the One seemed contrary to that goal. In practice, the preferred strategy in pursuing reunion with the One was to adhere to the Monad and everything it represented in contrast to the Dyad. The Monad was considered by Pythagoreans to be good, while the Dyad was – if not exactly evil – at least fraught with ambivalence, and thus somewhat suspect.

Through Plato's adoption of Pythagorean sensibilities, this focus on the inequality between the Monad and the Dyad led to the establishment of dogmatic, monotheistic, judgmental religions that tended to treat the Dyad as "sinful," and punish it accordingly. Although the doctrinal sources and the terminology were very different in the East, it should not be assumed that this attitude of inequality in the Realm of Two is endemic to the West. In China, for example, both Confucianism and Taoism espouse the virtues of harmony between yin and yang. Yet in practice, Confucianism tends to stress the importance of a patriarchal (yang) moral code, while Taoists tend to prefer a yin approach to life modeled after water (Oxtoby *Eastern* 369). When any religion, Eastern or Western, favors one side of any polarity and considers the other sinful or inferior, this is a failure to become fully conscious within the Realm of Two.

Within the Realm of Two, becoming is a matter of intentionally embracing the Other in all its many guises, learning what we can from it, and assimilating all that is different from us. In this way, we find a place of balance within ourselves, a center amidst the extremes. To do this requires that we consider everything the Dyad has to offer without judgment. Judgment leads only to separation, separation leads to suspicion, suspicion leads to intolerance, intolerance leads to conflict, and conflict leads to chaos – which takes us back to where we started in the Realm of Zero. Unfortunately, such a mindset seems endemic to the Pythagoreans, who did not hold the Dyad in the same light of reverence that they reserved for the Monad. It also appears to be endemic to most religions on the planet today, vociferously promoting their beliefs as superior to those that exist in contrast to them.

The number Two is potentially a source of ambivalence within this tradition of inequality, and the catastrophic backward slide toward the Abyss it precipitates. The discussion that follows focuses mostly on the roots of ambivalence that stem from Pythagorean principles,

◇◇◇

with the understanding that ambivalence is intrinsic to the Realm of Two, and at play wherever inequality prevails.

Two As an Echo of the Abyss

Our first clue to this sense of ambivalence within Pythagorean tradition comes from a passage in *The Theology of Arithmetic* (Waterfield 41):

> . . . the Dyad is all but contrasted to the nature of God in the sense that it is considered to be the cause of things changing and altering, while God is the cause of sameness and unchanging stability.

This contrast between stability and change is worth noting, given the context of our recent movement from the Realm of Zero into the Realm of One. Here we are. We have just managed to escape the Abyss, as our gods and goddesses of choice bring order to chaos. And now we are about to enter a Realm where forces exist that threaten to reintroduce disorder. Such forces must be eyed with some suspicion. Once we have projected a Creator god and pledged our allegiance to it as our ticket out of the Abyss, anything Other – in contrast to the One – can understandably be perceived as a counterforce in opposition to our god.

Yet, as the Pythagoreans also recognized, the Creation is not possible without a progression from One to Two. There can be no multiplicity of separate forms – no trees, rocks, birds, humans, rain clouds, or stars – without a dissipation of unity into diversity. "Hence the first conjunction of monad and dyad results in the first finite plurality, the element of things, which would be a triangle of quantities and numbers, both corporeal and incorporeal" (Waterfield 41). Two is a necessary first step from One to all the other numbers, along a path on which the evolution of consciousness and creation proceed. For this reason, the Pythagoreans also described Two with more affirmative epithets such as "daring," "movement," "generation," "change," and "Rhea," the mother of the gods in Greek mythology (Waterfield 41, 46).

The first source of Pythagorean ambivalence toward the Two is that it was both a necessary adjunct to the One and a deviation from it. Since the ultimate goal of human existence was considered by the Pythagoreans to be reunion with the One, this first necessary step would appear to be in the wrong direction, posing a huge dilemma for the true disciple of the way. This dilemma was solved by the adoption of a strict moral code involving dietary restrictions, periods of fasting and silence, abstinence from sex, minimal possessions, etc. – the idea being that such a way of life would bring one through the Realm of Two with a certain spiritual purity born of adherence to divine virtues. This sensibility was passed on to Plato, who – in elaboration of Pythagorean teachings – began to construe the body as a prison and a corruption of Spirit, and developed his moral code accordingly. In turn, Plato's ideas compounded a moralistic tone of inequality already present in the monotheistic religions – Judaism, Christianty, and Islam – that were influenced by him.

These religions eventually took the Pythagorean ambivalence toward the Dyad to a feverish pitch, projecting it as the counterbalancing force of Satan, who opposed God in an endless battle for human souls. Similar rabid opposition between good and evil became the rallying cry not just for religions influenced by Plato, but also for Zoroastrianism in ancient Persia (contemporary Iran), the Jain tradition of India, and the Omoto cult in 20th century Japan (Oxtoby *Eastern* 416-418). This battle has subsequently played itself out on many stages, including the Spanish Inquisition of the 15th and 16th centuries, which tortured and killed countless Jews who had supposedly committed heresies against a Christian god; the systematic genocide in the 17th and 18th centuries of indigenous peoples throughout North and South America, who had the audacity to worship gods other than Jesus Christ; and the terrorist war of holy jihad waged today against the Christian West and reciprocated by them[4]. Many examples of One at war with Two have marred the history of the human race.

Two As the Embodiment of the Shadow

As Jungian mythologist Marie-Louise von Franz pointed out, the duality inherent in the Realm of Two goes deeper than religious preferences, however violently defended. The ambivalence at the root of these preferences is likely as hardwired into our psyche as the inclination to project our gods in the Realm of One. This possibility is most clearly demonstrated on an archetypal level in the Two-Creator motif, mentioned in Chapter One, in which creation proceeds as a process of cooperation or competition between two gods. As Von Franz describes this motif (69-70, words in brackets are mine):

> All over the world there is the tendency to ascribe the act of creation to one figure who then retires and stays outside, while another figure steps into the act of creation. It seems to me quite obvious that this is an archetypal motif which describes the separation of individual consciousness from its unconscious background, the two together representing the pre-conscious totality. . . . The tendency toward consciousness [Two] would be the son of the pre-conscious totality [One], but at the same time the two exist in a natural union, because this tendency toward consciousness stems from the whole totality. . . .

At first, the two gods are identical. There may be a contest or a battle between them as one tries to prove himself superior, but it invariably winds up in a draw. Still, the intent of such a contest is to differentiate one god from the other, and eventually this comes to pass.

◇◇◇

In Von Franz' estimation, there are seven dimensions of this differentiation, some or all of which may be present in any given Two-Creator myth:

1) One is more active; the other more passive.
2) One knows more; the other knows less.
3) One is more human; the other less human.
4) One is a father type; the other is a son type.
5) One is male; one is female.
6) One is good; one is evil.
7) One is more towards life; one is more towards death.

There is an echo here of Aristotle's Table of Opposites albeit with a different set of polarities. Within von Franz' delineation, there is an additional sensibility that sheds light on the true nature of Two, for everything on the right (associated with Two in contrast to One) can be considered an attribute of the unconscious, while everything to the left can be understood as an embodiment of consciousness. This dichotomy between conscious and unconscious does not exist when everything is pre-conscious (or potential only) as it is in the Realm of One. It is only when we get to the Realm of Two that the unconscious has any meaning at all. When the unconscious becomes a separate creative force in its own right, then we have counterbalance to the conscious mind – one that often serves to undermine the conscious agenda, generate ambivalence, and produce results in stark contrast to the conscious goal.

In Jungian terms, this counterbalancing force is known as the shadow. When the shadow is engaged on a personal level – often through some relationship in which unconscious patterns are mirrored by one's partner – the Realm of Two becomes the arena in which resonance by wounding takes place. Resonance by wounding occurs within the Realm of Two as the necessity for dealing with the shadow. Whole societies also have their shadow, which is generally projected onto an enemy in times of war, and more generally onto various targets epitomizing evil – i.e., drugs, pornography, or terrorism. When the shadow is projected in the broadest possible sense on the collective level – as an embodiment of everything the Monad is not – it gives rise not to a Creator god, but to his arch nemesis.

This arch nemesis is usually a trickster god – depicted in indigenous cultures by an animal such as raven, coyote, hare, or spider, but also in more sophisticated anthropomorphic form in Greek, Chinese, Japanese and Hebraic traditions (Radin xxiii-xxiv). Although one might reasonably expect Satan to fall into this category, the Christian adoption of a shadow god is actually more complicated. Though mentioned in the Old Testament, Satan didn't really come into prominence as a counterforce to God until the Hellenistic period (c 200 BCE – 150 CE) (Lehmann & Myers 192). Before that, Yahweh himself at times became a trickster. According to Jung (Radin 196):

If we consider . . . the daemonic features exhibited by Yahweh in the Old Testament, we shall find in them not a few reminders of the unpredictable behavior of the trickster, of his pointless orgies of destruction and his self-appointed sufferings, together with the same gradual development into a saviour and his simultaneous humanization.

It was only when Yahweh was elevated in Hebrew thought to the status of a Creator god that Satan arose as a projection of Yahweh's trickster function. As religious scholar Willard Oxtoby wonders (*Western* 25), "Did a God now responsible for everything that happens need an adversary to whom people could assign the blame for what goes wrong?" If so, then typical of a trickster god, Satan becomes a scapegoat for everything within the resonant field of opposites that can't be comfortably contained in our concept of God.

As Jung tells us (*Archetypes* 270, 264):

The trickster is a collective shadow figure, a summation of all the inferior traits of character in individuals. . . . The trickster is inherently ambivalent, because it represents a primitive 'cosmic' being of divine-animal nature, on the one hand superior to man because of his superhuman qualities, and on the other hand inferior to him because of his unreason and unconsciousness.

We are simultaneously attracted to the trickster for what he has to teach us, and repelled by the mirror he holds up to our basest, most embarrassing incongruities. To fully enter the Realm of Two by embracing the path of equality, we are immediately challenged to step through the trickster's minefield of shadows, where every virtue we would claim is counterbalanced by a correspondent vice. The temptation is to keep the virtue and project the vice, but doing so immediately sets us on the path of inequality and back down into the Abyss. To step beyond this minefield onto the path of equality, we must own the shadow and find the secret light within it that dissolves the false barrier between virtue and vice altogether.

Two as the Gateway to the Fall

This predicament is depicted in the story of Adam and Eve in the Garden of Eden. In this myth, the serpent – a symbol for Mercurius, or Hermes, the Greek trickster god (Jung, *Archetypes* 312) – tempts the first humans with fruit from the Tree of the Knowledge of Good and Evil. Before this moment, there was no such thing as good and evil; only a neutral equality between everything on the left side of the Table of Opposites and everything on the right side. Tasting this fruit, everything shifted, and suddenly our primal ancestors found themselves on the path of inequality, where guilt and shame accompanied choices on

<><><><><><><><><><><><><><><><><><><><><><><><><><><><><><><><><><><><><><><><><>

the right side (dyadic choices), labor became necessary to dwell within creation, and birth became associated with pain and suffering.

Christians have interpreted this myth essentially to mean that entry into the Realm of Two was the source of original sin – that is to say, a fall from the natural state of grace that existed when the soul identified completely with the One. A more careful reading of the myth shows that the real fall is the embrace of the path of inequality, on which certain things, behaviors, and ideas were now considered good and others were construed as evil. Before the knowledge of good and evil, such a distinction would not be possible. With this knowledge, half of creation is suddenly cast in shadow. The Pythagorean solution – central to the religious agenda – is that in order to return to God, one must move out of shadow and back into light. Ignored in this injunction is the deeper recognition that shadow and light exist only relation to each other, and that to embrace the Wholeness that is God, one must reject neither.

Striking out in what first appears to be a different direction, Richard Heath argues that (6):

> The fruit that hangs on this tree is knowledge of the numerical relationships between planets that both initiates an understanding of number and reveals that creation is indeed structured in a numerical way. The numerical culture would eventually measure the size of Earth itself and "become as one of us (the gods) to know good and evil." But when humans become knowledgeable of the cosmic, it angers a God that would keep them within a created paradise or, alternatively, forces a newly self-conscious couple to leave the garden paradise and make their own way in the world.

Implied in Heath's statement is recognition of the awesome power of number – the knowledge of which – potentially gives us mere mortals, the power of the gods. Setting aside for the moment the metaphysical implications of this knowledge, it is not hard to see how numbers – the root of all mathematics, and thus the language in which science expresses itself – have given the human race the power over life (e.g., life-saving medicine and the ability to clone other living things) and death (e.g., weapons of mass destruction and toxic chemicals). Applied use of numbers has also given us a magical Pandora's box of inventions that allow us to do everything the gods can do – e.g., defy gravity (with airplanes and space shuttles), ignore the limitations of space (through the Internet and wireless phone technology), and suspend time (using recording devices such as cameras and camcorders, digital recorders and hard drives). All of these inventions – ultimately made possible only through our knowledge of numbers – have given us the power of the gods, or at least the semblance of their power.

The question remains, what is good or evil about this knowledge? It is not the knowledge itself that is good or evil, but the use to which we put the knowledge. To the extent this knowledge of number – of the archetypal power underlying creation – is pressed into service to promote equality between the opposites, a certain creative synergy takes place that ultimately furthers the evolutionary process of becoming. To the extent that this knowledge is used to promote inequality between the opposites, then we have a devolutionary force that undoes creation through war, genocide, escalation of microbial and viral diseases with which we are also at war, and in general, a literal tearing of creation in two. The technology that arises from number is neutral. It is the intent behind our use of technology – and more fundamentally our attitude toward the Dyad out of which subsequent numbers spring – that determines whether our knowledge of number can be considered good or evil.

If our goal in moving through the numerical sequence – from potentiality in the One into a Creation blossoming in multiplicity – is to evolve as conscious beings living in a garden paradise, then using our knowledge of numbers to further this purpose along a path of equality might be considered good. Using this same knowledge to promote devolution through inequality might be considered evil. Ultimately, even this dichotomy is false, since devolution serves a legitimate purpose in the overall scheme of things, and is not more or less vital to that purpose than evolution. Once we have this knowledge, however, we must make a choice about how we will use it, and there will be a consequence to this choice. Put another way, consciousness entails responsibility, and to ignore this responsibility or to refuse it is to fall from grace.

Two as the Embodiment of Feminine Power

As suggested by the story of the Garden of Eden, the gift of consciousness and the necessity for responsible choice also require us to leave the Garden and enter the world. If the Garden can be understood to be a state of unconscious identification with the One, then the world must be understood as the Realm of Two, where navigation is essentially an exploration of the relationship between opposites – or the resonant field I refer to in *The Seven Gates of Soul* as soul space. The embodied soul – the unique fusion of Spirit and matter that makes us what we are – does not truly exist until the One leaves its state of pure potentiality and takes on a body, or a plenitude of forms within the embodied world.

Entering this dyadic realm of form was an additional source of ambivalence for the Pythagoreans, because it entails embracing a distinctly Feminine way of being. The Dyad is Feminine because in necessitates entry into matter (from the Latin, *mater*, meaning mother), changeability (associated with the Feminine Moon moving through its phases), and plurality (through the Feminine process of giving birth).

The Monad, considered by the Pythagoreans to be an essentially Masculine force, contains all potentiality within itself, and is essentially self-contained. To enter the world of

◇◇

form, not as a potential, but as an actuality, a birth is necessary. This comes only through the Feminine. In a literal way, it is through the womb of the mother that we enter this life[5]. In a more cosmic sense, it has always been the archetypal Feminine that represents the power to give birth, the power to bring Spirit down into a life conditioned by form, matter, time and space – into existence as an embodied soul.

The ambivalence that the Pythagoreans projected toward the Dyad – through which the manifest creation came into being – also extended to the Feminine. This ambivalence was shared by others – the Jews and Christians, in whose primal myth of creation, Eve was the temptress offering the apple to Adam; the early Greeks, for whom Pandora's box was the source of all the suffering in the world; and in many indigenous and shamanic cultures, where menstruation is viewed as a source of ritual contamination (Lehmann 296-297).

Without embodiment, no evolution in consciousness is possible. Yet within a body, different rules apply than do within the domain of pure potential. We are no longer free to dwell unconsciously in Light; we must now grope our way through darkness. The movement through darkness is not up – where we project our gods in some transcendent heavenly abode – but down into the thick tangle of root and blood, bone and rock, where progress depends not upon adherence to the purity of some ideal, but upon the willingness to sweat and moan, suffer and cry out from the depths of being, touch the Earth and learn from everything that touches us. This movement downward shatters our self-containment, fragments our internal unity, and forces us to reclaim it through relationship with everything foreign to our nature.

Such a path made the Pythagoreans uncomfortable, because it defied the logic – the Logos – of all that they revered. Reunion with the One was their goal, but the path to that goal was apparently down into a darkness populated by an overwhelming messy diversity of seemingly separate forms. Even though they called the Dyad, "the Goal, because She is that to which the Monad proceeds" (Opsopaus, Part II), it made no rational sense because it was to the Monad that they felt themselves returning.

As deep ecologist and feminist scholar Joan Halifax reminds us, this downward Feminine movement that gives birth to the life of the soul has it own logic, its own purpose that can only be understood through the experience of it (137):

> *We go into darkness, we seek initiation, in order to know directly how the roots of all beings are tied together: how we are related to all things, how this relationship expresses itself in terms of interdependence, and finally how all phenomena abide within one another. Yes, the roots of all living things are tied together. Deep in the ground of being, they tangle and embrace. This understanding is expressed in the term non-duality. If we look deeply, we find that we do not have a separate self-*

identity, a self that does not include sun and wind, earth and water, creatures and plants, and one another. We cannot exist without the presence and support of the interconnecting circles of creation – the geosphere, the biosphere, the hydrosphere, the atmosphere, and the sphere of our sun. All are related to us: we depend on each of these spheres for our very existence.

This interdependence is what allows us to exist as souls – splinters of the fragmented One – within seemingly separate bodies. The separation – feared by Pythagoreans as a necessary evil in moving evolution forward – is as illusory as the absolute authority of the reference point from which we project our gods and goddesses in the Realm of One. For as we consent to participate in the opportunity that the Realm of Two offers us – the formation and exploration of relationships – we learn that we are not separate. We share the same human spirit as those who appear radically different than us. We belong to the same biosphere as every other living creature on this planet. Our bodies are composed of the same elements as mountains, oceans and stars. We are a unique mix of ingredients, occupying a singular niche in the cosmic food chain, and serving an individual purpose that we can call our own. Yet, ours is not a solitary dance, and within the Realm of Two, we get to explore the larger choreography.

This is essentially a Feminine exploration, because women are biologically conditioned not just to give birth, but also to foster life. Relationship is the vehicle for this task, which begins with the sacred relationship between a mother and her breast-feeding child – an act that is the epitome of physical nourishment, emotional bonding, and psychological security. To contain the mother-child relationship, family provides a larger network of support; family extends to community; and community expands to include all living things and the ecological interdependence that sustains them in an all-inclusive network of Life.

Within the Realm of Two, it is the Feminine embrace of interdependence that allows becoming to proceed. This is no longer just a solitary becoming, although it is that – since we are still growing as individuals within the context of our relationships. It is also a collective becoming, since interdependence by definition is a quality of the many. While our chosen gods and goddesses provide a role model as we seek to evolve as individual souls within the Realm of One, the relationships between them that are part of any polytheistic pantheon, provide the role model for our interactions with each other within the Realm of Two.

The monotheistic religions and those that adhere to a dualistic framework pitting good against evil, fail to provide this more pluralistic role model, since the rich diversity of polytheism is subsumed by the Monad against which the Dyad is considered inferior or sinful. In keeping with Pythagorean ambivalence toward the Dyad, the Feminine expression of divine power survives only as a virgin birth that is incidental to the main attraction, or

◇◇

a consort goddess who plays a secondary supportive role to some male deity, or a goddess cult that remains a marginalized echo of what it once was in its matriarchal heyday. This is changing slowly, though most denominations still do not ordain women priests, God is still predominantly addressed as He, and patriarchal attitudes still limit the power and authority of women in the Church. Though Eastern religions have, in general, been more accommodating of female deities, the actual treatment of women in the religious societies of the East has paralleled or lagged behind that of the West.

Our Collective Need for the Balancing Medicine of the Feminine

Aside from the politics of inequality, in which women are treated unfairly, we pay a steep price spiritually for our ambivalence toward the Feminine. This temporary life we live as spirits housed in bodies is a dance of balance between the Masculine and the Feminine – each of which is necessary if the dance is not to be a blind stumble back into the Abyss. The Masculine is not the sole prerogative of men, nor is the Feminine a possession only of women. Quite the contrary, life requires us to embrace and integrate both, regardless of our physical gender. This integration – called the *hieros gamos* by Jung – is a necessary dimension of the quest for Wholeness, which is another word for the Pythagorean desire for a Return to the One. The Return does not happen, except through *hieros gamos* – through an integration of the Masculine and the Feminine. Given the continuing dominance of the Masculine within contemporary society, *hieros gamos* necessarily means making a larger space for a re-integration of Feminine values.

Re-integration of the Feminine does not just mean an affirmation of women's rights, nor the return to a matriarchal society. It requires a much more fundamental shift in consciousness than that, toward a culture of true equality, in which an awareness that the "*the roots of all living things are tied together*" (Halifax 137) serves as the basis for every decision we make – as a culture, as a political body, as a global society. Within the Realm of Two, we must find a more humble place within the natural order, where our presence contributes not just to our own well-being, but to the well-being of every other life form with whom we share this planet.

The Masculine seeks to dominate, and according to this rule, we have assumed dominion over the Earth – forgetting how much more durable the Earth is than we are. We cavalierly raze forests that are older than us, older than the most ancient ancestor we could possible claim relation to. We blast through mountains that pre-date the human species as though they were put there just to challenge our technological prowess. Harvard biologist Edward O. Wilson estimates that by the year 2100, half of all plant and animal species on the planet will be gone (Whitty 38). Global warming; toxic chemicals in our air, food and water; the litany goes on and on. All in a day's work for the unbridled Masculine.

Care of the Feminine requires us to realize that whatever damage we do to the Earth we do to ourselves. Her roots and ours are inextricably bound. In the end, it is we who will suffer, our children and our children's children. Future generations will live in a world that no longer quite so easily supports or sustains human life. Some may die prematurely. Many will become sick, with weakened immune systems, diminished capacity to procreate, no air to breathe that is not contaminated, no food to eat that is not toxic, no water to drink that has not been polluted. It is already starting to happen.

The only antidote to such a future is to create a balanced culture in which both side of the polarity are honored – the Masculine drive toward progress and the Feminine capacity for caring about how progress affects our culture, our children, and the larger web of life on which we depend for our survival and quality of life. At various times in our history, we have reached for equality, though within the last 25 years or so, we have gotten severely out of balance. We seem to care less as a culture than we did a short time ago. In this age of austere budget cuts for social welfare programs counterbalanced by bloated military spending, caring itself is often seen as a sign of weakness – as an antiquated vestige of liberalism. This is a dishonoring of the Feminine, and women in power are no less at fault than men. What must change is not so much the percentage of women who rise to the top, but the culture in which the opportunity to rise is given. This requires a massive collective shift in consciousness, within the Realm of Two, from an attitude of inequality to one of equality.

The Relationship Between the Realm of Two and the Second Chakra

In *Tracking the Soul* I consider the second and fourth *chakras* as Feminine states of consciousness in which the interconnectedness of all life is honored, relationship is emphasized, and a caring life is the epitome of conscious choice. In the second *chakra* (see p. xxiv), we discover what gives us pleasure and pay attention to the quality of life – to what Thomas Moore calls "care of the soul." This approach to life is distinctly Feminine, because it requires us to become receptive to all that the world has to offer – its beauty, its pleasure, the miracle of its very existence. In an imbalanced world where the Masculine rules, all of this is taken for granted, and lost in the rush toward bigger, better, and more. Yet as Moore reminds us (286):

> *The emptiness that many people complain dominates their lives comes in part from a failure to let the world in, to perceive it and engage it fully. Naturally, we'll feel empty if everything we do slides past without sticking. . . . Soul cannot thrive in a fast-paced life because being affected, taking things in and chewing on them, requires time. Living artfully therefore, might require something as simple as pausing.*

CHAPTER TWO

◇◇

In the second *chakra*, we learn to live artfully – to pause – so that we can discover what is true for us on the most basic level. We become receptive, learn to feel and listen and intuit our way into a relationship to life that is sensual, emotional, and filled with resonance. We develop conscious preferences – hot rather than cold, soft rather than hard, slow rather than fast, thick rather than thin – and create a life of quality that encompasses these preferences. At first blush, this may seem like a path of inequality, but such preferences are not formed without experiencing their opposite, and not indulged without the necessity for moderation. When preference becomes addictive or obsessive or rigidly compulsive, an imbalance born of excess will create resonance by wounding that leads back to resonance by contrast. The preferences and pleasures cultivated in the second *chakra* are a self-regulating mechanism because too much of anything will lead to pain and require its opposite as an antidote.

Coming to full awareness in the second *chakra* means first becoming aware of our preferences – which are often more a matter of unconscious conditioning than choice – experiencing contrast to our preferences, and then rising to a higher level of flexibility with regard to our choices. Ultimately, in a fully awakened second *chakra*, everything can be appreciated for what it is, and we are able to respond creatively to each moment from a broad repertoire of optional behaviors. Life becomes rich and full, because we are free to be who we are and to enjoy a smorgasbord of choices within a resonant field that is itself enriched by diversity, the encouragement of spontaneity, and freedom of choice. On the way to such freedom, we are inevitably challenged to move beyond our comfort zone and embrace the unfamiliar.

While, life itself will see to it that we are constantly exposed to new experiences, this is also often an intentional teaching strategy adopted by spiritual teachers who fit the trickster mold. As Prier Winkle writes in an insightful article about such teachers (15-16), the Russian mystic Gurdjieff presents a prime example:

> His invariable method of teaching (interpreted by his expounders and apologists as designed to break down restricting habits and habitual attitudes, and so to allow the pupil to 'wake') was to set each one doing some task to which he was not accustomed. Men who had led sedentary lives or worked at typewriters were set to carting loads of stones from one end of a garden to the other, or to digging large holes with primitive shovels, while others who had led an active life were made to sit still. Society women were made to wash dishes and to share rooms with others to deprive them of privacy. . . . Later on, when he no longer taught in an organized fashion at his own Institute, the same pattern appeared instead at his teaching dinner parties. These were held in rooms which could hold twenty people or so at a pinch into which from forty to sixty were crammed. Special morsels were handed out by Gurdjieff to particular guests, meat to vegetarians, fiendishly hot spices to the unprepared, vodka to those unaccustomed to alcohol, etc.

⟡⟡

Those who survived this unorthodox forced exposure to everything foreign, invariably came to appreciate that their preferences were largely a matter of conditioned response, and that they could, in fact, choose again. The whole point of the exercise, seen in its most enlightened framework, was to reduce all relative truths to the power of Ain Soph – the power to choose anew in each moment – and most importantly within the Realm of Two from a place of non-judgment with regard to choices. From this perspective, working at a typewriter is not better or worse than carting stones; washing dishes is not an inferior experience to being served with them; abstaining from meat or alcohol is not morally preferable to indulging them. It is the power of this realization – and of non-judgmental choice – that marks the primary evolutionary step onto the path of equality within the second *chakra*.

When this step is taken collectively – if we ever do take it – society as a whole will be profoundly transformed by tolerance, a global willingness to learn from our differences, and co-creative synergy that spans all possible divides. As a culture, a balanced second *chakra* means making broad allowances for individual differences – ethnic diversity, local culture, societal encouragement for the development of unique individual styles. It means educating people to make intelligent choices, instead of indoctrinating them to adhere to imposed standards. It means celebrating the precious individuality of each soul, in the knowledge that we all have our essential parts to play in a whole that would suffer without our conscious, creative individuality.

In a world where the second *chakra* is feared, commercialized monoculture predominates. Every town looks and feels the same, with all the same national chains, local radio stations that pump out the same twenty hits, local papers that subscribe to the same wire services, food grown by the same factory farms, packaged by the same giant corporations, and shipped from the same distribution centers. Individuality is a myth sold on television as a ploy to create brand allegiance, while true individuality is eyed with suspicion and often punished. The kind of quality that comes from slow, careful attention to the details of everyday life holds no currency unless it can be sold for consumption. Conspicuous wealth is held forth as the gold standard to which everyone aspires, yet true contentment and satisfaction with the quality of life is as rare as gold.

Within the model of spiritual development posed by the *chakra* system, true individuality arises within the third *chakra*. This is also typically where we vie with other individuals for territory – for our sense of place in the world, and for the space to demonstrate our competence, develop our talents and skills and make our contribution to the whole. An imbalanced society, traveling the path of inequality through embrace of the Masculine at the expense of the Feminine, promotes a competitive atmosphere within the third *chakra*, without encouraging those who compete to appreciate the fact that we are also bound together as parts of the same whole. To remember this is to understand that cooperation among truly individuated souls

◇◇

facilitates a higher level of functionality than competition. Before this becomes possible, we must – as a society – learn to embrace both the discovery of true individuality and the broad tolerance for individual differences that arises within a healthy second *chakra*.

The Relationship Between the Realm of Two and the Fourth Chakra

In the fourth *chakra* (see p. xxiv), we cultivate relationships with others based on shared affinities, the potential for balance, and unconscious contracts to work together toward re-triggering and healing of dovetailing issues. On a personal level, our most important relationships contain all three elements, and provide ample ongoing opportunities for growth through sharing what we enjoy, discovering new pleasures in contrast to our preferences, and revisiting old wounds that are opened again by those we love. On a collective level, these same relationships contain the potential for improving the overall quality of life for everyone, learning from each other, and healing the longstanding wounds that pit brother against brother, brother against sister, rich against poor, black against white, and human against the other species with whom we share this planet.

The healing of relationships is an embrace of the Feminine at the level of the fourth *chakra*. From this healing comes the cooperative atmosphere of co-creativity that we suggested was possible at the level of the second *chakra*. It also gives us the power to address such monumental collective issues as the long-term damaging effects of war, genocide, species extinctions, global warming, displaced refugee populations, widespread hunger, infectious pandemic diseases, economic inequality, and essentially every other issue plaguing us today. Until we open our hearts to a compassionate embrace of what the Buddhists call right relationship with everything that is, we lack the spiritual will to fully acknowledge the existence of these problems, much less address them.

In the West, we are largely ineffective in addressing global problems, because we are conditioned by our capitalist heritage to look for personal advantage in every situation. This is true for individuals and it is true for nations. We do not think of ourselves as part of an interconnected web of life; we think of ourselves as independent agents vying with other independent agents for limited resources. From a spiritual perspective, this is the third *chakra* run amok, encouraged by a competitive system that rewards those who kick and claw their way to the top of the pile, trampling those beneath them with impunity. From within the context of our present discussion, this is the path of inequality through the Realm of Two, entrenched and institutionalized as an expression of the imbalanced denial of the Feminine. Until we care as much about each other as we do about ourselves, this will not change.

⬦⬦

Until we make a collective shift as a culture from preoccupation with the third *chakra* – where competitive advantage reigns – to an integration of the third and fourth *chakra* – where compassion fosters true cooperation – we will continue to function from a win-lose paradigm that gets us nowhere. One person's misfortune will be another's opportunity; one person's gain will be another's loss; every triumph of the human spirit will be borne on the backs of those who suffer because of it. While certain individuals will prosper under such a system – indeed the history of what we call progress in this world is the story of those who have – collectively, there will continue to be no net gain. Until we collectively integrate the third *chakra* and the fourth in the Realm of Two, our world will be one in which inequality remains a source of intense and traumatic suffering for us all.

Beloved American mystic Ram Dass puts it quite simply (223):

> *Common to all those habits which hinder us is a sense of separateness; we are divided within ourselves and cut off from others. Common to all those moments and actions which truly seem to help, however, is the experience of unity; the mind and heart work in harmony, and barriers between us dissolve.*

> *Separateness and unity. How interesting that these root causes, revealed in the experience of helping, turn out to be what most spiritual traditions define as the fundamental issue of life itself. Awakening from our sense of separateness is what we are called to do in all things....*

It is in the fourth *chakra* that we awaken to a sense of unity of heart and mind, and of compassionate interconnectedness. This realization is taken to its ultimate in the Buddhist vow of the *bodhisattva*, in which the decision is made to postpone one's own enlightenment in order to work toward the enlightenment of all sentient beings. This does not mean proselytizing others, or converting them to a particular vision of the truth. It simply means caring for them as you would care for yourself, because in truth, you *are* caring for your Self. This act of caring from a place of interconnectedness is what transforms the Realm of Two from a war zone into a cooperative eco-system existing in peace and harmony, where true evolution is possible.

Caring harnesses everything we have achieved in the third *chakra* – all our personal competencies, our talents, our skills, our capacity for contribution – to a larger vision of service to the whole. It is this vision, born of an integration of the third and fourth *chakras* in the Realm of Two that enables the process of becoming to proceed as an integrated, cooperative venture on the path of equality. Through this venture and only through this venture can we begin to alleviate the suffering that we impose upon ourselves and each other, as we seek to distance ourselves from the Abyss.

CHAPTER TWO

Endnotes

1. I introduce the concept of resonance in *The Seven Gates* (Landwehr 193-197), and elaborate it to encompass three kinds of resonance in *Tracking the Soul* (Landwehr 57-59). I refer the reader to these earlier books for a more complete discussion of these ideas.

2. Heath associates the Moon with the number Three for reasons that I will explore in Part Two.

3. This is true, even in gay and lesbian relationships, since it is male and female principles we are talking about here – yin and yang – and not merely body gender. In order for attraction to exist at all between two people, there must be some polarity within the relationship.

4. This statement is not meant to minimize the role that other factors – such as control of oil – plays in Middle Eastern tensions. Whether religious differences are merely the justification for a more complex set of variables contributing to war, or the central focus of the war, they fuel a strong sense of separation and dichotomy in the Realm of Two, which ultimately promotes a path of inequality through it.

5. Even though patriarchal medicine has attempted to bypass the womb through techniques of *in vitro* fertilization, the eggs that are the carriers of life, are still unavoidably of the Feminine.

Chapter Three: Into Conscious Power

If the Pythagoreans were ambivalent about the Dyad, their faith in the divine order of things, evolving from the Monad, was restored when they got to the number Three. Says Iamblichus (Waterfield 49):

> The triad has a special beauty and fairness beyond all numbers, primarily because it is the very first to make actual the potentialities of the monad – oddness, perfection, proportionality, unification, limit.

This appreciation is further elaborated in the same text by Nicomachus, who continues to extol the sense of completion it brings to the promise of the One (Waterfield 51):

> The triad is the source of actuality in number, which is by definition a system of monads. . . . The triad is the first to be a system, of monad and dyad. . . . It is also the very first which admits of end, middle and beginning, which are the causes of all completion and perfection being attained.

It is almost as though the resolve inherent in the intentions of the One, tested by movement through the Realm of the Dyad, reasserts itself with Three, but now no longer as a mere potentiality. Having met and clothed itself in matter in the Realm of Two, Three becomes real in a way that One is not. It is this reality - touchable in the form and flesh of the embodied world - that suggests the possibility of a divine order, not just in the abstract, but in the actuality of a full-bodied experience.

Three As the Essence of Order

The divine order revealed by Three within the manifest world becomes evident everywhere. The early arithmologists converging in Iamblichus' seminal work point toward: the three "so-called true means" - arithmetic, geometry, and harmony; the three faces of time - past, present, and future; the three rectilinear angles - acute, obtuse, and right; the three types of odd number - prime and incomposite, secondary and composite, and mixed; the three types of triangle - acute-angled, obtuse-angled, and scalene; the three configurations of the moon - waxing, full, and waning; the three kinds of planetary motion - direct, retrograde, and stationary; the three circles of the zodiacal plane - the tropic of Cancer, the tropic of Capricorn, and the ecliptic; and the three kinds of living creature - land, winged, and water (Waterfield 52). With the benefit of modern science, one could argue that in some cases the arithmologists overstated or oversimplified. Still, there is also a bit of intuitive truth

◇◇

to their assertion that is echoed in the words of Homer: "All was divided into three" (Illiad 15:189, qtd in Waterfield 53).

To this list, we might add the triune God of Christian iconography – manifest as the Father, the Son, and the Holy Ghost; a parallel designation in Hindu mythology – taking form as Brahma the Creator, Shiva the Destroyer, and Vishnu the Sustainer; the three gunas of yogic philosophy – *rajasic* (active), *tamasic* (passive), and *sattvic* (pure and balanced); the three doshas of Ayurvedic medicine – *vata*, *pitta*, and *kapha*; the three subtle nerve channels in yogic practice– *ida*, *pingala*, and *sushumna*; the three parts of the psyche in Freudian psychology – ego, id, and superego; the alternate scheme of Transactional Analysis – Parent, Child, and Adult; the three styles of Greek column – Doric, Ionic, and Corinthian; the three primary particles in an atom – proton, neutron, and electron; the three types of color – primary, secondary, and tertiary; the three basic parts of a sentence – subject, verb, and object; the three Furies, three Fates, and three Graces of Greek mythology; the Hegelian dialectic of thesis, antithesis, and synthesis; the three states of matter – solid, liquid, and gas; the three wise men; the three blind mice; the three wishes given by a genie; the priest, the minister, and the rabbi of countless jokes; and the rules of baseball – three strikes and you're out, three outs to an inning, and three times three innings to a game (Eck).

Whenever we want to make something accessible to conscious understanding, we begin by dividing the object of our scrutiny into threes. Whether this is a reflection of the actual construction of the embodied world, or of our own need for manageability, we can associate the number Three with the process of bringing order to chaos.

In this function, the number Three echoes the number One, which is the source of a Creator god who brings order to the chaos of the Abyss. With Three, however, we are no longer talking about a Creator god, but Creation – the ordering as opposed to the One who orders it. This ordering is a reflection of the Creator god's immanence within Creation, so that through the number Three, the One is revealed beneath the veil of appearances.

More central still is the possibility of discerning the order within Creation, which is a property of the number Three accessible to humans. With the capacity to observe the natural order within the embodied world, we use the gift of consciousness provided by the Creator god to become more godlike ourselves. If we are able to discern the natural order, then we can also participate more consciously within it, and this participation makes all the difference in the quality and depth of our experience.

Discerning the natural order within the manifest world was the impetus for the development of civilization. Tracking solar and lunar cycles meant that early humans could plan their hunting, food-gathering, and planting activities in harmony with the seasons, and be more productive. Later, experiential traditions arose – such as early Taoism, shamanism, alchemy, and Hermetic magic – that attempted to mimic natural order so that actions could be infused with the power inherent in that order. Expanding on this concept, empirical scientists sought to discern the natural order within everything they studied, so that we could have more control over our own destiny as a species. Today, everything from stock market analysis to weather forecasting to complex medical diagnoses depends upon our ability to discern the patterns that order our world. Ultimately, these activities and the quest for an understanding of order that motivates them are a reflection of our relationship to the number Three.

Three and the Power of Consciousness

In Chapter Two, consciousness was described as an awareness born of comparison and contrast in the Realm of Two. Comparison and contrast in turn are predicated on the participation of the object to which we apply our consciousness in one or more polarities. If Object A is hotter than Object B, we have a basis for this comparison because of a continuum of contrasting possibilities between hot and cold.

In the Realm of Three, this basic awareness of polarity is taken one step further. Now we have an observer to observe the polarity, and to change it through his/or her observation. If we know how Object A is different than Object B, we also know how to alter Object A so that it is like Object B. If Object A is cold and Object B is hot, then we simply apply heat to Object A in order to change it. In this way, conscious participation within the Realm of Three not only allows us to observe, but also infuses our observations with the creative power to initiate change. In Chapter One, we spoke of the necessity for taking responsibility for being the change we hoped our gods and goddesses of choice could make for us. It is the power of consciousness awakened to its creative capacity to shift any object along any continuum we can observe within the Realm of Three that makes this responsibility possible.

In the material world, we witness this capacity when the observations of science are transmuted into the inventions of engineering. An observation of the difference in polarity between an anion and a cation makes possible the movement of electricity through a wire that then allows us to operate everything from a waffle iron to a computer. An observation of the aerodynamic principles that make a bird lighter than air leads us to invent airplanes that fly. An observation of the behavior of healthy cells as compared to cancerous cells allows us to experiment with various treatments. All of this is made possible within the Realm of Three.

The number Three becomes even more exciting once we enter the domain of spiritual psychology, which is the subject of this book. The potential for personal growth is predicated on our ability to observe ourselves in relation to our own patterns, and to make decisions about how we want to shift those patterns. In *Tracking the Soul*, I explore the possibility for observing ourselves along a number of continuums – hard and soft, wet and dry, rough and

◇◇

smooth, and so on. Each of these sensory polarities encompasses a broad range of attitudes toward life, propensities, desires, and likely behaviors to which they are related by metaphor.

Someone who is closer to the hard end of the hard-soft continuum, for example, might be more of a disciplinarian to her children than someone who is soft; work longer hours; and be less expressive of feelings. Someone who is closer to the dry end of the wet-dry scale might be more of a loner than someone who is wet; approach life with a more linear, more rational and more practical bent; and be more sensitive to heat. Someone who is closer to the rough end of the rough-smooth scale might tend to be emotionally insensitive; not very good with details; and sloppy with regard to matters of personal hygiene.

Within the Realm of Two, these affinities are simply taken for granted, and often exercised unconsciously. We are who we are and our expression of who we are revolves around resonance by affinity. Where affinity leads to excess, as it often does, we feel pain and begin to draw to ourselves experiences that force us to become more conscious of the source of our pain. I call this kind of experience resonance by wounding, because it is our wounds, born of excess and/or deficiency that cause us pain. In the Realm of Two, we are aware of pain, and perhaps even of the external source of that pain, but we lack the perspective to see how we are responsible for our suffering. It is only in the Realm of Three that we gain the objective distance from ourselves to recognize our excesses and our deficiencies for what they are. With this recognition comes the possibility for change.

Our hard person might realize that her workaholic tendencies were creating problems, and be motivated to cut back her hours at work and create more time for herself and her family. Our dry person might experience the pain of loneliness and be motivated to reach out more deliberately to others. Our rough person might experience a few costly business mistakes due to lack of attention to details, and start to pay closer attention or hire an accountant who will. Through the awareness cultivated in the Realm of Three, each of these people will begin to gravitate toward experiences that embody qualities and attributes that are somewhat foreign to their nature – experiences evolving through what I call resonance by contrast. Resonance by contrast, in turn, facilitates an expanded range of options as well as the personal freedom and growth that comes through such an expanded range. In this way, the Realm of Three becomes the place where awareness and perspective lead to growth.

A Personal Example of Resonance By Contrast in the Realm of Three

Before sitting down to write this chapter, I found myself responding to an email from a new friend going through a difficult patch. As it happened, this correspondence wound up taking an inordinate amount of my day, and my working agenda fell by the wayside. I was truly enjoying my communication with this friend, and more than happy to be there for her however I could. Nor am I so rigid in my adherence to schedules that I can't break them when necessary or even just because I want to. But in spending the entire day on this email, I began to feel a bit imbalanced – a signal I have learned to take as evidence that some polarity on which I function has been tipped to excess.

Within the Realm of Two, I would have simply noted my discomfort and moved on, probably blowing off the rest of my day as lost and scrambling to make up for lost time the next day. Discussing the Realm of Three, however, I am motivated to recognize my awareness of imbalance as an opportunity for new growth. In Part Two of this two-volume set, I will revisit this exercise to reveal the hidden astrological knowledge that guides me. Here, I will merely share my process in order to demonstrate how the exercise of self-awareness in the Realm of Three contributes to the possibility for growth.

When I think about the various spectrums on which this sense of imbalance might sit (using my knowledge of astrology as an intuitive springboard), what comes first to mind is the polarity tight and loose. If I had to place myself along this spectrum, I would put myself closer to the tight end. What this means is that most of the issues that arise as a consequence of my identification with the tight end will arise because in some sense I am too tight. If I take the time to think about tightness as a metaphor for my life – consciously moving my awareness from the Realm of Two (a sense of discomfort) into the Realm of Three (increasing awareness about the source of my discomfort) – a list begins to emerge:

> I am tight in the sense that normally I do not reveal much about myself to others. It is only as I begin to feel comfortable with them, that I open up and share. As the expression goes, I play my emotional cards pretty close to my chest.

> I tend to keep close tabs on my time and my money. I don't like wasting time, and normally I don't spend money unnecessarily or frivolously.

> I tend to have few close friends, rather than a broad social network. My tolerance for social interaction is limited, and I can be something of a loner.

> I generally try to keep only those material possessions that I actually use. I am constantly weeding things out, reorganizing, and optimizing my personal space. I do not like clutter or disorganization.

> I prefer the familiar, the tried and true to new, unfamiliar, untested experiences. Even in matters of entertainment, I gravitate to certain authors,

◇◇

actors, musicians, and would rather read, watch, or listen to everything they have done than try something new on a whim. I have a small collection of favorite DVDs that I watch over and over again without tiring of them.

I am fairly disciplined about my work life, able to set priorities quite easily, am organized and efficient in approaching the work I have decided takes priority, and generally stay focused on the task at hand.

I am loyal and monogamous by nature. Even in my wild and reckless youth, during the Sixties at the height of the sexual revolution, it was difficult for me to get emotionally or sexually involved with any woman without imagining a full life together until death do us part.

While I recognize that my creativity depends upon my ability to innovate, experiment, and explore new ideas in new ways, I prefer to lay a solid foundation for this creativity. My first book, for example, was heavily dependent on research into the ideas of others, which then became the license I felt I would need to explore my own ideas.

While I care deeply what is going on in the world, I have had to limit the amount of news I am exposed to. I distrust most news that comes through traditional media channels, and tend to get mine through a few select publications I do trust, and the newsletters and/or magazines of the non-profits to which I contribute.

I prefer to work by myself and for myself. I have inherited enough business savvy from my father and grandfather to live an entrepreneurial work life, but I have always stopped short of hiring employees, and can no longer possibly think in terms of working for someone else.

I am increasingly disenchanted with the culture and the political climate that seems to be evolving in the US. I don't watch TV, try to minimize my interaction with unfeeling corporations, and feel only a tenuous connection to the local community. In some ways, as the larger world becomes more interconnected, my world appears to be shrinking. As one of my t-shirts reads: "I live in my own little world, but it's ok. They know me here."

Early in life, I dropped out of the Catholic Church where I was raised and became disillusioned with religion in general. In my 20s and 30s, I explored Eastern religions and pursued a number of spiritual experiences within the

New Age community, but became increasingly disillusioned with most of them as well. Now, my relationship to Spirit, which is still critically important to me, comes through my own meditation and private attempts to discern the will of the divine within my own heart and soul.

I call these attitudes toward life "tight," because they pull me inward, they tend to contract my soul space (the psychic arena in which I live my life), and they facilitate a more limited and in some sense more manageable focus. The word "tight" might mean other things to other people and have different connotations in different circumstances. Nor does this list exhaust the way tightness manifests in my life. It does illustrate how this sensory metaphor applies to my experience, and describes an attractive force in a field of resonance by affinity. Because I am "tight" by nature, I will tend to attract to myself experiences that reflect and reinforce this dimension of my being.

Though the quality of "tightness" might be judged within the Realm of Two by those to whom it represents a reflection of their shadow – associated with such derogatory epithets as "uptight," "tightwad," or "tight-ass" – ultimately such judgments projected onto another simply take us down the path of inequality back into the Realm of Zero. On the path of equality in the Realm of Two, the quality associated with tightness is neutral. It has its place, just as looseness does. To be "tight" as a matter of preference, or resonance by affinity, is no better or worse than to be "loose," or to be anywhere on the spectrum between "tight" and "loose." What matters in relation to tightness, or to any other quality we might use as a sensory metaphor for experience, is our subjective affinity to it. Where such affinity exists, to be "tight" is to live in harmony with one's nature, and that can only be a good thing, regardless of how our "tightness" might be judged by others.

While the thought of being "tight" might seem undesirable or even abhorrent to some, for the most part, it works for me. Since I have a preference for tightness, however, I also inherently have a tendency to become too tight. When this happens, I generally get an emotional signal that I have crossed a line into excess. When I withdraw into my own little world, for example, stop reaching out to others, and have no meaningful conversations with anyone for any length of time, I become too tight, and I begin to feel lonely. When I keep too tight a control on my money, live the life of a Spartan ascetic, and become provincial in my tastes, I become too tight, and start to feel bored and restless. Despite my penchant for discipline, organization, and thoroughness, there are times when these qualities – which I value – become excessive and squeeze the very life out of my creativity. When this happens, I start to feel dull, tired, and depressed.

All of these painful emotional signals are telling me that I have indulged my resonance by affinity with tightness to the point of excess. The antidote to this excess is moving away from the tight end of the tight-loose scale toward the loose end. To make such a move, however, requires me to experiment with behavior that is somewhat foreign to my nature, and my unconscious tendency will be to believe that more of the same is the appropriate solution. I will do this, even when I know, or some part of me knows, this is not true.

I have been around the block enough times to understand that adding fuel to fire makes it burn hotter, longer, and more dangerously. But to some extent, the experience of pain initiates a vicious cycle. With pain comes stress, and under conditions of stress, our natural tendency is to cling to what is familiar. In terms of our model of resonance, this means that when some affinity leads to excess, we will attempt to indulge more of our affinity as a comfort, hoping it will alleviate our stress. As we all know, it doesn't. It just intensifies our pain. If we continue to ignore our pain, past a certain point of reason, our pain becomes chronic, and begins to crystallize around a wound that then becomes a magnet for resonance by wounding.

The good news here from a spiritual perspective is that pain will intensify to the point where it gets our attention. Eventually, we must address our wounds or they will kill us – often in a physical sense, but more seriously through a dwindling spiral of slow soul death. With attention, and the commitment to honor our awareness with action, we move into the Realm of Three, where change is possible.

The Possibility for Mitigation of Excess in the Realm of Three

Just before meeting my new friend, the walls of tightness were beginning to close in on me. Living alone in my writer's sanctuary here in the Ozarks, after having just ended a 13-year relationship, I was beginning to feel lonely. I had been reaching out to friends here and there, both locally and at a distance, but there was a certain amount of inertia that brought back the echoing void as soon as I stopped making the effort. Not getting any worldly news, going into town once a week for supplies, living simply with my books, my music and the luscious green forest that surrounds me has been unquestionably idyllic, but I had also begun to feel increasingly out of touch, bored, and restless. Even my writing had begun to feel a bit stale – suffocated for lack of fresh air – and my enthusiasm for work that had previously excited me was starting to wane. I was, in fact, depressed and not sure how to break out of my self-imposed predicament.

This was a familiar scenario. I had lived this very same life before, and had to recognize my emotional pain as a reflection of some wounded part of me that was attracting it. By indulging my natural preferences through resonance by affinity, I had created a life that despite its external appeal was the byproduct of resonance by wounding. My life had become too tight, and the pain of excess was becoming a source of motivation toward change.

Aside from being motivated by our pain, the imbalances in our lives also attract people, ideas, and experiences that model the possibility of balance. We are attracted to these possibilities through resonance by contrast so that we might learn from them, and find a way back to our center. As we draw from these people, ideas, and experiences the quality opposite to that which we indulge in excess, we are restored to balance, and any chronic wounds we have suffered through excess (and/or deficiency), are assuaged and gradually healed.

My new friend was healing me through her presence in my life, and the deeply intimate correspondence we were sharing. In fact, though circumstances did not allow us to be together, it felt like we were falling in love. As we all know, there is nothing like new love to motivate us to break old patterns, and it felt very much like this was what was happening to me. In terms of the model of resonance we are exploring here, there was a strong sense of resonance by affinity, as we seemed to share so many things in common. There appeared to be very little resonance by wounding to complicate this relationship, although certainly time would tell whether or not that was true or just wishful thinking. Less noticeably, but no less an important part of the attraction between us was a subtle undercurrent of resonance by contrast. There were certain ways that we were different from each other, and so far at least, that also seemed to be part of our attraction. But more fundamental than that, my love for my new friend was somehow triggering a shift in me that had been imperceptible until now.

To describe this shift in a word, I was slowly gravitating toward the loose end of the spectrum. Having felt comfortable with her immediately, I was revealing a great deal about myself. I was spending more and more time with each passing week, lost in the delicious reverie of our communication, and whatever work schedule I had been trying to keep for myself was coming undone. I had recently spent my money freely to celebrate her birthday, and adopted her orphaned cat, and that felt good. I was fantasizing with her about living together in some exotic place – writing together, playing together, creating a business together, and sharing every aspect of our lives. My little world was beginning to expand, if nowhere else but in my imagination. All of this was a healing balm to the excess tightness that had previously described my wounding.

Though I am articulating this shift in some detail now, it was imperceptible to me until I crossed a certain line. The attraction itself was a reflection of a powerful resonance within the Realm of Two. What shifted this attraction into the Realm of Three - where conscious choice and intentional movement toward change becomes possible - was the awareness that tightness was giving way to looseness. This awareness only occurred as some part of this process began to feel too loose. It was only as I began to feel the pull back toward a more disciplined work life that I knew something in me had shifted. Exactly what had shifted was not clear until I sat down to write what I have just shared with you. As I took a step backward and began to look more closely at what was going on, I shifted into the Realm of Three.

CHAPTER THREE

<><><><><><><><><><><><><><><><><><><><><><><><><><><><><><><><><><><><><><>

The Realm of Three
and the Potential For Changing Collective Awareness

Within the Realm of Three, these personal shifts form the basis for a life that gradu-ally reflects increasing self-awareness and opportunity for personal growth. But the Realm of Three is more than just an arena for personal growth. It is the stage upon which his-tory provides collective lessons that we would all like to believe gradually move us toward a more humane, more conscious, more progressive culture. In my more despairing moments – watching the hijacked jets of 911 crash into the Twin Towers; watching our country arro-gantly flex its muscles in shock and awe at a small, powerless country that had done us no wrong; watching our liberties quietly stripped away and our moral code loosened to encom-pass torture in the name of national security – I have had to question just how much we have learned from history.

As a German-American growing up in the lingering shadow of World War II, I found myself obsessed with the horrific rise to power of Adolf Hitler and the subsequent atrocities that were committed in his name. I could not possibly imagine how such a thing had hap-pened, much less that in my adult life I would witness similar atrocities being committed in Bosnia, Rwanda, and Sudan. When we blindly follow a path of inequality in the Realm of Two, history repeats itself in the Realm of Three and we learn nothing from it, except per-haps how to fool ourselves more skillfully.

Yet, there is also reason for hope. As our world becomes too hot, in all the ways in which that metaphor has meaning, we begin to search more earnestly for solutions. On a literal level, Al Gore's award-winning documentary *An Inconvenient Truth* raised awareness about global warming to the point where those who would still insist on denial have increas-ingly less credibility with the media and with the public at large. Unlikely champions of this cause – such as Republican Governor of California, Arnold Swarzenegger; Chicago Mayor, Richard M, Daley, and corporations like Goldman Sachs, Citigroup, and John Hancock Insurance – emerged to lead a bandwagon that is likely to result in real change within the foreseeable future. Though the United States – under the regressive policies of recalcitrant old school oil baron G.W. Bush – refused to endorse the Kyoto Protocol, close to 400 cities in every state plus Washington, DC signed the US Mayor's Climate Protection Agreement, while other global players such as Japan and many of the countries of Europe have taken significant steps toward reduction of greenhouse gases in recent years (Snell 74).

This represents a positive opening within the Realm of Three that stems ultimately from increased pain through resonance by wounding on a global scale, as we continue blindly down the path of inequality in the Realm of Two. Although evidence for global warming has been accumulating for decades now, the increase of extremes in temperature and weather patterns across the globe, the painful and expensive aftermath of storms such as Hurricane Katrina, and the irrefutable evidence of melting polar ice caps have all raised our awareness to a certain threshold where we begin gravitating toward contrast – in this case, cultivating strategies to take us down the hot-cold scale toward the cooler end of the spectrum.

Wisdom, Denial and the Terrible Plunge Back into the Abyss

As with hot-cold, so too with all the polarities through which Spirit expresses itself within the manifest world. Within the Realm of Three, both individually and collectively, we have the opportunity to learn from our mistakes – to transmute our wounds into wisdom, and to shift our world in a positive direction. Regardless of how troublesome life appears, there is no problem that cannot be alleviated through focused awareness. It is this possibility that makes the Realm of Three a potential pathway beyond the Abyss. In the Realm of Three, we apply our Monadic powers of consciousness to the ambivalence inherent in the Realm of Two on a path of learning and growth. Thus, the arithmologists noted (Waterfield 51):

> The triad is called 'prudence' and 'wisdom' – that is, when people act correctly as regards the present, look ahead to the future, and gain experience from what has already happened in the past: so wisdom surveys the Three parts of time, and consequently knowledge falls under the triad.

Conversely, when we fail to survey the three parts of time – when we have knowledge, but do not use it; when we see our mistakes, but do not correct them; when we live through history, but do not take it seriously – we stumble blindly through the Realm of Three. In psychological terms, this is called denial.

When we live in denial, individually or collectively, we refuse the divine gift of consciousness. If there is such a thing as the wrath of God, it will invariably be evoked as we turn our back to it, for we cannot evoke a god and then ignore its teaching, without expecting dire consequences. If this god is (potentially at least) our pathway out of the Abyss, the subsequent denial of this god will be a return to the Abyss, but with a vengeance. Our return to the Abyss on the coattails of denial will be a dramatic experience of crash and burn.

It is interesting, in this regard, that the arithmologists "call the triad 'piety': hence the name 'triad' is derived from 'terror' – that is fear and caution" (Waterfield 51). In a footnote, Waterfield suggests that Iamblichus was linking *trias* (triad) with *trein* (to be afraid). The implication is that when we do not live our lives as an act of piety – that is to say, in allegiance to whatever god or gods we have chosen to be our guide – our failure to do that is cause for fear.

◇◇

It is no great revelation to note that in the early 21st century, we live in a political atmosphere that is charged by fear. Events of September 11, 2001 and its aftermath have greatly contributed to this atmosphere, as have dubious wars in Afghanistan and Iraq, and the ongoing threat of war with other countries like Iran, whose political intentions are suspect. Yet the fear is rooted more deeply than that – in century-old policies of colonialism and corporate neo-colonialism that fuel hatred of Americans around the world, and our willful refusal to take those policies into account as a major cause of terrorism; in duplicitous behavior that espouses a desire for peace, yet is willing to arm both sides of any conflict to the teeth; in promoting global economic prosperity at the same time our policies are dependent upon the cheap labor of Third World sweat shops, immigrants, and refugees pressed into virtual slavery; in official rhetoric that claims the moral high ground as champions of democracy abroad, while simultaneously dismantling it at home through underhanded fast-tracking of omnibus bills that no one reads, with invisible appropriations riders that no one debates; in an attitude of denial toward the cumulative environmental damage we have done, and continue to do in the name of business as usual; and countless other areas in which we say one thing, and do another, working insidiously against the very gods we evoke in our public declarations of intent. All this makes a mockery of whatever pretension to piety we might assume, and contributes toward an atmosphere of fear and terror in the Realm of Three.

The Sad Fate of Truth on a Path of Inequality in the Realm of Three

Consider, the fate of those who dare to tell the truth – commonly known as whistleblowers – in the face of this denial (Schulman). When Army Specialist Joseph Darby turned over to the media photos showing the graphic abuse of military prisoners at Abu Ghraib, he was publicly praised by former Defense Secretary Donald Rumsfeld, yet simultaneously warned by the Army not to return to his hometown of Cumberland, Maryland, where he was commonly regarded as a traitor. Called "a walking dead man . . . with a bull's eye on his head" by his wife (Wikipedia), he was shunned by friends and neighbors, his property was vandalized, and he and his family was put under protective custody at an undisclosed location. For all intents and purposes, his life was ruined – all for telling the truth, for rising to the level of piety required by the Realm of Three in a political climate stuck on the path of inequality in the Realm of Two.

In a different arena, Bogdan Dzakovic led an FAA red team that probed weaknesses in airport security and, 90% of the time, managed to breach the system. Testifying before the 911-commission, Dzakovic said, "We were ordered not to write up our findings in some cases and not to retest airports where we found particularly egregious vulnerabilities to see if the problems had been fixed" (Schulman 55). In the wake of this testimony, the FAA started providing advance warning to airports about what would be tested and when, so they would more consistently pass their tests. Dzakovic was demoted from his position commanding an elite security force with the FAA to a file clerk position with the Transportation Security Administration. The reward for telling the truth in a culture of denial is excommunication.

Lastly (though certainly not exhaustive of the cases of this type), we have Sandillo Gonzalez, who blew the whistle on a paid informant for the Bureau of Immigrations and Customs Enforcement who was implicated in a series of murders, showing that the agency had foreknowledge of these murders yet failed to intervene. Years earlier, he had also reported the suspicious disappearance of 10 kilos of cocaine from a government raid. Gonzalez was told by the authorities to which he reported these serious violations of law by government agencies, that the OSC (Office of Special Counsel, charged with the task of investigating whistleblower complaints) "receives a large number of matters concerning disclosures of information" and that "cases are generally processed in the order in which they are received" (Schulman 57). For his trouble, Gonzalez was threatened with a choice of early retirement or a downgrade on a crucial performance review, while none of his violations reports was ever investigated.

To the extent that we continue to live in denial in the Realm of Three, we will live in a world that is not safe, not free, and not conducive to the evolution of consciousness necessary to follow in the footsteps of the gods that we deny. Our moral standards will continue to deteriorate; our vulnerability to terrorist attacks will increase; the corruption of our government will continue to make a mockery of everything we stand for publicly; we will continue to poison our air, our food, our water, and ourselves through empty rhetoric designed to allow business as usual while creating the mere appearance of change. The path of denial within the Realm of Three has only one possible destination. On this path, we will crash and burn, then plummet back into the Abyss. Our only hope of redemption – the exercise of conscious awareness toward growth – will have been squandered.

Three and the Power of Mediation

Beyond the exercise of consciousness in relation to the polarities of the Realm of Two, within the Realm of Three we are offered a second set of opportunities for movement onto a more productive path of becoming. These opportunities are hinted at by the arithmologists when they suggest (Waterfield 53) that . . .

> . . . *"All was divided into three,"* given that we also find that the virtues are means
> between two vicious states which are opposed to both each other and to virtue, and

◇◇◇

there is no disagreement with the notion that the virtues fall under the monad and are something definite and knowable and real wisdom – for the mean is one – which the vices fall under the dyad and are indefinite, unknowable and senseless.

They call it 'friendship' and 'peace' and further 'harmony' and 'unanimity': for these are all cohesive and unificatory of opposites and dissimilars. Hence they also call it 'marriage'.

In a footnote, Iamblichus goes on to explain that according to Artistotle, virtues were "each a mean between two vices at the extremes, one of which was excessive, the other defective" (Waterfield 53). Within these passages we now have a very rich prescription for understanding not only how Three mediates all polarities within the Realm of Two, but also a deeper understanding of how it makes change possible, once awareness is applied to a condition of excess or deficiency. There are three possible scenarios here, which I will consider in turn.

Three and the Resolution of Conflict on the Path of Inequality

In discussing the Realm of Two in our last chapter, I suggested that there were basically two pathways through it: the path of equality and the path of inequality. The path of equality, the more enlightened choice, leads to the possibility of creative synthesis, love, mutual respect, and compassionate tolerance of differences, an *hieros gamos* through which we are empowered to evolve, individually and collectively, in the footsteps of our chosen gods and goddesses. The path of inequality leads to conflict, hatred, disrespect, and cruel intolerance through an unleashing of devolutionary forces capable of derailing our progress through the numbers into the promise of becoming.

Assuming for the moment that we have chosen the path of inequality, and consequently entered a state of conflict – essentially a war between two extremes of a given polarity or set of polarities – within the Realm of Three, a neutral perspective outside the polarity offers the opportunity for resolution of conflict. Between the two extremes, we have a vantage point, from which it is possible to perceive that the two extremes exist not only in opposition to each other, but also along a continuum that connects them and makes them part of the same reality. Understanding this continuum, it is then possible to find our way from conflict to common ground.

In the late 1980s, I worked briefly for a non-profit organization called The Trinity Forum for International Security and Conflict Resolution. The goal of this organization was to provide mediation services between warring factions in a series of conflicts impacting international security. As I wrote in a Project Handbook for one of our conferences:

◇◇

The process emphasizes the building of relationships between people of diverse view-points. Often people who fundamentally disagree on important issues assume they have little in common with those on the other side of the political fence. People who already agree tend to talk amongst themselves, and exclude those with whom they have basic differences. The dialogue process encourages quality interaction among those who normally would not talk to each other and creates an atmosphere in which it is possible to go beneath obvious differences to look for common ground. Although the process takes a considerable amount of time and patience, the results often reflect solid thinking that encompasses the genuine needs and interests of all sides.

Such a notion, of course, is not new, although in today's climate of belligerence toward manufactured enemies, where an exercise of pre-emptive military force is often preferred to more moderate pathways of diplomacy, it is good to be reminded. In my role of Project Coordinator at the Trinity Forum, I witnessed on several occasions, the power of mediation to reach below whatever external conflict was raging to the place where both sides shared a common bond of humanity and fundamental human concerns. One conference sponsored by The Trinity Forum, for example, brought together a Sandinista representative, a Contra leader, and two US ambassadors of different political views to dialogue around the Nicaraguan conflict that became a focal point of controversy within the Reagan administration.

Though the process was grueling at times, it shed some genuine light on issues that had previously been hopelessly polarized. Though much disagreement remained on the table by the end of the five-day conference, one participant thought the process extremely helpful in "stripping specific issues down to the bone to what is essential to be dealt with." Stripping away the rhetoric and posturing typical of any polarized conflict, in other words, was conducive to seeing more clearly a continuum of possibilities stretching between the polarities.

A similar result occurs in situations where ordinary Israelis and Palestinians sit down to discuss their hopes, fears, and needs (Wikipedia "Projects"); when gang members are able to talk outside of the peer pressures induced by their respective gangs (Advancement Project Los Angeles); when environmentalists and loggers bravely come to the table to explore common ground (Bright Future); or when any couple exits the traditional divorce process and undergoes mediation. Conflict is in no one's interest, except perhaps those who in some way profit perversely by another's misfortune, or in the international arena, from selling weapons to both sides. Where conflict is actively pursued by government, one can reasonably assume that there is a secondary agenda beneath the rhetoric that is somehow worth more than the lives of those sacrificed, and the vast amount of resources that conflict consumes.

The cost of the Iraqi war as of September, 2010, for example, was estimated at $747 billion (National Priorities Project)[1], not counting the incalculable cost of human lives and the

◇◇

immense suffering of those who have managed to survive. The US Department of Energy estimates that Iraq's total oil reserves could reach as high as 400 million barrels (Paul "Oil in Iraq")). In 2004, the value of this oil was estimated to range from $582 billion to $8,960 billion. Since the high end of this estimate was based on a price of $41 per barrel, which has long since been surpassed, this high-end estimate must now be considered low[2]. To put these figures in perspective, the combined worldwide profits of the five largest oil companies was only $35 billion in 2002 (Paul "Iraq Oil Bonanza) – the year before the war began. For those who think solely in terms of business opportunities, the cost of this war is an attractive investment, with the potential to earn a massive profit. For everyone else, it is a dismal refusal of the power of Three to mediate extremes that is creating hell on earth.

Three and the Impetus Toward Creative Synthesis on the Path of Equality

The second possibility for mediation in the Realm of Three exists whenever we choose the path of equality in the Realm of Two. As discussed earlier, this path leads to creative synthesis, love, mutual respect, and compassionate tolerance of differences, effecting a *hieros gamos* or integration of polarities that lies at the root of these more enlightened possibilities. Within this scenario, Three becomes the agent of synthesis and/or the embodiment of something new that draws upon resources from both ends of the polarity it is synthesizing.

If I can walk a path, for example, upon which tight and loose are understood to be of equal value, each in its own place within a balanced set of preferences that includes them both, the potential exists for me to redefine myself in terms that are neither tight nor loose, but some more versatile combination of the two. It is reasonable to assume that on the path of equality, there will be experiences within the Realm of Three that point me in this direction and serve as a catalyst toward synthesis. Through a lifetime of learning related to this polarity, this has been characteristic of my journey.

This book, for example, is a catalyst to synthesis on the path between tight and loose. Previously, I mentioned that in my preference for tightness, I do not normally reveal much about myself to strangers, and that I prefer to lay a solid foundation for my creativity on which I might more freely expound my own ideas. On the basis of these preferences, one might assume that I would write a book that was factual and relatively impersonal – that is to say, fairly tight in its organizational structure, control and presentation of content, and style. That is not what has happened. As I entered more deeply into this book, it became apparent to me that it required a kind of free-wheeling prose that can dip into the intimately personal, then soar above for an aerial view of the collective, then burrow more deeply beneath the surface to extract whatever nuggets of philosophical wisdom might be buried there.

◇◇

While the nature of this book might seem to relegate it to the loose end of the spectrum, in contrast to my natural affinity for tightness, this would only be true on the most superficial level. Despite its free-wheeling nature, there is a cohesive structure that propels me in linear fashion through the expected sequence of numbers, showing how each is related to a continuum of numbers, as well as to various dimensions of truth – arithmological, mythological, philosophical, psychological, personal, cultural and political. Within my discussion of each number, I am free to meander loosely among these various dimensions, but the structure itself contains my meandering. Meanwhile, each station on my meandering journey must necessarily be built upon facts, persuasive arguments, and relevant examples,

In this book, I have evolved a process that is neither tight nor loose, but both simultaneously, and an end result that reflects the synthesis between tight and loose. What shall we call it? Too-loose Lautrec? Lucy in her Tights with Diamonds? This is the nature of creativity. We invent nothing new until we synthesize old polarities, but when we do, everything changes. This becomes possible only as we follow a path of equality in the Realm of Two into the Realm of Three with conscious intent.

Consider Albert Einstein's famous equation: $E = MC^2$, which broke the mold posed by the old Newtonian continuum on which matter and energy were polar opposites. Now, post quantum theory, we live in a world where reality is neither matter nor energy, but some strange shape-shifting amalgam of the two. The full implications of this theory have yet to be explored, either scientifically or philosophically, but our world has been and will continue to be utterly transformed because of it. On a mundane level, quantum theory has given us nuclear weapons, fiber optics, and GPS satellite systems, as well as CD players, DVDs and wireless Internet.

On a deeper, perhaps more subliminal level, quantum theory has thrown conventional wisdom on its head, and ushered in an era of post-modern ambivalence. For "if we take the quantum nature of all physical systems into account, the statement that a certain specific event q 'has happened' (or, equivalently that a certain variable has or has not taken the value q) can be true and not-true at the same time" (Stanford Encyclopedia), suggesting that very little in this life can be taken for granted or affirmed with absolute authority.

The same creative synthesis of matter and energy has fueled renewed interest in old Theosophical speculations about the laws of manifestation, in which the power of intention is harnessed to create something out of nothing. One popular version of this paradigm shift in our thinking is the controversial video, *The Secret*. Proponents claim it to be a key to well-being and a life of creative fulfillment, while critics call it "potentially psychologically harmful, ethically deplorable and scientifically nonsense" (Adler qtd in Killoran). Regardless of which side of this controversy you land on, the debate itself is predicated on the collapse of the old order in which matter and energy were once at opposite ends of a spectrum. This

◇◇

collapse in turn was facilitated by the advent of quantum theory as a perspective in the Realm of Three – offering a vantage point on the continuum between matter and energy on which a creative synthesis of the two became possible in our collective imagination.

Three as a Catalyst to Growth

The third way in which the number Three facilitates the possibility of becoming is through alleviating the conditions of excess and deficiency that lead to resonance by wounding. Whenever excess or deficiency exists, the antidote becomes moving toward the end of the continuum most likely to restore balance – e.g. toward the loose end of the spectrum between tight and loose in order to alleviate conditions of excess tightness. Yet one need not wait until tight shoes make it impossible to walk, before removing them and letting tired feet breathe. Knowing that excess tightness is an ever-present possibility for someone with an affinity for tightness, one can choose instead to compensate on the fly. Within the life currents tending toward tightness, one can keep a steady counter-flow of looseness bubbling through. As with the assistance that Three provides toward synthesis on the path of equality in the Realm of Two, developing such a counter-flow transforms the continuum between opposites into something entirely new.

As the correspondence between my new friend and I heats up, for example, our deepening love automatically generates its own counter-flow. I seem to be getting the same amount of work done as before, but in less time and with more ease. After reading an unbearably erotic email from her one morning, I had to break my routine and go down to the river, where I hoped to cool myself down enough to focus. I swam and lounged upon a large rock jutting out of the river, watching dragonflies mate by my feet, and Cooper's hawks lazily ride the thermals overhead. The day was hot and languid, and time stood still. I had all the time in the world, and took it. It was deliciously loose.

When I felt satiated and refreshed, I ambled up the hill to my cabin in the woods, and went back to work. Back home, I was able to focus, without a sense of strain or pressure, and enjoyed my work in the same way that I enjoyed the river earlier in my day. There was now an undercurrent of looseness bubbling through and oxygenating even those areas of my life where tightness would ordinarily prevail. Yet none of the advantage that tightness gave me had been lost. There was no sense at all of deprivation or of not being able to indulge my affinities, and no need to deliberately try to push myself toward one end of the spectrum or the other. All was just tight enough to allow the luxury of looseness to deepen my sense of balance, rather than detract from it.

Love is a wonderful catalyst to this balancing dance between flow and counter-flow in the Realm of Three. But I have also experienced this same dance in other moments. When my father died, the grieving process brought intermittent tears and a watery looseness that I resisted at first, but that gradually became soothing to me as counterpoint to a tightly scheduled life. On vacation in Germany, the newness of everything around me required me to hold my planned itinerary loosely, so that serendipity and spontaneous discovery became an unpredictable part of my adventure. Every workshop I have taught, every talk I have given, every astrological consultation I have done has been an exercise in allowing my tight preparation to breathe and to fibrillate in response to feedback in real time from my students and my clients. In the best of these presentations, the loose unpredictability of that feedback allows my tightness to be the cocoon out of which some new species of butterfly emerges to the amazement of everyone involved. In each of these moments, allowing an undercurrent of looseness to flow through my affinity for tightness, I am free to leave the entire continuum behind in a flight along some new trajectory, as yet undefined.

On a collective level, we can observe the same process at work at various points of our history – the influx of women into the workforce during and following World War II; the desegregation of schools in the 1950s and 60s; the intermingling of East and West Germans when the Berlin Wall came down in 1989; and the more recent influx of Mexican immigrants across the border into the workforce of the US – to name a few. In each case, the dominant paradigm offered tremendous resistance at first, but gradually as the counter-flow of the unfamiliar became more commonplace, fear subsided and something new and more integrated was born out of the confusion. These historical trends are still unfolding, and will be for some time, largely because wherever the path of inequality prevails – which is to say, throughout much of our culture most of the time – this balancing of flow and counter-flow is perceived as an undermining force by the powers that be and those who have vested interests in inequality. Still, once set in motion, counter-flow is not easily stopped, and at the ground level, serves to erode the continuum along which inequality was previously made possible.

Take the controversy surrounding immigration, for example. Those who are opposed to immigration generally fear the loss of jobs to foreign workers, the introduction of public health issues through the potential spread of infectious diseases brought by the immigrating population, the potential threat of terrorism and other national security issues, cultural contamination by foreign ways of life, population increases that stress existing resources and infrastructures, and dilution of national identity. Those who are in favor have become habituated to cheap labor, and have discovered through daily contact that their xenophobia was largely misplaced. They see how free-trade agreements have decimated local economies from which immigration flows and carry with them a certain ethical obligation to those who have been displaced by our policies. There would seem to be reasonable arguments on both sides of this issue, and it is not that easy for anyone who has seriously considered these arguments to land squarely on one side of the issue or the other.

◇◇

Whether one is overall in favor, or opposed, the flow of US jobs overseas that began in the 1980s and the counter-flow of immigration into this country in the early 21st century appears to be a form of rebalancing in the wake of globalization that is unstoppable. This rebalancing is changing the nature of the old continuum between citizen and foreigner – releasing something new, a kind of nomadic global worker, for whom the entire planet is the job market. We think of immigration as a precarious strategy of low-level farm or factory workers, but the same phenomena is happening in every strata of the global economy. Americans often work overseas; foreigners increasingly work here; the Internet increasingly facilitates collaboration across geopolitical boundaries. The barriers between formerly separate nation states are becoming more permeable, even as governments scramble to fortify them.

Just as earlier waves of immigration – from Africa in the early 19th century; from Western Europe in the early 20th century; from Asia in the 1980s; and from Eastern Europe in the 1990s – have forced awareness about the differences and the commonalities between ethnic groups of all possible variety, so too does this continuing wave from the South now provide further opportunity for positive changes in the way that we relate to one another. Whether we take full advantage of these opportunities or not, they are being offered within the Realm of Three through the flow and counter-flow of people across borders once considered less permeable.

The Realm of Three and the Integration of the Third and Sixth Chakras

In the Realm of Two, choosing the path of equality is a matter of cultivating an essentially Feminine way of being – open to the pleasures and responsibilities of a soul-centered life in the second *chakra*, and to an underlying sense of interconnectedness fed by caring and compassion in the fourth *chakra*. To continue this path within the Realm of Three, where consciousness serves as the vehicle for change, we must cultivate a similar relationship with the sixth *chakra*. In the sixth *chakra* we learn to hold our beliefs lightly, observe reality with fresh eyes, and adapt ourselves accordingly. It is this supreme flexibility that allows us to see new solutions to old conflicts, to creatively synthesize polarities, and to establish the flow and counter-flow that leads to purposeful change.

Making the necessary collective shift in consciousness to take full advantage of the opportunity for growth in the Realm of Three requires us to forge a psychic connection between the third and sixth *chakras*. In *Tracking the Soul*, I describe this intrachakra connection as a relationship of power. This is not the power of dominance, or what pagan eco-activist Starhawk calls "power over," but power underwritten by consciousness, or what she calls "power-from-within" (1-14). Power-from-within is the capacity to live from a place of awareness that sees how everything fits together with everything else, and how everything has its essential place within the whole. It is easy to get caught up in the third *chakra*, believing that ours is the only truth, and then using brute force – often in denial of evidence to the contrary – in order to force our worldview on others. It is only when the sixth *chakra* is open and functioning clearly, that we can see the relative nature of our point of view, and modify our behavior accordingly.

Whatever we think is going on is incidental to the process of Life unfolding all around us. When we are able to step back and discern our place within that unfolding, we gain the level of detachment necessary to initiate and orchestrate meaningful change – resolving conflict on the path of inequality, becoming midwives to creative synthesis on the path of equality, and setting appropriate counter-flows in motion. When the third and sixth *chakras* are working in tandem, these small acts of intervention become homeopathic in nature, adding minimal amounts of like energy to like in order to move something along that is already in motion. It is this homeopathic approach to creative involvement in the world that constitutes the highest possibility within the Realm of Three.

Endnotes

1. According to the National Priorities Project, "The numbers include military and non-military spending, such as reconstruction. Spending only includes incremental costs, additional funds that are expended due to the war. For example, soldiers' regular pay is not included, but combat pay is included. Potential future costs, such as future medical care for soldiers and veterans wounded in the war, are not included. It is also not clear whether the current funding will cover all military wear and tear. It also does not account for the Iraq War being deficit-financed and that taxpayers will need to make additional interest payments on the national debt due to those deficits."

    Barack Obama declared an official end to the war on August 31, 2010. But 50,000 troops remain in Iraq (as of mid September, 2010), incurring incremental costs, and the Department of Defense has requested an additional $51.1 billion for Iraq in FY 2011. Thus, the total cost of the war will continue to rise.

2. The price of oil in mid September, 2010 was about $77 per barrel with a one-year forecast of $89 per barrel, after reaching an all-time high of about $130 per barrel two years previously (Oil-price.net).

Figure 3: The Rune Nauthiz

Constraint, necessity, the limitations of time and space

Chapter Four: Into Manifestation

Entering the Realm of Four, we see the completion of a process that was only a potential when we crawled out of the Abyss onto dry land. Conjuring our Creator god of choice in the Realm of One, we were the ones suddenly faced with choice in the Realm of Two – to see as the Creator does, from a place of neutrality where everything contributes to a larger interconnected whole; or to be blinded by unnecessary judgments about right and wrong, good and bad, ok and not ok. Whichever way we chose, we then were challenged, in the Realm of Three, to rectify our imbalances and rise to a higher level of creative resolution, using the consciousness infused in us through our relationship to our Creator god of choice. How we respond to this challenge, in the Realm of Two and the Realm of Three, determines what sort of creation will prevail. We make our bed on these more fundamental levels, and then we live in it – in the Realm of Four. This is where consciousness solidifies into what we commonly call hard reality – the consensus reality upon which our subjective image of the world is projected.

The apparent solidity of this world is nowhere more convincing than in the Realm of Four. As Iamblichus discusses it (Waterfield 55):

> Everything in the universe turns out to be completed in the natural progression up to the tetrad, in general and in particular, as does everything numerical – in short, everything whatever its nature. The fact that the decad . . . is consummated by the tetrad along with the numbers that precede it, it is special and particularly important for the harmony which completion brings; so is the fact that it provides the limit of corporeality and three-dimensionality. For the pyramid, which is the minimal solid and the one which first appears, is obviously contained by a tetrad, either of angles or of faces, just as what is perceptible as a result of matter and form, which is a complete result in three dimensions, exists in four terms. . . .

> Because the tetrad is like this, people used to swear by Pythagoras on account of it, obviously because they were astounded at his discovery and addressed him with devotion for it . . . .

The Decad (the number Ten, containing all numbers) is completed by the Tetrad, because it is visually composed of four layers of dots, with four at its base, and each preceding number above it. This figure – called the *Tetraktys* – was held in great reverence by the Pythagoreans, since it showed how this apparently solid universe was constructed from 10 numbers arranged in 4 groupings, and is a visual representation of the truth that is encoded in it. It is not until we reach the Realm of Four that this revelation becomes possible.

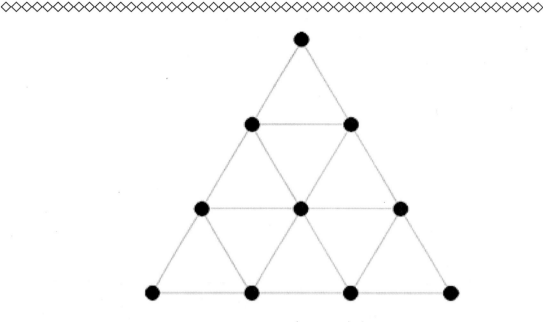

Figure 4: The Tetraktys

The Tetraktys and the Ascent to the One

The Pythagoreans also valued the *Tetraktys*, because it showed the way of Ascent from the fallen state of the embodied world back to a mystical identification with the One. Moving from the bottom of the pyramid up, you have a depiction of the individual soul rising above its entrapment in matter through various stages of refinement until it reaches the pinnacle of possibility, where a return to the One is achieved.

As pointed out in *The Seven Gates of Soul*, this desire to escape the material realm that motivates the Pythagorean Ascent became the mindset of inequality that came to serve as the basis for Judeo-Christian and Islamic moral codes, which denigrated the body as an impediment to spiritual achievement. A similar development took place in the East, where concepts such as *karma*, *samsara*, and reincarnation were understood to ensure a form of entrapment in the embodied state, to be transcended through various forms of enlightenment.

In *The Seven Gates of Soul*, I propose that a more useful spiritual path is one that seeks not to escape this manifest world but to bring as much consciousness as possible into it. Such a path evolves through a deepening recognition of equality and of the immanence of the One in all things in the Realm of Two, and through making conscious choices in light of this recognition in the Realm of Three. Despite the Pythagorean desire to escape the embodied world, it is possible to understand the *Tetraktys* as a description of this conscious movement through the Realm of Four in the opposite direction - the Descent of Spirit[1] more deeply into embodiment. In this direction, moving from the top of the pyramid down, you have a depiction of the Creator entering into and becoming immanent within Creation. Allowing a more liberal reinterpretation of Pythagorean principles, I will attempt to decode the *Tetraktys* in this way.

Four and the Construction of the Physical Universe

Just as the number Three figured prominently in our attempt to make sense of the creation, so too does the number Four figure prominently within the creation itself. If the number Three shows how we attempt to understand the manifest world, the number Four is built into the very fabric of the manifest world. Iamblichus makes his case throughout the chapter on the Tetrad in *The Theology of Arithmetic*, extolling with wide-eyed hyperbole the wonders of the four "foundations of wisdom" - arithmetic, music, geometry, and astronomy; the "four sources of the universe," which he leaves unspecified; the four basic categories of entity within the universe - angels, daemons, animals, and plants; the four kinds of planetary movements - progression, retrogression, direct, and retrograde stations; four kinds of plants - trees, shrubs, vegetables, and herbs; four kinds of virtues - wisdom, moderation, courage, strength, and power, and justice, beauty, and friendship (with a little obvious finagling to fit eight virtues into the space of four); the four parts of the human body - head, trunk, legs, and arms; the "four sources of a rational creature" - brain, heart, navel, and genitals; the four seasons of the year - spring, summer, autumn, and winter; the four seasons of man - childhood, youth, adulthood, and old age; the four "measures of general change" - eternity, time, critical time, and passing time; and so forth (Waterfield 56-59).

As with Iamblichus' groupings of Three, we in our postmodern wisdom may puzzle over some of his designations. The intent behind these groupings was to show how the manifest universe was constructed - not conceptually, as with the number Three - but as a tangible reality amenable to the senses. Of these categorizations, none was so basic - or so pertinent to our understanding of the spiritual implications of Four as a path to the full embodiment of the One - as the four elements: Earth, Air, Fire, and Water[2].

The Four Elements

Again, given our knowledge of modern chemistry, it is easy for us to laugh at the simplicity of this concept. Yet understood metaphorically, the four elements endure as one of the cornerstones of contemporary metaphysical understanding. In Part Two, I discuss the

◇◇◇

astrological implications of this fourfold division. Here I will briefly explore the more general metaphorical implications, noting that a great deal has been written about the elements elsewhere throughout the metaphysical literature.

Despite the deceptively crude division of matter into four categories that would beg for more sophisticated analysis in any modern chemical laboratory, the four elements were an early conceptual depiction of physical matter. Aside from any additional metaphorical associations they might have had, earth, water, and air corresponded roughly to solid, liquid, and gas; while fire was understood to be the agent of transformation – essentially energy – that allowed one state to mutate into another. These physical states and their metaphorical correlates in turn served as an early conceptual template for human physiology that gradually evolved to encompass psychology as well. This template – which divides the full range of human experience into four multi-level domains – is still useful today.

Earth represents that which requires our practical attention – everything from the health of the physical body; to the job, vocation, and career path that puts food on the table and a roof over our heads; to the decisions that we make about where and how we are going to live; to deeper questions about our inherent mortality and the fate of the soul after death.

Water symbolizes the emotional realm – encompassing our relationships with parents, children, friends, and all those with whom we share a common bond of the heart; the feelings that these relationship elicit; the desires that draw us more deeply into this life; the wounds we suffer as a consequence of our involvements; the memories around which our sense of identity coagulates; and the deeper, less easily defined spiritual longings that propel us along the soul's path.

Air encompasses the life of the mind, and all of its ramifications for our participation in the embodied world – the realm of thoughts, ideas, and beliefs; our need to communicate with others; our participation in the commerce of our collective life; our creative inventiveness, problem-solving abilities, and capacity to use our talents and skills in making a contribution to the greater whole of which we are part; our social life; our exposure to cultural sources of knowledge and conditioning; and our capacity to understand and articulate our experiences.

Lastly, Fire can be metaphorically associated with our capacity for growth and for identification with the spiritual dimension of our being – the power to initiate change; the underlying sense of purpose, motivation, and ambition that drives us forward in this life; our identification with causes, ideals, and values; our ability to transcend our experience, to gain perspective, to reach for vision, and a more enlightened understanding of the mundane, as well as a deeper sense of connection to Spirit.

◇◇◇

Phase One: The Opportunity for Awakening

Understood in this metaphorical way, the four elements provide a foundation – the bottom row of the *Tetraktys* – on which a four-phase Ascent in the Pythagorean sense must begin. According to the Pythagoreans, "the Ascent begins in the material realm with the contemplation of the Divine in the objects of sense; this is the Awakening" Phase (Opsopaus).

From our perspective, the awakening of consciousness through the senses is not the first step toward a contemplation of the Divine, so much as it is an ongoing discovery of the presence of the Divine within the material realm. The reunion with the One sought by the Pythagoreans, in other words, is less an Ascent from the material realm, but rather a meeting within. The four elements on the bottom rung of the *Tetraktys* provide the raw material through which this meeting – or awareness of immanence takes place. From the perspective of the soul, the opportunity for awakening to the omnipresence of Spirit provided by the four elements is the spiritual essence of the embodied world. This awakening takes place through the sensory and emotional experiences associated with each element, as well as through participation in the domains they govern.

Phase Two: Purification By Wounding

The second stage of the Ascent – represented by the three dots on the second tier of the *Tetraktys* – was known by the Pythagoreans as the Purification Phase (Opsopaus). This was where the ascending soul sought to "calm the non-reasoning mind," the sensory self, filled with desire, moved by emotion, and driven by instinct. Once the senses and emotions were unleashed during the Awakening Phase, the ascending soul had to make sure that they did not lead it astray onto a path of sin and self-delusion. Purification of the senses and the emotions was necessary to assure the soul would continue ascending.

As pointed out in *The Seven Gates of Soul,* the instinctual body wisdom of the senses and the emotions was deeply mistrusted by the Pythagoreans, by Plato and the other members of the Pythagorean Succession, by the religions that sprang up in the Mediterranean during the Hellenistic era, by the early empirical scientists of the 17th and 18th centuries, and by many of the early psychologists of the 19th and 20th centuries. Yet, from the soul's perspective, it is exactly this "non-reasoning mind" that allows us to make direct contact with an immanent god.

As we sense, feel, and follow our instincts we begin to explore what the path of soul is for us. This individual process of self-discovery takes place within the Realm of Four as we encounter the four elements through the "non-reasoning mind" of the Pythagoreans, more accurately understood from the soul's perspective as the wisdom of the body. Mistakes will

◇◇

be made as we explore the wisdom of the body, and we will be wounded. In the Descent of Spirit into matter, mistakes and wounding afford an opportunity for the soul to learn. Where learning takes place in a non-judgmental atmosphere, the soul evolves in its capacity to consciously embody Spirit and to draw one's deity of choice more deeply into the embodied world.

A second task on the third tier of the *Tetraktys*, was considered by the Pythagoreans to involve "quieting the inner discourse" of the reasoning mind (Opsopaus). It was thought that quieting the mind would allow the soul to hear the thoughts of God[3], which were "neither discursive nor sequential in time." God does not speak to us in words, as in ordinary conversation, nor does God's message necessarily unfold from one moment to the next.

As discussed in both *The Seven Gates of Soul* and *Tracking the Soul*, the lessons we are here to learn – which is Spirit speaking to us – are more easily seen through windows that open during critical stations of various astrological cycles that unfold over the course of a lifetime. In these previous books, I demonstrated how to track these messages through a technique called the cyclical history, in which we attempt to see repeating patterns within our memories of these stations. I will return to this technique in Part Two, when I speak about the astrology related to the number Four. For now, suffice it to say that in order to see Spirit within our lives, we must step backwards and take a broad cyclical view of time - in much the same way that we might have to climb higher to get an aerial view of a giant footprint we couldn't see as we were standing in the middle of it.

Paradoxically, the other way in which we might more clearly hear Spirit speaking to us is by coming more completely into the present moment. As discussed in *The Seven Gates of Soul*, most of the world's mystical traditions speak of this prized awareness of Spirit's presence as a matter of focused attention in the here and now. Most meditation techniques involve practices designed to train the mind to attain this focus at will. When the Pythagoreans speak of "quieting the inner discourse," this is essentially what they are talking about.

While I would agree with the Pythagoreans about the necessity for doing this, I disagree with their assertion that quieting the mind brings one closer to God. What I would say instead is that it creates a space for Spirit – manifest as consciousness – to enter more deeply into the embodied experience. When we are focused – so absorbed in the moment that the moment is all there is – we become inhabited by a Fully Present Awareness that is intelligent, creative, and clear. We cannot function at this level of awareness - or experience the presence of Spirit within us - when there is static on the line. By quieting the mind, we reduce this static.

The third and final task on the third tier of the *Tetraktys*, as envisioned by the Pythagoreans, involves a "non-willful surrender" to the Divine Nous, or the spark of the Divine that exists in each soul. For the Pythagoreans, this surrender allows a direct, intuitive perception of the Ideals (Truth, Beauty, Courage, Justice, Strength, etc.) that lie beneath the outer appearances we perceive as reality. Once the non-reasoning mind and the reasoning mind have been quieted, what remains is the noetic mind, through which direct apprehension of God becomes possible.

Again, I would agree with the Pythagoreans that the awakening of intuition is a desirable goal, but present a slightly different understanding of the noetic mind. From my perspective, it is not the quieting of the mind that creates the possibility of intuition, but the focusing of body wisdom (sensation, feeling, and instinct) within the moment, balanced by a broad awareness of cycles studied in retrospect, that allows the perspective necessary to see life as a whole.

According to the Pythagoreans, the quieting of the non-reasoning mind and the reasoning mind so that the noetic mind could function more clearly was a matter of cultivating the Kathartic Excellences (Opsopaus) – virtues associated with control and suppression of the body. The goal in all matters Pythagorean was to facilitate the soul's Ascent from matter. To the extent that the soul's purpose is construed instead as bringing Spirit – or one's chosen deities – more deeply into the embodied life, then catharsis is not an Ascent from matter at all, but a Descent into it, armed only with consciousness and the power of choice. Like spelunkers entering a cave with a flashlight, each of us is charged with the task of seeing and making our way through the darkness. Inevitably, we will stumble and fall, but this universal experience – of wounding, and of learning to negotiate the issues at the heart of our wounding with greater compassion and clarity of consciousness – will ultimately prove to be the catharsis through which we are most deeply motivated to conjure a better, clearer, brighter god or goddess and invite our deity of choice to dwell more solidly within us.

For, as pointed out in *The Seven Gates of Soul*, it is our suffering – not our desire to escape suffering – that provides the greatest opportunity for spiritual growth in this life. If we pay attention to our body wisdom, from moment to moment to moment, we will become aware when we have crossed a certain internal line from affinity to excess, or begin to suffer from a deficiency potentially rectified by contrast. If we then catalogue these moments in relation to the larger patterns unfolding through cyclical time, we gain the perspective necessary to transmute our issues so that they no longer have the power to wound us and throw us out of balance with ourselves.

This is not a matter of escaping our issues, or of permanently curing ourselves of their sting, but of bringing more consciousness to bear upon them, as we live with them and learn from them. Like an alcoholic staying sober one day at a time, these core issues are the demons through which we invoke our Higher Power – in whatever shape or form that takes for us. The invocation begins with body wisdom and deepens with growing awareness

◇◇

of the larger patterns of wounding, creative adaptation, and release that weave through the embodied life.

Body wisdom begins to evolve in the Realm of Three as we become conscious of the various polarities on which we are engaged, and our relationship to those polarities. In the Realm of Four, this awareness, or body wisdom, begins to anchor itself more specifically to various circumstances in our lives – circumstances that evolve through cyclical time to reflect our growth, or lack of growth, in relation to them. Our experience of life, and particularly of the suffering that life entails, becomes the mechanism for evoking a higher level of love, creativity and conscious application of our god-given talents and abilities in response. Or put another way, it is our core issues – the most intractable sources of suffering within each life – that serve as the attractive force that draws our deity of choice more deeply into us.

In *Tracking the Soul*, I introduced a concept from yogic philosophy called the five *koshas* – or five sheaths – to measure how deeply the embodiment of Spirit goes. The five *koshas* describe a continuum ranging from pure matter at one end to pure Spirit at the other. Together they can provide a conceptual scheme within the Realm of Three for increasing awareness of where we are in relation to a given polarity within the Realm of Two. In the Realm of Four, we can begin to experience the reality behind the concept as each of the *koshas* correlates with one or more of the four elements. *Annamaya kosha* – or the physical body itself – is associated with Earth, in both its literal and its metaphorical implications. *Pranamaya kosha* – or the level of energy dynamics – is associated with Fire; and *manomaya kosha* – or the emotional, psychological level – is associated with Water. *Vijnanamaya kosha* – or the imagistic, symbolic level – reflects a combination of Water and Air; while *anandamaya kosha* – or the level on which Spirit fully penetrates matter – reflects a combination of either Water and Fire, Fire and Earth, or Earth and Air.

The last four elemental combinations are called disparate, because their integration requires a certain amount of spiritual growth. Water does not combine with Air, on an internal alchemical level, until we gain access to the unconscious mind and become skilled at consciously integrating its contents. Water does not combine with Fire until we have transcended our individuality enough to experience the interconnectedness of all life, and learned to go about our personal business with care and compassion for others. Fire and Earth grate against each other until the soul understands that its most painful and difficult experiences are also its greatest opportunity for growth. Air and Earth do not work well together until we can discern the spiritual implications of our mundane affairs and negotiate our way through them with clarity of spiritual intent, commitment and integrity. It is because these disparate combinations provide the greatest challenges, that they become the primary source of purification on the second tier of the *Tetraktys* in the Realm of Four.

◇◇

Phrase Three: Illumination Through Elemental Alchemy

Phase Three of the Ascent, as understood by the Pythagoreans, involves the falling away of what the Hindus call *maya*, or illusion, and the ability to see that all the apparent separate forms in which God is housed are God in disguise. This ability is the natural function of the noetic mind, or Divine Nous – the spark of God that dwells at the center of every living soul. The Pythagoreans sought identification with the Divine Nous as a stepping stone to identification with the Divine, believing that once the soul was able to quiet the non-reasoning mind and the reasoning mind, and center itself within the noetic mind, it would commune with the gods and goddesses and essentially become one of them. From within the conceptual scheme I am introducing in this book, such an attainment is tantamount to becoming the god or goddess in whose footsteps one has been following.

Our focus here is on the Descent of Spirit into matter – experienced by the embodied soul as increased awareness. Becoming a god or goddess does not mean leaving the Earth plane and transcending to some mythical plane on a par with Mount Olympus or Vahalla. It means that through our willingness to address our core issues consciously, as a spiritual opportunity on the second tier of the *Tetraktys*, we attain a certain godlike flexibility in relation to them on the third tier.

Instead of living passively at the receiving end of the alchemy of our wounds, we become alchemists, able to change the lead of unconscious suffering into the gold of conscious self-redemption. Instead of being victimized and wounded by our excesses and our deficiencies, we gain the power to shift them. We do this not only through conflict resolution, creative synthesis and the intentional generation of counter-flow – measures available in the Realm of Three – but now also increasingly as a process of creating elemental balance.

Within the Realm of Four, we are now moving – both literally and metaphorically – toward a more elemental exercise of consciousness. This is so because any polarity we could possibly imagine in the Realm of Two or seek to balance in the Realm of Three is a manifestation of some elemental relationship.

In Chapter Three, I revealed my own personal relationship to the tightness-looseness scale. Here this scale can be understood as a relationship between Earth and Water. The greater the proportion of Earth, the farther toward the tight end of the scale we move. The greater the proportion of Water, the farther toward the loose end of the scale we move. Excess tightness essentially means too much Earth, too little Water; excess looseness means too much Water, too little Earth. Adjusting the tightness-looseness scale then can be envisioned as a matter of mixing Earth and Water in the same way that one might mix concrete to lay the foundation of a house, or create a slurry that would separate a heavier metal from the dross.

◇◇

Seeing it this way, while knowing something about the nature of the elements, allows us to approach any dilemma on the tightness-looseness scale with a simple clarity that is nonetheless quite powerful. When I was concerned, in Chapter Three, that the new love that had entered my life was loosening my schedule, I instinctively went down to the river to settle myself down. My concern arose from a condition of excess tightness, now revealed to be excess Earth. The antidote to that condition was Water. In other situations in which my excess tightness became problematical I have also sought Water in less literal forms. In moments of abject loneliness and social isolation, I have cried, and somehow my tears helped move me back to state of balance. Often when I am stuck in my creative process, I will get up and make myself a cup of tea or grab some juice from the refrigerator. One of my favorite things to do when feeling disconnected from the larger world in which I live or the alien culture that surrounds me is to walk into a bookstore magazine section and let myself be drawn to various magazines – not as a rational (earthy) decision, but responding to subtle cues in color, design, or headline that catch my peripheral vision (a much more intuitive, or watery approach).

Just as with tightness and looseness, so too can any other polarity be understood as the interplay of two elements. Hot and cold, for example, can be construed as an interaction between Fire and Earth, or perhaps in certain contexts between Fire and Water. Wet and dry can be imagined as either a Water-Fire dynamic or a Water-Earth dynamic. These designations are common to classical astrology, but within an astropoetic mindset, the sensory attributes need not stop there. Rough and smooth might be envisioned as another Earth-Water dynamic; young and old as Fire-Earth; male and female as Fire-Water; thick and thin as Earth-Air; wild and tame as Fire-Earth; fast and slow as Air-Earth; in and out as Water-Fire; high and low as Air-Earth; bitter and sweet as Fire-Air; and so on. Each elemental dynamic gives birth to multiple sensory polarities; many sensory polarities can be represented by more than one elemental dynamic; and the designations we use are somewhat subjective and context-dependent[4]. Still, the exercise itself is valuable, since it allows us to approach the illumination phase of the *Tetraktys* with a language that is simple, intuitively obvious, and limited in application only by the poetic imagination.

Secondary Elements

We gain flexibility in our elemental alchemy by positing the formation of secondary elements when any two elements combine. Fire and Earth can be understood to produce ash and smoke; Fire and Water, steam; Fire and Air, wind; Earth and Water, mud, clay, or slurry; Earth and Air, dust or pollen; Water and Air, foam or fog. These and other possible secondary elements then serve as a new set of metaphors for understanding what happens on an elemental level, when we encounter any situation in life begging for creative transformation.

◇◇◇

I alluded earlier, for example, to the possibility of mixing Earth and Water to make concrete or slurry in addressing my issues along the tightness-looseness scale. These secondary elements (concrete is a form of mud) become valuable metaphors in their own right, as I contemplate my options along this continuum.

Thinking about mud, for example, in relation to my tendency to want to anchor creativity on a solid foundation of thorough research, I am reminded instantly of two things. First is a memory of a book I read when I was getting ready to build my house, called *The Owner-Built Home* by Ken Kern. Kern was a maverick builder who liked to experiment with unconventional techniques and building materials – an approach that appealed to me in theory, if not in practice. One of his ideas was to shape the walls of a house not in right angles or squares, but rather in sinuous curves, using chicken wire as a base and ferrocement over that to form the walls. Ferrocement is a form of mud.

What I realize in contemplating this image as a metaphor for my creative process, is that this is what I am essentially doing in this book. My discussion of numbers here, and the Pythagorean theory about numbers, forms a frame of chicken wire, while free-ranging discussion of everything from the war in Iraq to ferrocement as a building material is the mix of mud (Earth-Water) that I am using to give substance to the structure.

My second memory is of finger-painting in grade school, and how free that made me feel – the slippery feel of the paint on my hands, the sheer wild joy of swishing it around the paper, and the carefree abandon of not feeling compelled to come up with a masterpiece at the end of the day for all my effortless effort.

This memory immediately triggers another of a workshop I gave several years ago around the astrology of the inner child. As part of that workshop, I made a space for a friend and student of mine to teach a module on mask making. Participants each made a mold of their faces, and then formed their mask to the mold, after which they dried, painted and decorated them. Somewhere in the gleeful madness that ensued, the students together decided they wanted to paint my face. Though I resisted at first, I let them, and wound up thoroughly enjoying myself. I felt loved, free to be silly, and filled with wonder at how the exercise had somehow bonded us as a group. The work after that was deep and powerful, and gone was any sense of embarrassment or reticence in the sharing of intimate truths.

Within the context of these secondary elements, an intuitive connection can be made between finger-paint and slurry – a combination of Water and Earth, with more Water than Earth. On the tightness-looseness scale, slurry is more loose than tight. Finger-painting and allowing myself to be painted moved me farther toward the loose end of the spectrum than I am normally comfortable with, but doing that proved to be a liberation. Paradoxically, adding a bit more water to the mix allowed something within the group experience to gel.

◇◇

Remembering that, and allowing myself to slosh a bit the next time I find myself in slurry, gives me a wider range of alchemical freedom. No longer am I quite so bound by a definition of myself as registering 6.5 on the tightness-looseness scale, or whatever other imaginary number I would assign to indicate more tight than loose, but not too tight. With this freedom, I can move as I choose - up or down the scale. What is this freedom ultimately, but the versatility of a Creator god, able to create or re-create the world any way It wants?

Phase Four: Union With the One

The final phase of the Ascent, as understood by the Pythagoreans, and as represented by the top tier of the *Tetraktys* - a single dot - was reunion with the One. This was the ultimate goal of the Pythagorean School, and it was achieved when even the Divine Nous lost its identity in flowing back to the Source. If on the third tier, one attained equality with the gods, on the Fourth tier one surrendered the final illusion that there was anything to be equal to. On the way up the number scale from One to Four, One precedes the notion of a choice between equality or inequality, because choice between any two options belongs to the Realm of Two. On the way down the number scale, from Four to One - ironically called an *Ascent* by the Pythagoreans - one must surrender the notion of equality at the gate. To make the final shift - back to Ain Soph, the Source of All Things - the soul must lose even the "I" that might experience its own return. Or as Pythagorean scholar John Opsopaus has put it, the soul "must rise above Form and Idea in the Inebriation of Love".

In the Inebriation of Love, all sense of separation dissolves. In terms of the Ascent, this means complete identification with Spirit, discussed in relation to the *chakra* system in *Tracking the Soul* as the province of the seventh *chakra*, and to some extent within each *chakra* penetrated to the level of *anandamaya kosha* (pure Spirit on the continuum from matter to Spirit).

In terms of the Descent - that is to say, the journey of the embodied soul - this means essentially becoming an agent of Divine will here on Earth, with no interference whatsoever from the conditioned personality, the fear-based ego, or the small identity limited by space and time. These lesser vehicles don't completely disappear once the fourth tier of the *Tetraktys* is reached, but they do become transparent in their ability to allow Spirit to function unimpeded through them.

At this level, one can assume any position on any continuum relevant to the situation at hand, serve as an agent of conflict resolution on the path of inequality, an agent of creative synthesis on the path of equality, and a catalyst to counter-flow where balance is needed. On the fourth tier, one can also shift any element up or down the intensity scale, adjust the blend of elements, or trigger the creation of any secondary element that might be necessary to further the outworking of a divine plan.

◇◇

Following a Personal Path Through the Realm Four

What is this plan? It is an image of the embodied world – the Creation – that reflects the intent of the Creator god in whose footsteps we follow. Obviously, this image will vary depending upon our god or goddess of choice. Yet any deity worth the name will lead us from the Abyss onto a path of equality, where judgments are necessarily suspended in the experience of a truth that transcends good and evil, light and dark, right or wrong, to encompass both.

Along this path, our chosen deity will teach us how to resolve internal conflict, how to balance the various polarities within our nature, how to shape-shift at will in order to move any situation toward resolution, creative synthesis, and/or awareness of a broader range of possibilities. Lastly, by following this deity, we are empowered to create a life for ourselves that reflects the life of the deity itself in some way. The mythology related to the deity will become the template for our lives. As we fill in the template with the choices that we make at each juncture along the path, the plan reveals itself. The world we inhabit shapes itself in the image of the myths that inform it.

In following an Odinic path, for example, I am naturally led on a quest for runic wisdom. This wisdom is obtained only by turning everything – especially conventional wisdom – upside down. As I do this, I begin to understand a certain level of paradox built into the fabric of the embodied world, and more pertinently, into my own nature. The experience of paradox makes no sense from the standpoint of the rational mind, but it makes perfect sense along an Odinic path, for Odin was a god who contradicted himself at every turn (Gundarsson 225):

> Thus, by the mystery and paradox of Odhinn's nature, he holds all things within himself. He builds the walls to ward Midgardhr and Asgardhr, but wanders outside at will. He is a lord of oaths and of betrayal, of making and unmaking. As a god of war, he makes his heroes invincible till he himself comes to take them in battle. It is said, and rightly, that Odhinn enjoys stirring up strife: only through struggle can one be tested and grow, and only by the clash of opposites can synthesis be achieved. Every step of the Odian path is a battle of some sort, whether external or internal. He is not a comfortable patron and was seldom loved by the folk as a whole.

My Odinic path began with a childhood game of king of the mountain, a battle for dominance of an ash pile in my grandfather's backyard. It continued through endless battles on the playground of grade school, in which I was confronted not only by my fellow students, but also by the school authorities and my own parents. In the Vietnam War, I became a conscientious objector, while fighting angry external battles against the government and an-

◇◇

gry internal battles against my own demons, who were always demanding more of me than I seemed able to give. In the land cooperative where I now live, I have fought endless political battles with my neighbors over large issues, affecting the entire forest in which the cooperative is housed and small issues, affecting nothing but the petty egos of those compelled to fight. Fighting these battles has not endeared me to any one, yet refusing to fight them would have been a denial of my nature.

I look around me now as I write this chapter and I see endless battles raging – between an American government attempting to secure Mideastern oil and those countries that own it; between Israelis and Palestinians over the right to exist; between nationalist Chinese and Taiwanese over control of identity; between Sudanese refugees and the government-backed Janjaweed seeking to exterminate them – to name just a few of the ongoing battles that currently rage across the planet. Following the Odinic path, I see these battles clearly as exercises in futility along a path of inequality in the Realm of Two. Yet I cannot contemplate any of these battles without choosing sides, and becoming emotionally attached to the outcome – that is to say, without contributing to the very inequality that I recognize and abhor as the root cause of global ensnarement within the Abyss.

One could simply say that inequality is built into the fabric of the embodied world, and let it go at that, but as a follower in the footsteps of Odin, I do not have that luxury. Instead I must constantly wrestle with the part of myself that can be provoked to battle – not suppressing it completely, which would be an act of self-denial, but also not giving it free reign, which would be an act of self-destruction. I must pick and choose my battles carefully, and in that choice, find the balance between equality and inequality within myself that renders even this dichotomy meaningless. For where I am drawn to battle, there is invariably some inequality or imbalance within myself that can only be equalized by externalizing it. Where I am not drawn to battle, there is an equality within myself that contradicts and at times can help alleviate the inequality that rages around me. In the end, there is only the path of the Odinic warrior that I must follow without judgment – sometimes through battle and sometimes through abstention from battle. When I can do this with clear intent, I become an Odinic shaman in the Realm of Three, able to shift elemental balances within the Realm of Four at will.

Such shamanic ability on the Odinic path was embodied by the berserkers – a fierce group of warriors, able to transmute themselves into bears and wolves, "said to be in a kind of ecstatic trance, a holy rage, when they rode into battle, howling eerily, disdaining shields, and inspiring terror in their enemies" (Metzner 75). While there were definitely external battles for these legendary warriors to fight, "in her paper 'The Transformed Beserk,' Jungian scholar Marie-Louise von Franz suggested that the berserker trance was a kind of visionary state, an out-of-body experience in which the soul of the warrior, sometimes in animal form,

◇◇

raged in battle, while the physical body lay as if asleep" (Metzner 76). This implies that when conscious intention is established within the Realm of Three, peace and balanced tranquility among the elements is possible within the Realm of Four. It is to this high state of intentionality – also embodied in the Taoist notion of *wu-wei* or doing without doing – that I aspire on this Odinic path. At times, this quest has thrown my personal world within the Realm of Four – as well as the world of consensus reality around me – into chaos. When I can stop the battle or transmute the necessity for it through the ferocity of my intention, I will have achieved my goal.

The Realm of Four as a Horizontal Compass

Beyond the vertical model of Ascent through the *Tetraktys* and its counterpart in the conscious Descent of Spirit into matter, explored in the first part of this chapter, we might also postulate a horizontal dimension to the Realm of Four that can be described through the metaphor of the compass. Essential to the compass are the four cardinal directions, which serve as a navigational tool for orientation within the embodied world. Not only is the soul seeking to embody Spirit. It is also seeking to do the work of Spirit within Creation, an ongoing challenge that requires it to have a sense of direction and purpose within a world made solid on an elemental level. It is through an understanding of the Realm of Four as the source of this compass that this sense of direction can be found.

Nowhere have I found the metaphor of the compass more clearly delineated than in the Native American concept of the Medicine Wheel – an understanding of the compass that honors the sacred nature of the art of navigation. As Seneca medicine teacher Jamie Sams describes the concept (*Medicine Cards* 21):

> All space is sacred space. . . . The Medicine Wheel is a physical expression of this knowledge, and can be used to set up sacred ceremonial space. . . . The Medicine Wheel is used to gather together the energies of all the animals or creature beings, the Stone People, Mother Earth, Father Sky, Grandmother Moon, the Sky World or Star Nation, the Subterraneans, the Standing People or trees, the Two Leggeds or humans, the Sky Brothers and Sisters and the Thunder Beings. These we consider to be all our relations in the Native teachings. . . . The Medicine Wheel is a symbol for the wheel of life, which is forever evolving and bringing new lessons and truths to the walking of the path.

The Medicine Wheel, in other words, is a recognition that all of life is for learning within the sacred arena of resonant soul space, and that within this learning process, we have the power to evoke appropriate deities from all four directions to deal with whatever issues, life

◇◇◇

circumstance or lesson we might be facing within the Realm of Four. Perspective is gathered and intentions set within the Realm of Three, but then we must move and act within the Realm of Four. On this level, it is the Medicine Wheel that provides direction.

Navigating the Medicine Wheel

This sense of direction can be understood in two ways – as a sense of momentum that comes from within, and as an awareness of the flow of life around us that quite naturally dictates the appropriate response. These two different senses of the Medicine Wheel can be mirror images of each other, when the soul is in harmony with itself. They can also present a seeming contradiction that must be reconciled when the soul is not in harmony, before outer movement within the Realm of Four becomes possible.

As a sense of momentum that comes from within, each of the four directions points toward a different source of motivation and internal drive. When we are driven by a sense of vision, or a strong internal image that we have projected onto the embodied world – whether through belief or conditioning, or through a divine revelation from the deity in whose footsteps we are following – we are said to be coming from the East. When the source of motivation is a strong emotion, or movement of the heart, toward involvement with another person or cause or toward a specific place or set of circumstances, we are said to be coming from the South. In the West, we are moved to act because of some breakdown in the ordinary fabric of life as usual, a death of sorts, large or small, which puts us face to face with the ever-present Abyss. In the North, our motivation comes through family or spiritual lineage and a sense of duty to parents, children, or community.

As a sense of awareness about how life is flowing through the soul space we inhabit, the four directions can be understood through the metaphor of wind. As Jamie Sams describes this association, "The powers (talents and lessons) of the Four Directions can be immediate answers when sent by the Wind Spirits. Traditional Native American teachers always teach children of their Tribe to feel the Wind so that they will know what to do if they are lost or afraid" (*Sacred Path Cards* 84-85).

When the wind blows from the East, the soul is being asked to use common sense in addressing the situation at hand, to weight the pros and cons, to penetrate the veils of illusion with mental clarity, to make a decision. When the wind blows from the South, the answers to the question at hand lies in the heart, and in the deepest desires currently moving on the emotional level. From the West, the wind requires a letting go, and a deeper level of courage and trust in facing the unknown that always follows death. From the North, the wind suggests seeking help from those we respect and from our family, biological or spiritual. Here we can also ask, "What would Jesus (or Odin or Aphrodite) do?" and then act accordingly.

Ultimately, navigating the Medicine Wheel is a matter of simultaneously seeking to understand the source of our motivation or internal drive, and looking to see which way the wind blows. Movement within the sacred space of the Medicine Wheel is then a matter of reconciling the two directions, often by beginning in the place of motivation and then moving with the wind.

As I began writing this book, I was just ending a relationship of 13 years, feeling the death of everything I hoped and longed for in that relationship; letting go not only of the person I loved, but of all the dreams that I had attached to this relationship. Despite the love we had shared, I felt myself teetering on the edge of the Abyss yet again, thinking myself a failure at the delicate art of relationship, resigning myself to terminal bachelorhood, hoping to ride off into the sunset on a solitary horse, having made my peace with my lot in life as a single man approaching elderhood.

Yet, within a very short time (less than six months after this ending in the West), it appeared the prevailing winds had started blowing from the South. I met someone new – the new friend mentioned earlier – quite unexpectedly, and began to feel the wondrous churning of the heart that accompanies new love. For a while, I experienced a strange juxtaposition of realities as the letting go continued, but my heart was also opening to new possibilities. This allowed a broader range of feelings to wash through me than if I had merely stayed in the West, where the emotional climate is often one of sorrow, grief, and despair. Those feelings were still occasionally there, but increasingly displaced by spontaneous joy, erotic longing, and a renewal of enthusiasm for life. Soon I was nearly out of the West entirely and residing in the South. In this case, it was rather easy to align myself with the prevailing winds, although I still needed to move through a transition. In other instances, there could potentially be more resistance, and consequently a harder, longer transition.

The Medicine Wheel and the Elements

Each of the four cardinal directions of the Medicine Wheel can be associated with an element. While the designations vary somewhat from tradition to tradition, what makes most sense to me is the following: East = Air (associated with the mind); South = Water (related to feelings); West = Fire (the element most often correlated with transmutation of existing circumstances); and North = Earth (rooted in biological lineage). In Part Two, I will demonstrate how this arrangement makes astro-logical sense, although not in the way that astrology is traditionally practiced.

Although the Medicine Wheel is a navigational tool in the Realm of Four, with this arrangement of elements and directions, we can also begin to see how the possibilities it offers have additional implications rooted in the Realms of Two and Three. If, for example,

◇◇◇

we consider the relationship between any two directions that are across from each other on the wheel, i.e. East and West, or North and South, we find we are attempting to reconcile elements that are essentially compatible with each other – Air and Fire, or Earth and Water. Such reconciliations are well suited to the path of equality, although as with any symbolic scheme like this, how the alchemy actually plays out depends upon the consciousness that is brought to bear upon it. In the absence of major resistance, reconciliation between East and West, or North and South will be relatively easy to negotiate along a path of equality in the Realm of Two. If our experience is rooted in the West, for example, and the prevailing winds are blowing from the East, there is a natural meeting of energies in the middle that normally results in a relatively easy blending, more conducive to creative synthesis and easy establishment of counter-flow, than to a need to mediate conflict, in the Realm of Three.

If, on the other hand, we consider the relationship between adjacent directions on the wheel, i.e. East and South, South and West, West and North, or North and East, there is greater potential for a sense of inequality in the Realm of Two and the need for conflict resolution in the Realm of Three. This is so because the elements associated with these directional combinations are disparate, that is to say, not easily reconciled on an alchemical level. Air and Water (East and South); Water and Fire (South and West), Fire and Earth (West and North), and Earth and Air (North and East) all represent more difficult challenges to the soul seeking alchemical blending.

Again, this is not to say that an individual soul cannot transcend these obstacles through the choices that she makes, just that in the absence of conscious intent, the default position posed by these directional combinations tends toward conflict. In a very graphic way, we can see that any prevailing wind from one of these directions will be flowing at apparent cross-purposes to the prevailing wind of a direction 90 degrees to either side. This sense of cross-purpose then tends to evoke resistance to change in general, and more specifically to the shift in momentum required by the Medicine Wheel dynamic in play.

Having said that, I must consider why this has not been true for me moving from West to South. My sense of intellectual honesty demands it, for I am putting forth a theory here, and yet my own experience does not match the theory. There must be a reason. I believe it is this: when I first met my new friend, she was going through a messy divorce that entailed not just the ending of a relationship, but also the loss of a piece of land that she loved and a business she was instrumental in creating with her ex-partner. From the perspective of the Medicine Wheel, she was even more deeply anchored in the West that I was. This made it easy for me to leave the West in order to offer her some solace.

My intention, in responding to her dilemma, was to shift East – a relatively easy transition from the West – and offer her astrological insight. I did do a bit of that, but since her time of birth was in question, it took me awhile for an Eastern response to her Western dilemma to become a primary option. In the meantime, I shifted South and offered her the comfort of a friend, and the simple empathy that one human being offers another when they can identify with their suffering. There are astrological reasons why this was natural for me to do, which I won't go into now. Suffice it to say, she was grateful for my Southern hospitality, and I was glad to have something besides the ending of my own relationship in the West to obsess about. Gradually her gratitude evolved into love and trust, while my caring mutated into longing, and my compassion into passion, which I was delighted to find met by her in equal measure. Together, it seems, we had moved out of the West into the South, our respective resistances melted away first by the warmth, and then by the heat of our connection.

A more difficult transition from West to South followed, as we each desired to remain friends with our ex-partners. When we met, my new love had yet to go through a difficult divorce, and it was too soon for her to think about friendship with the man she left. My ex-partner and I had not spoken since I chose to end the relationship, and although I made overtures and expressed my interest to stay in touch, she did not respond. Naturally, when someone is hurt so badly that their world is shattered (which is generally what happens in the West), they become angry (evoking a fire within to match the raging fire of destruction without), and forgiveness, compassion and love (all Southern emotions) are difficult to come by.

As discussed in Chapter Zero, moving from the West to the South is no less a monumental task than finding compassion for those who have courted the Abyss and dragged you kicking and screaming into the Abyss with them. This transition on the Medicine Wheel is depicted by Christ turning the other cheek and advising us to "love thy enemy."

It is Nelson Mandela inviting his captors to his inauguration party, then establishing the Truth and Reconciliation Commission. It is soldiers in Vietnam falling in love with Vietnamese women. It is Hillary Clinton forgiving her husband, ex-president Bill for his infidelity after several years of mortifying public humiliation. It is in fact, a transition so rare, that historical examples are hard to come by. Seldom in this world, do those devastated by loss, suffering and death of loved ones, open their hearts to the perpetrators of the devastation. Yet that is exactly what the true movement from West to South requires.

In a similar way, movement from South to East entails gaining perspective on feelings and psychological processes; movement from East to North requires anchoring a vision on the solid ground of material manifestation; and movement from North to West requires letting go of our accomplishments, our loved ones, and our investment in the manifest world in order to face death in all of its many literal and metaphorical disguises.

Movement is also possible in the opposite direction around the wheel. Shifting from West to North generally involves a restoration of what has been lost or damaged, such as the

◇◇◇

rebuilding efforts in the wake of Hurricane Katrina. Movement North to East means stepping back from our established mindset and thinking outside the box, perhaps the kind of thinking required to establish a more sustainable energy economy in a post-oil world. Movement from East to South means allowing the emotional reality of our experience to sink in, shifting from abstraction – such as that encoded in statistics – to the actual human-scale implications of those abstractions and the living stories behind the numbers. Movement from South to West means letting go of that which is dearest to us, as each chapter of our lives comes to an end, and entering a place of uncertainty and not-knowing in relation to the future – the kind of shift that Victor Frankl experienced, for example, in entering a concentration camp during World War II.

The Implications of Realm of Four Dynamics Within the Chakra System

None of these transitions is easy. All of them require a quantum leap in consciousness – an application of awareness in the Realm of Three, as well as a major shift in circumstance involving some form of elemental rebalancing in the Realm of Four. All of them are catalysts for finding and cultivating extraordinary resources within ourselves in the face of extraordinary circumstances. In my previous book *Tracking the Soul* (Landwehr 97-98), I speak of the seventh *chakra* as a place of orchestration, where we are empowered to view our circumstances from a divine perspective as opportunities for growth and transformation, and then re-enter them accordingly in ways that allow us to cooperate more consciously with the process. Thus, when negotiating any cross-purpose dynamic on the Medicine Wheel, the challenge is to rise to the seventh *chakra* for the necessary perspective, and then move back down again to the *chakra* being triggered by the circumstances at hand in order to make the necessary change on the elemental level in the Realm of Four. Ultimately this process will entail integrating the seventh *chakra* and the first, which is where the elemental level in the Realm of Four is anchored. In *Tracking the Soul*, I speak of this combination of *chakras* as the Alpha and Omega (Landwehr 108-109), where body and Spirit, heaven and Earth, the divine and the elemental come together in synergistic harmony.

It is worth reiterating here that each numerical realm challenges us to effect a major leap in consciousness – whether collectively or individually. These leaps can be mapped to the *chakras* as follows:

> In the Realm of Zero: from the 1st *chakra* to the 4th and 6th.
> In the Realm of One: from the 4th and 6th *chakras* to the 5th.
> In the Realm of Two: from the 3rd *chakra* to the 2nd and 4th.

◇◇◇

In the Realm of Three: the integration of 3rd and 6th chakras.
In the Realm of Four: the integration of 1st and 7th chakras.

Accomplishing the tasks set forth in the Realms of Zero and One allows us to begin following in the footsteps of our chosen gods or goddesses. Accomplishing the tasks set forth in the Realms of Two and Three enables a shift in consciousness from a patriarchal mindset, in which every dog must fight for its place in a world that is marked by conflict, to a more egalitarian mindset that encompasses the Feminine, where the interconnectedness of all life serves as the primary basis for choice. Accomplishing the task set forth in the Realm of Four allows the soul to become the co-creative agent of its chosen deity within the manifest realm, an elemental alchemist, and a skilled navigator of resonant soul space.

In discussing the Realm of Five in the next chapter, I will show how using these consummate skills with clear intent becomes the basis for a humane and compassionate global culture.

Endnotes

1. I use the word "Spirit" in this chapter to describe the immanent Creative Intelligence that pervades this embodied world, and everything in it, including the individual human soul. The immanent god that is Spirit is not generally recognized by most monotheistic religions, though it was often taken for granted among polytheistic indigenous peoples around the world, pagan culture, mystical, shamanistic, and animistic traditions, and certain Eastern religions such as Taoism and Shinto.

2. When used in the archetypal sense, the elements will be capitalized – i.e. Earth, Air, Fire and Water. When Earth is used to refer to the planet on which we live, it will be also be capitalized. When referring to the actual physical substance, or to a specific instance of expression – i.e. the fire of destruction – the elements will be not be capitalized.

3. I use the word "God" in this chapter, not in its strictly religious sense, but as an epithet for the One, which was the Pythagorean deity of choice. God and its referrants – such as the Divine – will be capitalized when referring to the One, and left uncapitalized when referring to a deity, i.e. one among many possibilities.

4. The elements involved in any sensory dynamic can be tied more specifically to the patterns in a birthchart that evoke them. I will explore this in some detail in Part Two of this two-volume set.

Figure 5: The Rune Berkana

Fertility, the emerging Life Force, growth in consciousness

Chapter Five: Into Life

The Pythagoreans made much of the fact that the number Five lay midway between the Monad and the Nonad – that is to say, provided the pivot point of the entire range of numbers they were exploring – and that it comprised exactly half of the Decad, in which the work of the Monad was fully realized (10 being a higher manifestation of 1: $1 + 0 = 1$). In this function, they saw Five as essentially a preferred, new and improved, version of the suspect Dyad. According to Anatolius, one of Iamblichus' teachers (Waterfield 66):

> *Reciprocally, we are able to see first in the pentad, compared with the greater limit, the principle of half, just as we see this principle first in the dyad, compared with the smaller limit: for 2 is double 1, and 5 is half 10.*

> *Hence the pentad is particularly comprehensive of the natural phenomena of the universe: it is a frequent assertion of ours that the whole universe is manifestly completed and enclosed by the decad, and seeded by the monad, and it gains movement thanks to the dyad and life thanks to the pentad, which is particularly and most appropriately and only a division of the decad, since the pentad necessarily entails equivalence, while the dyad entails ambivalence.*

Reading between the lines of Anatolius' discourse, it is easy to see that the Pythagoreans saw the Pentad as the opportunity to apply consciousness in the Realm of Three to the trouble raised by the inherent duality in the Realm of Two ($5 = 2 + 3$). Put another way, in the number Five, the Pythagoreans placed their hope that the path of inequality within the Realm of Two could be naturally and effortlessly transmuted into a path of equality – or to use their word, "equivalence." They ascribed terms to the Pentad such as "marriage" (creative synthesis), "lack of strife" (conflict resolution), and "justice" (creation of counter-flow) – all strategies for applying awareness within the Realm of Three to deal with the inherent ambivalence in the Realm of Two – as discussed in Chapter Three.

The Realm of Five and Consciousness

Within the Realm of Five we can see the alchemy of consciousness discussed in Chapters Three and Four at work – not just in isolated cases of intentional application, but as an intrinsic force within creation. What the Pythagoreans meant when they said that "*the pentad is particularly comprehensive of the natural phenomena of the universe*" is that this alchemy of consciousness is built into the fabric of the manifest world – as natural law. Natural law was

◇◇◇

understood, in part, as a revelation of the divine order of things, made possible through an expansion of the four elements to now include a fifth element – Aether. This fifth element is, on one level, Spirit Itself, the Monad, a Living Force now midway in its process of transformation into a Conscious Universe – symbolized by the Decad – through the evolution of natural law. On another level of understanding, the vehicle for this transformation emerges in the Realm of Five as the felt presence of an immanent Creator god, experienced by all sentient beings as consciousness. I first discussed consciousness in relation to the number Two, where it arises as a possibility. Here, in the Realm of Five, it becomes apparent that this possibility not only exists, but is characteristic of a manifest universe infused by the presence of an immanent Creator inhabiting that universe.

These are admittedly controversial statements, since not everyone is prepared to admit a connection between consciousness and an immanent Creator god, nor that the manifest Creation is conscious. To this point, I have been discussing consciousness as though these correlations were a given, but within the Realm of Five, it becomes crucial to look more closely at what we mean by consciousness – for it is within this realm, according to the Pythagoreans, that a previously inert creation becomes permeated by the Creator's presence, manifest as a conscious life force. Within the Realm of Five, Spirit is Life, and Life is Sentience or Conscious Intelligence.

As discussed in *The Seven Gates of Soul*, Pythagoras proposed these ideas at a time when the mythopoetic view of the world was still dominant. Within this worldview, it was taken for granted that gods and goddesses walked among us, and were still actively involved in the workings of the manifest world and in the affairs of human beings. As the emerging art of arithmology (Pythagorean number theory) evolved, it took this equation of Spirit = Life = Consciousness for granted, and saw within the number Five the metaphysical arena in which the truth of this equation was revealed.

How are we to understand this assertion today, in an age when science has all but explained away the mystery reverberating at the heart of your ability to read and comprehend this sentence, not to mention every other experience – of love, of creativity, of identity, of epiphany, of purpose – to which we might want to ascribe some spiritual meaning?

From the scientific standpoint, consciousness is little more than a neurochemical reaction in the brain, requiring no spiritual explanation at all. Yet, as scientists study consciousness, they are forced to acknowledge that it begs for a more sophisticated approach than neurochemistry is prepared to offer. In the words of J. Allan Hobson, Director of the Laboratory of Neurophysiology at Harvard Medical School (ix):

> As crucial as the traditional reductionistic approach of science is to our understanding of brain-mind unity, we are certain to be disappointed if we expect consciousness

◇◇◇

to reveal itself in an electron micrograph, an ion channel, a recording, or an electro-phoretic plot of brain proteins.

In addition to reductionism, we need three other –isms not always appreciated by scientists. We need emergentism to keep us aware that complex phenomena such as consciousness emerge at higher levels of system organization and cannot be discerned or analyzed at the level of the system's essential building blocks. We need holism to keep us aware that we must be conscious of something, and that something, be it the world, our bodies, or our selves, serves to structure consciousness every bit as much as the brain does. We need subjectivism because, since it is subjectivity that we wish to explain, we had better know – as precisely as possible – just what it is.

Between the lines of this scientific concession, it is possible to hear a Pythagorean sensibility at work, although no scientist that cares about his or her career would dare to claim such an affinity. Nonetheless, in the concept of emergentism, it is possible to recognize an acknowledgment of the forces of creative synthesis at work in the organ of consciousness. This admission is not at odds with our earlier proposal that this is what happens in the Realm of Three when consciousness is applied on a path of equality, nor with the Pythagorean designation of Five as "marriage." In the concept of holism, we hear an explanation of the sense of interconnectedness that must underlie the function of consciousness as discussed in relation to reclaiming the Feminine in Chapter Two - again a prerequisite to an emergent path of equality within the Realm of Five. Lastly - and perhaps most importantly - when Hobson calls for subjectivism in a scientific approach to an understanding of consciousness, he is cracking one of the primary barriers between science and the soul - discussed at length in *The Seven Gates of Soul* (Landwehr 137-150). Science is by definition a quest for objective truth, but a science of consciousness must also encompass subjectivity.

As science broadens its scope to encompass the subjective quest for a sense of its interconnectedness and the possibility of synthesizing apparent polar opposites within the manifest world, it begins to approach the same understanding of consciousness taken for granted by the Pythagoreans as being endemic to the Realm of Five. To be sure, Hobson's proposal is fairly radical given the current scientific paradigm, and he is careful to reassure his colleagues that it is not his intent to discard science for a return to the mythopoetic mindset out of which the Pythagoreans were operating (x):

At the same time, we must not let our commitment to the evidence offered by emergentism, by holism, and by subjectivism become a screen for romantic mysticism, a dodge from quantitative rigor, or an evasion of experimental hypothesis testing. I will therefore endeavor, always, to be of two minds at once: bold and modest, subjective and objective, psychological and physiological, brainful and mindful.

◇◇◇

Nonetheless, even this modest step forward is encouraging, as it suggests that ultimately consciousness must be understood as more than brain chemistry. To bring consciousness within the purview of science at all, it must be approached as the connecting link between the subjective soul and the objective consensus reality it inhabits. Though Hobson's concessions to the mythopoetic worldview of the Pythagoreans are qualified by his training in the demands of scientific rigor, his call for a consideration of subjectivity at least brings subject and object together in a participatory dialogue. Within the mythopoetic worldview shunned by science, a full participatory dialogue would require an interpenetrating fusion of subject and object that renders scientific objectivity impossible and irrelevant. Although Hobson is obviously not prepared to go that far, his desire to reach toward equality between subjectivity and objectivity at least nudges neuroscience toward a definition of consciousness not incompatible with Pythagorean sensibilities.

The Realm of Five and Participation Mystique

A continuous and fully participatory dialogue between subject and object was observed by anthropologists to exist throughout the indigenous cultures of the world. Participation mystique is a term borrowed by Jung from French anthropologist Lucien Levy Bruhl to describe the mystical connection, or identity between subject and object, that characterizes this dialogue. According to Jung (*Psychological Types* 441), participation mystique:

> . . . consists in the fact that the subject cannot clearly distinguish himself from the object but is bound to it by a direct relationship which amounts to partial identity. This identity results from an a priori oneness of subject and object. Participation mystique is a vestige of this primitive condition (456). . . It is also a characteristic of the mental state of early infancy, and finally, of the unconscious of the civilized adult, which, in so far as it has not become a content of consciousness, remains in a permanent state of identity with objects.

It is clear from this definition that Jung, who was trained as a scientist, considered the dissolution of the subject-object barrier to be a condition to be outgrown – as primitive human beings became civilized, as infants grew up, and as unconscious contents became conscious. Yet he also recognizes that the oneness of subject and object is an "*a priori* condition," meaning the original state out of which primitive human beings, infants, and the conscious mind all evolve. If this is so, then a return to the Source as it is understood in a spiritual sense – must inevitably involve a return to this *a priori* condition of participation mystique.

In Chapter Two, I discussed how consciousness was an emergent property within the Realm of Two, where the dichotomy between subject and object is first encountered. Within the Realm of One, no such dichotomy exists. Within the Realm of Five, the midpoint of the process and the beginning of the return (to be experienced in the Decad: $1 + 0 = 1$), the dichotomy exists, but now alongside the possibility for its erasure through participation mystique. Along the path of inequality, this would necessarily be understood as a devolutionary unconscious state posited in contrast to a preferred conscious awareness of clear boundaries between subject and object, between the conscious mind and the unconscious mind. Along the path of equality, however, participation mystique is nothing less than the creative synthesis of subject and object, unconscious and conscious minds – productive not of delusion or madness, but of a highly creative, spiritually evolved state of being.

Such a potential has been glimpsed by a few pioneers within the psychological community, such as former Assistant Professor of Psychiatry at Johns Hopkins University School of Medicine, Stanislav Grof. Grof speaks of a transpersonal experience he calls "dual unity" – which sounds like an oxymoron, but which I believe ultimately tends toward a functional state of participation mystique. Like participation mystique, dual unity involves a "loosening and melting of the boundaries of the bodyego and a sense of merging with another person into a state of unity and oneness. In spite of feeling fused with another, the subject retains awareness of his or her own identity" (46). According to Grof, this experience occurs most often in psychotherapeutic sessions in which an individual is reliving prenatal memories; during sexual union, particularly within the context of intentional Tantric practice; and in therapies conducted with psychotropic substances throughout the 1960s and 70s. However induced, "the experiences of dual unity are often accompanied by profound feelings of love and a sense of sacredness" (46).

What is this love and sacredness, if not an experience of the One obtained while still in an apparently separate body? If so, then dual unity represents a step forward, not a step backward, on the evolutionary trail. I consider it a point of entry into the Realm of Five - where the manifest universe established in the Realm of Four reveals itself to be conscious. I realize this is a radical statement from the standpoint of conventional understanding, which posits that it is we who are conscious of the world. But if there is ultimately no barrier between subject and object - which as Jung suggests is an *a priori* condition - then we might legitimately ask, "Is it we who become conscious of the world, or the world that reveals its true nature as a resonant field of consciousness when it engages us?"

Consciousness and the Implicate Order

That consciousness could be an attribute of the manifest world of objects and impersonal forces seems utterly antiquated when considered within the context of what most people believe to be the scientific worldview. Yet – as even theoretical physicists have begun to suggest

◇◇

in the wake of quantum theory – the old fragmented world order in which matter is acted upon by energy that exists apart from it is an illusion. So is the notion of an independent observer separate from what is being observed. This was apparent to me in my sophomore titration experiment (see pages 7–9) in which the observation of a chemical reaction led to an even more powerful transformation within me. Yet back then, in the late 1960s, the implications of quantum theory had not yet registered, and the dominant paradigm maintained the subject-object barrier with religious fervor. Now, 40 years later, that rigidity has begun to crumble.

Instead of the old world of separation between subject and object, quantum theory reveals a world in which everything is connected to everything else, and everything is in constant flux. Process matters within such a world, and process includes both subject and object. If this is so, then consciousness – the act of becoming aware of anything – is not something the subject does to the object, but rather an interaction between subject and object. Even this statement fails to capture the whole truth, which is that subject and object are one and One is consciousness. It is this awareness that begins to emerge within the Realm of Five, once we are able to look beyond the appearance of things as predicated by the subject-object split that is real in the Realm of Two.

Theoretical physicist David Bohm discusses this awareness – which is now a scientific perception as well as a metaphysical one – within the context of what he calls the implicate order. The implicate order is essentially the omnipresence of an integrated unity – we might boldly call it Spirit, though scientist Bohm must be more circumspect than us in his reference to matters spiritual. Implicate order is implied within the various parts within the whole that we routinely discuss in terms of subject-object relationships in the name of science. This discussion, which we take to be an accurate description of reality, is in fact, a matter of convenience that allows us – primarily within the Realm of Three – to impose what Bohm calls an explicate order on our observations of the manifest world (*Wholeness and the Implicate Order* 189-190).

> Generally speaking, the laws of physics have thus far referred mainly to the explicate order. Indeed, it may be said that the principle function of Cartesian coordinates is just to give a clear and precise description of explicate order. Now, we are proposing that in the formulation of the laws of physics, primary relevance is to be given to the implicate order, while the explicate order is to have a secondary kind of significance.

If Bohm's proposal is taken seriously, what this essentially means is that physics and metaphysics will once again converge, just as they did in the pre-scientific days when the mythopoetic worldview prevailed. In this view, life, consciousness, and the movement of gods and goddesses throughout the manifest world were all taken to be the underlying source of anything that could be observed. Bohm hints at this possibility, when he suggests (*Wholeness and the Implicate Order* 249) that:

> . . . the implicate order applies both to matter (living and non-living) and to consciousness, and . . . it can therefore make possible an understanding of the general relationship between the two, from which we may be able to come to some notion of a common ground of both.

It is refreshing to see scientists being encouraged to once again contemplate the manifest universe as a living, conscious entity brought to life through the omnipresence of Spirit within it, even if they are not free to discuss it in these terms. Although Bohm was shaped in his thinking by Krishnamurti and the Dalai Lama, he is still very much the scientist, attempting to coax his stodgier colleagues toward a profound paradigm shift in a language that must remain circumspect. Thus, the exciting Pythagorean implications in his theories are left for others to make.

Michael Talbot, commenting on Bohm's work in *The Holographic Universe* (271), for example, can afford to be a bit bolder than Bohm himself:

> Bohm believes that life and consciousness are enfolded deep in the generative order and are therefore present in varying degrees of unfoldment in all matter, including supposedly "inanimate" matter such as electrons or plasmas. He suggests that there is a "protointelligence" in matter, so that new evolutionary developments do not emerge in a random fashion but creatively as relatively integrated wholes from implicate levels of reality. The mystical connotations of Bohm's ideas are underlined by his remark that the implicate domain "could equally well be called Idealism, Spirit, Consciousness. The separation of the two ~ matter and spirit ~ is an abstraction. The ground is always one."

The Implicate Order and the Gaia Hypothesis

It is possible to see a further expression of this idea from quantum physics in biological terms, as presented in such theories as the Gaia Hypothesis. The Gaia hypothesis was originated by NASA scientist James Lovelock and elaborated by University of Massachusetts microbiologist Lynn Margulis. It proposes that living and non-living parts of the Earth form a complex interacting system that can be thought of as a single organism. Although Lovelock has been accused of teleological and quasi-mystical thinking (still taboo in scientific circles) and criticized, even by his supporters, for speaking in a distinctly non-scientific language, the theory itself has spawned two international conferences, and a body of research, some of

◇◇◇

which supports the original hypothesis and some of which does not. Margulis, among others, has been particularly instrumental in helping to gradually overcome scientific objections, so that a begrudging alternate hypothesis called Weak Gaia has become the starting point for scientific fine-tuning. That this theory is being considered at all by mainstream scientists attests to the fact that biological science is also edging toward a more Pythagorean view of creation, as experienced particularly within the Realm of Five.

According to Weak Gaia theory, "the Earth's biosphere effectively acts as if it is a self-organizing system, which works in such a way as to keep its systems in some kind of "meta-equilibrium" that is broadly conducive to life. The history of evolution, ecology, and climate show that the exact characteristics of this equilibrium intermittently have undergone rapid changes, which are believed to have caused extinctions and felled civilizations" (Wikipedia "Gaia hypothesis"). Science and Gaia theorists are united in their quest for the organizing principles pre-existent within nature. The Gaia hypothesis, even in its weak form, proposes that these organizing principles are a form of innate intelligence built into the system. Of course, the idea of intelligence existing within nature is just as much a heresy among biologists as it is among physicists, and Weak Gaia theorists are careful to avoid this kind of language when articulating their theories, although it is certainly implied.

Other theorists, not bound by old-school science, have taken the Gaia hypothesis into realms Iamblichus and other Pythagoreans would recognize more clearly as the Realm of Five. Yale Law School graduate James N. Garner and Senior Astronomer at the SETI Institute Seth Shostak, for example, have proposed the Selfish Biocosm Hypothesis. In their book, *Biocosm*, Garner and Shostak suggest "that there is a cycle of cosmic creation, in which highly evolved intelligences with a superior command of physics spawn one or more 'baby universes,' designed to be able to give birth to new, intelligent life. Thus, the ability of the present universe to support intelligent life as well as it does is not an accident, but the result of the evolution of a long succession of ever rmore 'bio-friendly' universes" (qtd in Wikipedia "Biocosm hypothesis"). Might we assume that these "highly evolved intelligences" are the gods and goddesses of the Realm of One, with whom we experience a kind of dual unity in the Realm of Five, as the creative capacity of our own consciousness begins to mirror theirs?

The Anthropic Principle and the Participatory Universe

The Anthropic Principle, first proposed by theoretical astrophysicist Brandon Carter at a 1973 Krakow Symposium honoring Copernicus' 500th birthday, states that the Universe we know is the only one we can know, because its laws are those that are discernible by us. According to University of Oxford mathematical physicist Sir Roger Penrose (qtd in Wikipedia "Anthropic Principle"), commenting on this theory:

The argument can be used to explain why the conditions happen to be just right for the existence of (intelligent) life on the Earth at the present time. For if they were not just right, then we should not have found ourselves to be here now, but somewhere else, at some other appropriate time. This principle was used very effectively by Brandon Carter and Robert Dicke to resolve an issue that had puzzled physicists for a good many years. The issue concerned various striking numerical relations that are observed to hold between the physical constants (the gravitational constant, the mass of the proton, the age of the universe, etc.). A puzzling aspect of this was that some of the relations hold only at the present epoch in the Earth's history, so we appear, coincidentally, to be living at a very special time (give or take a few million years!). This was later explained, by Carter and Dicke, by the fact that this epoch coincided with the lifetime of what are called main-sequence stars, such as the sun. At any other epoch, so the argument ran, there would be no intelligent life around in order to measure the physical constants in question – so the coincidence had to hold, simply because there would be intelligent life around only at the particular time that the coincidence did hold!

The universe we know, in other words, is knowable because its laws are those that are accessible to a human consciousness capable of knowing them. Other Anthropists have gone on to speculate the existence of multiple universes – co-existent with this one, built around other physical laws – that we can't comprehend because we are not equipped to sense them. American theoretical physicist (a later collaborator with Einstein) John Archibald Wheeler has proposed a "participatory universe" that requires conscious observers to bring it into being, a view that advocates claim can be inferred from quantum mechanics. Others have asserted that there exists only one possible universe, whose purpose it is to generate and sustain observers.

Intelligent Design Revisited

This latter adaptation of Anthropic theory has been seized by religious proponents of Intelligent Design, who claim that the creative intelligence woven through the universe supports the existence of God, and that because the universe was created by God it is infused with His Purpose. The originators of Anthropic theory have been quick to disavow their association with Intelligent Design, claiming that its teleological conclusions are unsubstantiated by science. Ever since the 17th century, when Thomas Hobbes, Francis Bacon and other empiricists successfully attacked and dismantled Aristotle's teleological worldview, science in general has systematically dismissed all suggestion that the material world (or the human psyche) could have a meaning or purpose. Since Intelligent Design presupposes a purposeful

◇◇

universe, it automatically becomes suspect on that basis alone. The US National Academy of Sciences has concurred that Intelligent Design is not science because it cannot be tested by experiment, does not generate any predictions, and proposes no new hypotheses of its own. The National Science Teachers Association and the American Association for the Advancement of Science have likewise declared it pseudoscience (Wikipedia "Intelligent Design").

Most proponents of Intelligent Design use the theory to promote the Biblical view of creation, which they take literally to be a seven-day process that defies the known laws of science, and which they insist ought to be taught alongside Darwinism as a cogent theory of the evolution of human life. Aside from its glaring disregard of scientific principles, those who advocate the teaching of this theory in schools are driven by a rather transparent religious agenda that is quite sectarian in nature, and an embarrassment to open-minded, free-thinking individuals who might otherwise entertain the possibility that the design of this universe, this embodied world in which we live, is – if not intelligent – at least ingenious in its precision and its hospitality to sentient life.

The Anthropic and Gaia hypothesists – who are by and large, highly credentialed scientists – have essentially come to this latter conclusion. Where they draw the line, however, is at a teleological explanation for this ingenuity. As discussed in *The Seven Gates of Soul*, teleological explanations were expunged from the scientific lexicon in the 17th century by early empiricists such as Francis Bacon and Thomas Hobbes, as a reaction to the then prevalent Aristotelian celebration of First Cause – the omnipresence of God or Spirit at the root of all observable phenomena. Distancing themselves from this idea was a necessary last step if science were to once and for all sever the umbilical cord from the mythopoetic worldview out of whose womb it was born, a worldview where divine purpose was implicit. If the manifest world could be considered devoid of purpose, then it could be studied as a merely scientific proposition, subject to quantifiable laws that were without metaphysical implications.

Yet, as scientists edge closer to the realization that this physical universe is designed in ways that mirror the function of the human mind – as Anthropists are beginning to do – and reflects an implicate order impossible to understand without reference to an omnipresent intelligence – as Bohm's followers and Gaia Hypothesists have – then perhaps they will reach a crossroads where it is impossible to go further without reconsidering their aversion to teleological explanations. If those who research human consciousness, such as J. Allan Hobson and Stanislav Grof, can embrace the idea that consciousness is, in part, a subjective phenomena – also formerly taboo within the scientific mindset – then perhaps theoretical physicists and other hard scientists, may one day embrace a form of neo-teleology. If and when that day comes, they will be entering the Realm of Five, as it was understood by the Pythagoreans.

A teleological universe is one in which meaning and purpose are no longer merely a matter of projection by the human mind, but rather intrinsic to the universe itself. Science has

◇◇◇

been slow to accept this possibility because it does not know how to do so without reinstating the Creator – otherwise known as Aristotle's First Cause – that it purged from the scientific lexicon 400 years ago. But as neo-teleologist (and astrologer) Richard Tarnas asks (35):

> . . . is it not an extraordinary act of human hubris – literally a hubris of cosmic proportions – to assume that the exclusive source of all meaning and purpose in the universe is ultimately centered in the human mind, which is therefore absolutely unique and special and in this sense superior to the entire cosmos?

To transcend this hubris and to bring mainstream science up to its own cutting edge, we must radically alter the way in which we think about our place within the cosmos. Do we stand apart from it as neutral observers? Or do we help create it through our participation in it? If the latter, then the intelligent design that is built into the fabric of the universe need not be understood as a narrow religious proposition, although its spiritual implications are profound. Instead we might simply postulate that intelligent design is what happens in a purposeful universe when consciousness informs matter with the capacity to learn from its experience.

If we approach our science in this way, then the entire cosmos becomes our teacher in the most profoundly spiritual sense of that possibility, as well as on an eminently practical level. This kind of science, for example, is already being practiced by the Biomimicry Institute:

> The core idea is that nature, imaginative by necessity, has already solved many of the problems we are grappling with. Animals, plants, and microbes are the consummate engineers. They have found what works, what is appropriate, and most important, what lasts here on Earth. This is the real news of biomimicry: After 3.8 billion years of research and development, failures are fossils, and what surrounds us is the secret to survival.

> Like the viceroy butterfly imitating the monarch, we humans are imitating the best-adapted organisms in our habitat. We are learning, for instance, how to harness energy like a leaf, grow food like a prairie, build ceramics like an abalone, self-medicate like a chimp, create color like a peacock, compute like a cell, and run a business like a hickory forest.

> The conscious emulation of life's genius is a survival strategy for the human race, a path to a sustainable future. The more our world functions like the natural world, the more likely we are to endure on this home that is ours, but not ours alone.

Such a language of intelligent design is not an anthropomorphic delusion – as mainstream science would currently have it. It is a necessity if we are to survive as a civilization and as a species. We can either continue to stand apart from the world we observe and in our institutionalized detachment from it, ensure its return to the Abyss. Or we can take our place within it as conscious participants, seeking to merge more seamlessly into the intelligent design that already exists everywhere around us.

Astrology as a Language of Intelligent Design

It should be noted here in passing that astrology is also, in essence, a language of intelligent design, not in an overtly religious sense, but in a spirit of inquiry unhampered by science's insistence on standing apart from a universe devoid of meaning. Astrologers routinely observe that the same patterns existing in the cosmos can be seen operating in various ways on various levels within terrestrial phenomena. Everything from earthquake activity to fluctuations in the stock market to historical trends to processes of individual human psychology can be mapped symbolically to a complex interplay of planetary cycles and timed to syncopated rhythms of waxing and waning intensity. Although we can note these planetary rhythms without ascribing any underlying intelligence to them, it is hard to study astrology for any length of time without marveling at the sophisticated simplicity of the source code behind anything we might observe. Science understands this source code as the observable laws of nature. Astrologers understand it as a language of multi-dimensional metaphor.

Within the Realm of Five, both scientists and astrologers must exercise consciousness to understand a reality that mirrors the structure of the world they are observing – essentially what Bohm calls implicate order. How this order came to be is a profound mystery that neither scientists nor astrologers can explain. Within the Realm of Five, astrology potentially provides a missing link between the observable patterns in nature studied by scientists and the intelligence that permeates the cosmos and infuses those patterns with discernable meaning and purpose.

Divine Intelligence and Human Civilization

Assuming for the moment that Pythagoreans were right, and the natural laws of the manifest world are an expression of the One's immanence within creation, then intelligence can be understood as the capacity to live in harmony with natural law. Put another way, intelligence can be understood - as I discussed it in Chapter One - to be the quest to follow in the footsteps of a Creator god. Whether we take this quest literally or metaphorically, following in the footsteps of a god or goddess is learning to live and create in harmony with natural law - to mimic nature in the process of human design. The two are one and the same, and nowhere does this become more obvious than in the Realm of Five, where the gods and goddesses inherent in nature reveal themselves to us as we comprehend the recurring patterns within creation. These patterns – understandable as natural law and in terms of astrological cycles – are the footsteps of the gods and goddesses we are following, a divine plan unfolding.

Those who chose the path of inequality in the Realm of Two, fail to perceive these patterns within the Realm of Five. Instead, they consider nature something to be conquered in order to make room for the often irrational and whimsical will of human desire to predominate. This was, and still largely is, the attitude promoted by mainstream science, and of the technological society that follows mainstream science as its god. In the footsteps of this god, entire mountains are obliterated for the minerals underneath, giant walls blocking migration routes are built to keep undesirable immigrants on their side of the line, entire ecosystems are placed in jeopardy so humans can enjoy relatively cheap transport, climate-controlled homes, and competitive edge in business and war.

We see the consequences of this path of inequality through the Realm of Five, everywhere we look – in rampant poverty and disproportionate distribution of wealth; in environmental destruction that contributes in turn to rising cancer rates and the proliferation of new cross-species diseases like AIDS, Avian Flu, Swine Flu and West Nile Virus; in growing refugee populations set in motion by environmental disaster, political bigotry, and economic necessity; and countless other horrific phenomena steeped in human suffering of tragic proportions. These chronic problems are the inevitable result of our systemic disregard for natural law – a disregard that does have consequences. The more out of alignment with natural law that we become, the more inhospitable our world is.

A Brief History of Natural Law

Since the term "natural law" has its own history within Western culture, I should take a moment to distinguish my use of this term from the more common understanding. The term was first used by Aristotle to distinguish *physis* (or nature) from *nomos* (or human law) on the other. This distinction implies a separation between humans and nature that later became the basis for the scientific insistence upon objective observation of nature. In the subsequent interpretation of Aristotle's thought on natural law – principally through Thomas Aquinas' application of the idea to ethics and jurisprudence – it also became the basis for what is now known as common law. At its best, common law was an appeal to common sense, human decency, and the idea that the commonwealth was a value to be considered at least on a par with the right of the individual to happiness.

The Aristotelian notion of common law was promoted and developed through the writings of 1st century BCE Roman philosopher Marcus Cicero, and later adopted by 17th century Elizabethan English jurist Sir Edward Coke, who exerted a profound 150-year influence

◇◇

on British Law. Cicero also influenced 18th century American statesmen like John Adams and Thomas Jefferson, and was written into the US Declaration of Independence and the Constitution. The fledging colony of America aspired to "the separate and equal station to which the Laws of Nature and of Nature's God entitle[d] them," and later evolved to become "a government of the people, by the people, for the people" all in accordance with natural law as it was outlined by Cicero.

Cicero also inspired early Church fathers, such as Paul of Tarsus and Saint Augustine, who equated natural law with the prelapsarian state of human beings before the Fall from grace in the Garden of Eden. As I suggested earlier, the Fall was the first step onto a path of inequality in the Realm of Two, in which Good was distinguished as the preferred alternative to Evil. In Augustine's view, the prelapsarian state in which natural law prevailed would have been one prior to knowledge of Good and Evil, and we might also postulate, in which participation mystique governed the conduct of humans in the natural world.

Natural law, in other words, was a state in which human beings were a part of nature, and acted accordingly. Correct behavior did not arise through some preconceived idea about right and wrong, but from an instinctual awareness that humans were not separate from the whole of nature in which the consequences of human actions would register. This understanding of natural law was perhaps articulated best in the famous quote attributed to 19th century Dkhw'Duw'Absh (Duwamish) chief Si'ahl (better known as Chief Seattle) (Zussy, Version 3[1]):

> This we know: The Earth does not belong to man; man belongs to the Earth. This we know.
>
> All things are connected like the blood which unites one family. All things are connected.
>
> Whatever befalls the Earth befalls the sons of the Earth.
>
> Man did not weave the web of life: he is merely a strand in it.
>
> Whatever he does to the web, he does to himself.
>
> Even the white man, whose God walks and talks with him as friend to friend, cannot be exempt from the common destiny.
>
> We may be brothers after all.
>
> We shall see.

This view of natural law was widespread among the indigenous peoples of the world, who remained in a state of participation mystique until it was drummed out of them by Christian missionaries, European colonialists, and a corporate culture that coveted the rich reservoir of natural resources harbored within indigenous territory. At its best, this concept of natural law was an attempt to pursue a path of equality through the Realm of Two, in which all human beings – men and women, blacks and whites, Muslims and Christians – as well as plants, animals, rivers, mountains, and stars were co-participants in a egalitarian dance of creative symbiosis. The consequence of this dance in the Realm of Five would have been a civilization built upon a widespread institutionalized understanding of natural law, as it was intuited by humans, trained from birth to participate in the mystique of the natural world.

Unfortunately, this proved to be the history neither of human civilization nor of the concept of natural law. Although Augustine, for example, recognized the *a priori* status of natural law, he also postulated that in the postlapsarian state – a life in harmony with the laws of nature was no longer possible – and that man instead should turn to divine law. Divine law was espoused primarily by Jesus Christ, but interpreted liberally throughout Christian history by diverse authorities such as 13th century scholastic Thomas Aquinas, 16th century Spanish Renaissance theologian Francisco de Vitoria, and 20th century Deist Ray Fontaine, who asserts on his web site that "What matters is that the truth about Nature's God enlighten the world like sunshine. It belongs to everyone." While "the truth about Nature's God" is self-evident to one who is not separated from nature, it becomes a matter of largely rational extrapolation of ideas from scripture in the hands of religious interpreters.

Within the history of the Church, natural law has provided justification for the domination of man over nature (Pelikan, *Genesis* 1:26) – quite the opposite of the kind of natural law that stems instinctively from man's participation in nature, as an inseparable part of it. In addition, the Church has also used natural law to justify all manner of inequalities, including missionary indoctrination of indigenous "heathens," torture and murder of "heretics," and condemnation of homosexuals (Nelson). The Islamic code of *shari'a* – translated in archaic Arabic as "path to the water hole" – represents the use of natural law to justify a profound and sometimes fatal attitude of inequality toward women. In the East, the natural law of the Tao (meaning "way" or "path") was used by Confucius to establish a hierarchical social system in which the authority of the state was assured dominance over its citizens. This political interpretation of natural law was later used as a justification for Mao Zedong's Cultural Revolution "a mixture of party purge and class warfare, during which radicalized students persecuted, humiliated, tortured, and even murdered alleged rightists or counter-revolutionaries" (Lüthi).

Interpretation of natural law in the West by 17th century political philosopher Thomas Hobbes has served as a similar foundation for all manner of Machiavellian political machinations. Hobbes believed that the unbridled governance of natural law would lead to a state of anarchy that he called a "'war of all against all' (*bellum omnium contra omnes*), and thus lives

◇◇

that are 'solitary, poor, nasty, brutish, and short'" (xiii). The antidote to this was to establish a social contract governed by a sovereign authority to which all individuals ceded their natural rights for protection by the state. Abuses by this authority were to be tolerated as the price of peace (Wikipedia, "Thomas Hobbes"). Hobbes had a strong influence on 18th century economist Adam Smith, whose idea that rational self-interest and competition could lead to economic prosperity and wellbeing became the basis for capitalism based on a free market economy.

Under the influence of Hobbes and Smith, natural law was distorted by fear of anarchy to become a social contract in which governments and corporations provide relative peace and prosperity at the expense of individual liberty. This institutionalized perversion of natural law in the indigenous sense leaves the natural world in great jeopardy and virtually ensures that inequality between rich and poor, powerful and powerless, privileged and exploited serve as the basis of a dysfunctional barren civilization, doomed ironically to the "war of all against all" that Hobbes had hoped to avoid.

An outside observer might be excused for concluding that the inhabitants of this civilization were devoid of intelligence, or – what we may now consider to be the same thing – a godless mob, devoid of any sense of this life as a sacred journey, or this life-sustaining planet as a sacred place. An outside observer who had access to our history as a species might conclude that this same divine intelligence or lack thereof was the pivot upon which human civilization has risen or fallen. Where humans have sought to live in harmony with natural law stemming from participation mystique, civilization has flourished. Whenever humans have become convinced in their hubris that they were somehow apart from nature and above its inherent law, then their pride generally went before a fall.

An Example of Culture Informed by Natural Law – The Maya

The ancient Mayan culture predominated throughout southeastern Mexico and northwestern Central America circa 2000 BCE – 250 CE. Although the factors contributing to the rise of Mayan civilization are complex and multivalent – including a supreme cultural diversity and adaptability to environmental conditions, well-developed trade networks, and military superiority – at the heart of the culture, and inherent in everything else they did was a cosmological system based on natural law. The worldview that gave rise to Mayan culture required them to live in harmony with this system and with the natural law that informed it. As Mayan scholar Robert J. Sharer notes (69-70):

> The ancient Maya made no distinction between the natural and supernatural
> worlds, as we do, and there can be no doubt that their ideology – their belief system,

◇◇

which explained the character and order of the world – was a significant factor in the development of their civilization. . . The Mayan cosmos was an animate, living system in which invisible powers governed all aspects of the visible world – all that was to be seen in the Earth and sky – and even the underworld hidden beneath. Each individual and social group had its role to play in this system, and the whole elaborate hierarchy of social classes, surmounted by the elite and ruling lords, existed simply to maintain this cosmological order . . . In ancient Maya society . . . kings were both political leaders and priests, and the ruling elite thereby came to direct all community activities – the giving of tribute, the building of temples and palaces, the maintenance of long-distance trade, the launching of military expeditions, and the performance of the complex of rituals that nourish and placate the gods – all as ordained by the cosmological order.

In our post-modern cynical sensibilities, we might look upon this cosmological order as superstition, or pass judgment about the so-called primitive mindset steeped in participation mystique that produced it. However, it is worth reminding ourselves that this civilization lasted for more than 2,000 years. To put this timeframe in perspective, it is helpful to imagine the Mayan culture ending today. If this were so, at its inception, North America would have been populated by indigenous tribes of Native Americans, just beginning to transition from hunter-gatherer society to a corn-based system of agriculture imported from Mesoamerica. In Europe, the ideas of Plato were just beginning to take root and intermingle with Judeo-Christian and Islamic culture throughout the Mediterranean region, after the conquests of Alexander the Great. In Asia, the Great Wall of China was just being built, and the teachings of the Buddha had not yet been written down.

Historians differ – sometimes widely – on their definition of civilization, their delineation of the various civilizations to appear and disappear on this planet, and their duration. By any account, what the Mayans achieved was remarkable. British historian Arnold Toynbee traces the rise and fall of some 21-23 civilizations in his magnus opus, *A Study of History*, all but four of which are now extinct (Wikipedia "A Study of History"). Egyptian civilization, broken down into eight separate dynasties by Michael Shermer, is more liberally thought to have lasted 2,820 years by historian S. E. Finer. Chinese civilization, broken down into nine dynasties and two republics by Shermer, is granted 2,133 years by Finer. At about 2,200 years, Mayan civilization is at least somewhere among the top three.

Shermer declares that since the fall of the Roman Empire, the average life span of a civilization (according to the tighter definition that he uses) is 304.5 years. Though Finer takes a broader view, he acknowledges various breakdowns of lesser severity than a final collapse. These sub-periods vary considerably from civilization to civilization, and even within the same civilization, but seem to average about one every 369.5 years (Cowen).

◇◇◇

An Example of Culture Threatened
By a Breakdown in Natural Law – 21st Century America

If we consider our own brief history here in America in this regard, then we might surmise that sometime during the late 21st century or early 22nd, we are likely to experience a breakdown in North American culture, as we experience it now. Indeed, as some are beginning to predict in the wake of recent economic events, it is perhaps already happening, slightly ahead of schedule (Hender). It is my contention that this breakdown is occurring, at least in part, because the United States has departed significantly from its original alignment with the spiritual impulse behind its inception, which in turn, was predicated on an alignment with natural law.

In theory, American democracy was conceptualized as a bold experiment – championing the rather egalitarian idea that each individual should have a say in how they were being governed. Before the birth of America, various forms of democracy existed in India, ancient Greece, and throughout the Roman Empire, although it is claimed by historian Jack Weatherford that the Founding Fathers borrowed their version of democracy from the Iroquois Federation (Wikipedia "History of Democracy"), known to themselves as the Haudenosaunee. Be that as it may, we might surmise that behind this political ideal was a natural law rooted in cosmology.

The Creation Myth Behind American Democracy
and Its Original Alignment With Natural Law

Haudenosaunee cosmology was formed through a combination of Earth-Diver and Dual-Creator myths. According to their account of creation, the Haudenosaunee came from the sky to the Earth at a time when it was covered with water. The Great Chief sent his daughter, Sky Woman to establish a patch of dry land in the watery world below, where his people could live. With the help of some animals, Sky Woman obtained enough mud from the bottom of the ocean to form land, where she settled and gave birth to twin boys named Flint and Sapling. Sapling, who was good and handsome, made the sun, straight mountains, and rivers. Flint, who was ugly and evil, made darkness, and turned the mountains jagged and the rivers crooked. Sapling made humans and taught them to plant corn; Flint created monsters, weeds, and vermin (Sutton 317).

Thus it can be seen that the Haudenosaunee were driven onto a path of inequality by their cosmology, often characteristic of Dual-Creator societies (see page 42). At the same time, they were thrown into fierce competition for land and resources with their neighbors, as we might speculate is characteristic of Earth-Diver cultures, where initially there is no land at all. Indeed this was the case, as before the Federation existed, there was considerable warfare among the five separate Iroquois tribes, as well as between the Haudenosaunee and their arch-rivals, the Huron.

This all began to shift, as two men – Deganawida (possibly a Huron who had joined the Mohwaks) and Haio'haw'tha (the eloquent legendary orator, immortalized by Henry Wadsworth Longfellow in his poem, *The Song of Hiawatha*) convinced the five tribes to accept the Great Law of Peace. Undoubtedly, as suggested by visionary Ozarks author Ken Carey (121-131), Haio'haw'tha reminded his people of natural law, and the necessity for living in accordance with a cycle of seasonal ceremonies called the Four Sacred Rituals (Sutton 323). Out of this return to natural law and the application of conflict resolution strategies on the path of inequality that had plagued the Haudenosaunee experience thus far, North American democracy was born and later adopted by the Founding Fathers.

America's Departure from Natural Law

Ironically, the US government has since broken up the Federation and forced the individual tribes to have separate governments. Pro- and anti-gambling factions now fight among themselves over what is left of Iroquois dignity, although a resurgence of interest in revitalizing the Federation has produced a bit of counter-flow (Sutton 124-125). Suffice it to say that if the Iroquois could once have contributed to a widely diverse North American indigenous civilization, they now suffer the same fate as other fragments of this culture at the hands of the white European invaders from across the ocean, who eventually coalesced in a mirror image of the original Haudenosaunee democracy.

Meanwhile, the original image of democracy adopted by the Founding Fathers is in serious trouble. It appears that democracy is now up for sale to the highest bidders, and that no one without connections to corporate and wealthy donor money makes it very far in politics today. The flow of information necessary to healthy democratic debate is increasingly controlled by the same corporate interests that buy and sell politicians. New technologies at the voting booth increase the likelihood of fraudulent or inaccurate returns, while those who do vote are becoming increasingly disenfranchised by a political system that is broken, and that few who benefit from the system seem in any hurry to fix. Large portions of the government now function behind the veil of non-accountability erected in the wake of terrorist attacks on September 11, 2001. Humongous Trojan horse bills like the US Patriot Act seek to revoke the rights of ordinary citizens, and other laws are increasingly passed by executive decree – with no democratic debate at all. The courts have been stacked to support administration policy, the differences between Democrats and Republicans have greatly eroded in the last 25

◇◇◇

years, and the entire notion of checks and balances – an important component of democracy – has apparently been sacrificed on the altar of political expediency.

At the very least, this is not the democracy the Founding Fathers envisioned, nor the democratic model adopted from the Haudenosaunee, based on principles firmly rooted on the path of equality, and sustained through adherence to natural law. Unless the US is fortunate to spawn its own Haio'haw'tha to remind us, a breakdown in the fabric of our civilization is not unlikely within the next 50-100 years. According to the more conservative historian Michael Shermer, this would likely signal the end of US dominance on the world stage, if not the end of the American experiment.

Some would say it has already started to happen, and that given our total disregard for natural law, this would not necessarily be a bad thing. Others say that Barack Obama is the Haio'haw'tha we have been waiting for. My sense is that unless we change our fundamental attitude and the culture as a whole to reflect a deepening sense of equality and participation in the natural world, it won't matter who leads us. We will continue our slide down the slippery slope of narrow self-interests toward an Abyss of our own making.

Natural Law and Consciousness

Within the Realm of Five, the fate of civilization depends upon our collective consciousness of natural law, and our willingness to live in accordance with it. The problem is that few of us – not urban sophisticates who have lost touch with nature, except perhaps for the occasional weekend excursion; or rural, blue-collar, fundamentalist rednecks, who rip down country roads in their monster trucks, tossing beer cans out the window as they go; or naïve environmentalist idealists, for whom wilderness is but an abstract concept – really know what natural law is. Consciousness of natural law – the first hinge on which the fate of our civilization relies – requires patience, a life that revolves around conscious participation in the natural world, and careful observation of natural process in the raw over long periods of time.

Some might argue that we have many dedicated scientists, who are engaged on a daily basis in just this kind of observation, many of whom have at least some input into public policy. As scientists become more familiar and comfortable with the new paradigm being promoted by pioneers such as Stanislav Grof, David Bohm, Lynn Margulis, and Janine Beynus of the Biomimicry Institute, we can be encouraged that scientific understanding of natural law will become a more accurate reflection of the divine intelligence built into the fabric of nature. Yet depending on scientists to make this leap for us is not enough.

In order to create a civilization based on natural law, the practice of observation and immersion in nature would need to be taught in our public schools alongside reading, writing

◇◇◇

and arithmetic – preferably before these more abstract ways of thinking. We would need to somehow rewire the entire public ethic in the US that places private gain above the collective good, and take corporate money out of the decision-making process, so that laws can be passed in harmony with natural law and for the collective good. I don't see any of these things happening soon. Until they do, our best hope against the impending fall of our civilization is for as many of us as possible to align ourselves with natural law, and withdraw our energies and support for those institutions, individuals, or corporations that refuse.

It is likely that even those of us who would support this intentional redirection of energies would argue amongst ourselves about natural law and its implication. There is, after all, no universally agreed upon definition. As an environmental activist working for rainforest preservation in the early 1990s, I can attest to the fact that the environmental movement is quite fractured, and plagued by internal competition for turf, public attention, and grant money. A new generation of eco-entrepreneurs seems to have embraced the notion that environmentally friendly enterprises – such as the sale of incandescent light bulbs, free-market coffee, windmills, carbon credits, and furniture from sustainably harvested redwoods – can do more good than encouragement of simplicity, less consumption, and more localized application of natural law. The jury remains out on this new mindset, I think. Meanwhile, there are a few basic principles of observation that I believe could benefit us all, regardless of our pet strategies.

Return to Observation of Natural Law

These seven principles are taken from an ecological systems design approach called permaculture, developed by Australian Bill Mollison, who has been hailed as "a new Schumacher, preaching smallness, with a touch of Emerson, for self-reliance, and Jefferson, for independence on the land" (*San Francisco Sunday Examiner & Chronicle* qtd in Mollison & Holmgren 128). Permaculture is essentially a hands-on realization of Bohm's concept of implicate order, based on a recognition of the unity and innate intelligence within nature. It is a well-developed and perpetually evolving worldview with both eminently practical and deeply philosophical implications – far more sophisticated than can be encapsulated here. Nonetheless, to give a taste of how permaculture or something like it might serve as the basis for a return to natural law, we might begin with the following principles (Heathcote), which can be understood both literally and metaphorically:

1. **Conservation: Use only what is necessary to achieve the desired outcome.**

There is a natural frugality in nature that does not generate waste. Plants and animals use what is available to them, but take no more than they need, while in turn giving them-

◇◇

selves to the rest of the ecosystem in which they live. Nothing in nature is wasted; everything is recycled.

Applying this principle to a life means giving back as much if not more than you take, and using the minimal amount of energy necessary to pursue your own agenda. Any condition of excess creates an imbalance, not just within you, but within the entire resonant soul space in which you live. Correcting the imbalance is best accomplished by recycling the excess in a way that benefits the whole system. The smallest change possible that will make a difference is generally the best option.

2. **Stacking functions: Every element within any system should ideally serve multiple purposes.**

Everything in nature contributes in multiple ways to the benefit of the whole. A forest provides shade, shelter and food for animals, fertilizes the soil, prevents erosion, serves as a windbreak, and yields lumber for building. The more integrated into the whole system something is, through the reach of these multiple functions, the more essential it becomes.

As a natural law applicable to human life, this principle recognizes that you are a multidimensional being, with multiple roles and functions to play. Society encourages specialization, especially with regard to career, but as a principle of conscious living, life requires you to develop as many of your talents and skills and interests as you can, and to find ways to express yourself on every level on which your spirit seems to demand expression. This well-rounded cultivation of self benefits not only you, but the entire resonant soul field in which you live, especially to the extent that the purpose of your life revolves around a desire to be of service.

3. **Repeating functions: Any need can be met in multiple ways.**

Not only does every element in nature serve multiple functions, but those functions are shared by multiple resources. The diets of most animals provide a simple example of this. Ravens along the coast of Maine, for example, often rely on seabirds and their eggs for food during hatching season. When nesting seabirds are scarce, ravens also feast on freshwater mussels from a nearby stream, or the entrails of deer or moose left behind by hunters, or the occasional mouse, or if need be, even beetles from a decaying tree (Heinrich 51-53). If ravens had to rely on only one of these sources, their survival would be placed in jeopardy.

Living in accordance with this principle essentially means becoming creative in the way you approach any need – developing multiple sources of income; exposing yourself to a wide range of information from a variety of sources; experimenting, exploring and pushing boundaries in every aspect of your life; cultivating a broad network of friends with whom you connect on multiple levels of interest. The more resourceful you can be, in meeting any need and desire, the more flexible you become, and the more fruitful a life of interconnectedness begins to weave itself around you.

4. **Reciprocity: Utilize all yields to meet existing needs.**

This principle is an elaboration of the first, in that within an efficient natural system, nothing is wasted. The leaf mold that builds beneath a tree in autumn serves as compost for next spring's growth. Every animal carcass serves as food for some other animal. The half-digested berries excreted by a coyote will be the perfect peat pot for new berries.

To apply this principle, you start to think about firing the trash collector – recycling everything you use, so that nothing you consume finds its way into a landfill. Beyond that, you can begin giving away what you don't need or can't use, bartering your skills in order to make them available to others who might not otherwise be in a position to afford them, and looking for ways to give of yourself – small acts of kindness, volunteering your time for some cause in which you believe, helping a friend in need. Instead of feeling sorry for yourself, reach out to cheer up someone else. Instead of bemoaning your wasted degree in English, volunteer to tutor those who can't read. Instead of blaming your spouse for something, find something to appreciate about her. If the energy you are putting out does not meet a need somewhere, the principle of reciprocity requires you to rechannel that output in a more productive way.

5. **Appropriate scale: What we design should be on a human scale and doable with the available time, skills, and money that we have.**

Our design systems are now adapted mostly to the needs of corporate interests, and only secondarily if at all to the people their designs are intended to serve, much less to the non-human components of the ecosystem affected by the design. Large monoculture farms make it easy for agribusiness companies to compete in global markets, but require huge outlays of capital for equipment, elaborate watering systems, chemical fertilizers, and toxic pesticides in order to compensate for the unnatural vulnerabilities of these cash crops to drought, disease or damage by insects. Because such ventures extend so far beyond a normal human scale, only the richest, largest, most voracious corporate farmers are able to assume the risk, and reap the financial rewards at the economies of scale these unwieldy systems require.

By contrast, nature designs in such a way that small concentrated pockets of complexity can produce more yield per unit of space, and simultaneously meet the needs of a variety of species co-existing with each other. Having observed an "edge effect," where two diverse ecosystems overlapping tend to produce a third region of increased diversity and biological abundance, permaculture designers now work this edge effect into landscapes, gardens and homesteads designed not for export, but to meet the needs of their inhabitants, human and non-human (Mollison & Holmgren 29).

CHAPTER FIVE

◇◇

To apply this principle more generally to a life, it is helpful to work cooperatively and co-creatively with others, to take a multi-disciplinary approach to any problem you might encounter, and to seek out interaction with those whose worldviews, backgrounds and values are different than your own. It is through the cross-fertilization of ideas drawn from many fields that the most interesting advances in our collective understanding are produced. Likewise, when resources are combined within a community of diverse individuals, each contributing something unique to the whole, edge effects are produced in which the whole becomes more than the sum of its parts. Competition among those who fear their differences produces inequality and its consequences; cooperation among those who remain curious about and willing to learn from their differences produces not just equality, but a vastly compounded resourcefulness that greatly enriches the quality of life for all involved.

6. **Diversity: We want to create resilience by utilizing many elements, all in symbiotic relationship with each other.**

Consider the multi-billion dollar pest control industry, which seeks to eliminate insects that threaten commercial crops, household gardens, and lawns. Insects pose a threat requiring such a dramatic invasive response, because the ecosystems being invaded have largely been reduced to monocultures, which are more vulnerable to major insect damage. In a healthy ecosystem, pests are understood to occupy an important niche within the system as a whole, which rarely allows any one component to dominate the entire system because the opportunity is just not there. Healthy plants planted in healthy soil, receiving plenty of water and sunlight but minimal soluble fertilizer or tillage, naturally resist insect attack. Symbiotic and antagonistic plant associations recorded in companion planting literature are capable of repelling insects and/or fostering the growth of competing insects. The proper arrangement of rocks, ponds and trees in relation to gardens also encourages animals such as lizards, frogs and birds that prey upon potentially troublesome insects (Mollison & Holmgren 32).

In application to a life, diversity ensures resilience by maximizing available resources for pleasure, production, wellbeing, and quality of life. If one avenue is blocked, another road can be taken to the same end. Variety in diet ensures adequate nutrition to support and sustain any endeavor. A multifaceted source of livelihood ensures a steady cash flow, even when any one source dries up. A broad network of friends, colleagues, health care providers and other service providers enables back-up support, when any one person upon whom you depend is unavailable. These same principles are applicable on any scale from the microcosm of an individual life to the macrocosm composing the infrastructure of an entire society.

7. **Give away the surplus: Create systems that are abundant and share the abundance rather than hoarding it for ourselves.**

◇◇◇

It is unconscionable that America, supposedly the richest nation on the face of the Earth, with a $2.77 trillion annual budget, geared heavily toward military spending and beefing up the Department of Homeland Security, cannot seem to afford to take care of its own people. It is estimated that 38 million Americans live in households that struggle to put food on the table (Learner). The vast majority of these live below the poverty level (National Poverty Center). Approximately 43.3 million people under the age of 65 are without health care insurance (National Center for Health Statistics). Over a five-year period, 5-8 million people will experience at least one night of homelessness (Substance Abuse & Mental Heath Services Administration). While the political sensibility in this country has drifted since the Reagan era toward a less compassionate response to those in need, we have increasingly chosen to give away our surplus to wasteful inefficient corporations, who care about nothing but their own bottom line.

Nature, by contrast, creates a natural abundance, which is distributed automatically to those who participate in any given ecosystem. There is sometimes competition for resources in areas where more than one species depends upon the same food source, but in general, all except man take according to their need, and leave the rest for others. A scarcity of resources often leads to competition, but where nature exists in a state of balance, sharing of abundance is the rule rather than the exception.

As a general principle of life, sharing also makes sense. Not only does it create goodwill and enhance whatever natural bonds of interconnectedness already exist, but it also reduces the possibility of conflict, while increasingly the opportunity for co-creativity and a synergistic approach to common problems.

When I built my cabin here, my neighbors came through the woods to help me erect the walls and attend to the larger details of construction beyond the capability of one person, and then in return, when they needed my help, I made myself available. It was not so very long ago that this kind of neighbor-to-neighbor ethic provided the social glue that held America together as a meaningful whole, though in recent decades this seems to have broken down, except perhaps through formalized programs like Habitat for Humanity. The same is true on an international level, though it seems the US has bent over backwards in recent years to offend former allies, and flaunt a self-assumed superiority in the face of international law. This attitude, which flies in the face of natural law, is not conducive to abundance, balance, or cooperative synergy in the Realm of Five.

The Implications of the Realm of Five Within the Chakra System

The spiritual task within the Realm of Five, for both individuals and the collective, is to discern the creative intelligence at the heart of the cosmos through a careful observation of

◇◇

the laws of nature, and to align oneself with this intelligence by applying these laws in one's every action. Such a task should be differentiated from the religious task of imposing one's point of view on others, otherwise known as proselytizing, for this will necessarily entail willfully denying the laws of nature when it seems convenient or expedient to do so. The laws of nature are an expression of the intelligence that our Creator gods have invested in creation. Our intent in following in their footsteps is to live in harmony with creation as conscious embodiments of and participatory contributors to its intelligent design.

Just as the Pythagoreans understood the number Five as the midpoint between the Monad and the Decad, in which the intelligence at the heart of the Monad was realized in physical form, so too might we postulate that the challenge at the heart of the Realm of Five involves taking the synergy of the extremes to the midpoint. In Chapter Four, I suggested that the spiritual task in the Realm of Four was integrating the first and seventh *chakras*. The task within the Realm of Five is bringing that integration back into the heart center, or the fourth *chakra*, midpoint of all seven *chakras*.

As discussed in *Tracking the Soul*, the fourth *chakra* is where life becomes a resonant field of relationships, and we become the central protagonist navigating that field. When we do this as agents of natural law, then we bring divine intelligence (as revealed in the seventh *chakra*), down to earth (encompassed by the first *chakra*), in a characteristically human way. It is here that the potential exists for the human species to take their rightful place within creation, not as an overzealous weed always threatening the delicate balance of the garden ecosystem, but as a supreme expression of the intelligence with which the entire garden is permeated. It is our job within the Realm of Five, to consciously and intentionally weave this intelligence throughout the manifest world, so that human culture serves to enhance, draw forth, and celebrate the natural order.

Magic – White and Black

Within the domain of hermetic philosophy, this capacity to intentionally create in the image of the natural order was known as magic, and was symbolized by the pentagram, or Five-pointed star. The pentagram brought the element of Aether into the material realm of the Tetrad, and with it the power to work consciously and intelligently in harmony with divine intelligence, thereby rising to this level of intelligence and co-creating the material realm in the image of the divine. That this was considered magic, suggests it to be rare. Indeed, the historical discussion of natural law has been intent upon bending it to unnatural agendas.

The capacity to live in harmony with a sense of natural law that evolves through conscious participation in nature comes through the open heart, which is where energy moving through the Realm of Five converges in co-creative, inclusive, cooperative synergy. When

◇◇◇

this movement meets blockages formed by wounding and fails to dissolve them, the Realm of Five becomes perverted by a closed and damaged heart. The hermetic philosophers considered this perversion – when pursued intentionally along a path of inequality toward personal power-over – to be the essence of black magic. Symbolized as an inverted pentagram, known as the "sign of the cloven hoof," this path was one that followed not in the footsteps of a Creator god, but of the Devil (Hall CIV).

Though this understanding derives from a dualistic mentality along the path of inequality, we can also recognize the cloven hoof of the inverse pentagram as symbolic of a progression through the numbers: from a choice of inequality in the Realm of Two, through a state of denial in the Realm of Three, through a state of fear and desire to dominate in the Realm of Four, through a refusal of divine intelligence – or willful ignorance – in the Realm of Five. In each Realm, we have the opportunity to choose a path that is conducive to becoming – co-creating civilization in the image of a Creator god worthy of our emulation. When we consistently refuse to do that, what we experience within the Realm of Five is essentially black magic, a dangerous denial of soul that invariably backfires as a catastrophic plunge back into the Abyss.

When instead we choose equality, learn to observe and apply consciousness toward conflict resolution, creative synthesis, and counter-flow within the Realm of Three, begin the integration of spirit and matter and find our internal compass in the Realm of Four, then more actively begin to exercise the divine intelligence that lives in us in harmony with natural law within the Realm of Five, what emerges on this level through the vehicle of the human heart, can be understood as magic, in the most sublime sense of that word.

Endnotes

1. The origin of this quote is a matter of great controversy, since at the time it was given in 1854, there was apparently no written transcription. Only after the speech, did several versions appear. The most popular version of the speech, often quoted by environmentalists is now attributed to Ted Perry, the screenwriter for *Home*, a 1972 film about ecology. Whether or not this speech is authentic, the sentiment behind it is nonetheless an articulation of natural law as I am discussing it in this book.

Figure 6: The Rune Thurisaz

Thorn, gateway, the rune of soul work on core issues

Chapter Six: Soul Work, Healing, and the Quest for Wholeness

I n the number Six, the Pythagoreans saw the completion and perfection of all that was begun in the Realm of Five. As the manifest creation established in the Realm of Four was infused with Divine intelligence in the Realm of Five, a process was set in motion that culminated in the Realm of Six as all things came into a state of natural balance and harmony with each other. Says Iamblichus (Waterfield 75-76):

> After the pentad, they used naturally to praise the number 6 in very vivid eulogies, concluding from unequivocal evidence that the universe is ensouled and harmonized by it and, thanks to it, comes by both wholeness and permanence, and perfect health, as regards both living creatures and plants in their intercourse and increase, and beauty and excellence, and so on and so forth.

The idea that the universe is "ensouled" by the number Six is particularly intriguing, since it is soul that we are most interested in here. In *The Seven Gates of Soul*, I describe the soul as a fusion of Spirit and matter that allows consciousness to coalesce around a sense of identity, evolve a life of meaning and purpose, and gravitate slowly toward recognition of its true unity with Spirit. All the major ingredients are in place by the time we reach the Realm of Five, since this is where consciousness makes possible the soul's journey. What remains, however, is for the soul to actually take that journey, and this is what happens in the Realm of Six. As Iamblichus describes this soul-making property of the Realm of Six (Waterfield 77):

> . . . if the soul gives articulation and composition to the body, just as soul at large does to formless matter, and if no number whatsoever can be more suited to the soul than the hexad, then no other number could be said to be the articulation of the universe, since the hexad is found stably to be maker of soul and causer of the condition of life. . . . For when soul is present, the opposites which have been admitted by the living creature are reconciled and ordered and tuned as well as possible, as they yield and correspond to each other and hence cause health in the compound . . .

This reconciliation, ordering and tuning of the opposites in relation to each other is the work of the soul, on its way to *hieros gamos* that in its highest form constitutes a reunion with Spirit while still in the body. This work takes place within the Realm of Six, producing a sense of articulation or purpose, and a sense of completion as the soul comes into balance through resonance by affinity, wounding, and contrast. Out of this work evolves a deeper

◇◇

sense of wholeness, psychological integration, emotional wellbeing, and a healthy body. The progression from soul work to completion, wholeness, healing and health is the natural province of the Realm of Six.

The Realm of Six and Soul Work

Despite the infusion of the embodied world with the availability of conscious intelligence, the soul in the Realm of Five remains a potential only until it is actualized in the Realm of Six. More specifically, the soul must exercise conscious intelligence to heal what does not appear to be whole, by reconciling the opposites that proved to be problematic or divisive in the Realm of Two. What makes this soul work both possible and necessary is the arithmological fact that within Six, Two and Three come together as factors to be synthesized – Two referring to the psychic rift between opposites in need of healing, Three to the methodology.

The arithmologists seemed to distinguish between factors that were added together to compose a higher number (e.g. 2 + 3 = 5) and factors that were multiplied (e.g. 2 x 3 = 6). Addition generally meant that these factors were present, but not necessarily integrated, while multiplication meant that there was some synergistic fusion of factors.

Thus if they saw the Pentad as the opportunity to apply consciousness in the Realm of Three to the trouble raised by the inherent duality in the Realm of Two, in the Realm of Six, they witnessed the emergence of soul in response to that opportunity. If Five was associated with consciousness, the soul that emerged in the Realm of Six arose through the application of consciousness to a specific state of imbalance needing attention. In the Realm of Six, we are compelled to solve problems, to address the core issues that generate and perpetuate a state of imbalance. Consciousness is still the vehicle for this, but through the actual doing – through intentionally addressing our core issues, and implementing a strategy for their rebalancing – consciousness is embodied by a living soul. This embodiment, as spelled out in some detail in *The Seven Gates of Soul*, is the soul's work.

My Descent into the Abyss

In Chapter One, I described my initiation onto a path following in the footsteps of the Teutonic storm god Odin, experienced as a cessation of breathing when the neighborhood kids ganged up on me and stuffed ashes in my mouth. I further outlined this path as a series of activities – learning to play saxophone, studying *pranayama* in a yogic *ashram* for seven years, seeking out the wind in the canyons and deserts of the Southwest – through which I sought to heal my lungs and bring them into balance. I concluded by sharing that I had recently experienced a recurrence of difficulty breathing, and then asked, "What would Odin do?" The unexpected answer was a trip to Havasu Falls and Chaco Canyon. I wish now to explore this trip as an application of consciousness toward the resolution of a problem – in this case, my difficulty breathing – that marks the essential evolutionary step forward in the Realm of Six.

Before I do this, it might be worthwhile to trace my process through each of the preceding Realms, so that we can begin to see how the various layers of numerical activation work together as a seamless whole. My journey through the number Realms began with my inability to breathe in the midst of the ash pile incident – an event that cast me headlong into the Underworld of the Abyss. Without breath, life as we know it is not possible, and at age 3, while choking on the ashes that had been stuffed into my mouth and up my nose, I entered the Abyss abruptly and encountered the terror of my own mortality. In that moment, I found a deeper will to live and stepped unconsciously onto my spiritual path. It would not be until years later, that I consciously felt the primal wound that was set in motion on that day, and only much later that I would recognize it as an initiation into the soul work I am compelled to do now. But the beginning of my soul's journey can nonetheless be traced to this unexpected plummet into the Abyss.

Discovering Odin as a Pathway through the Abyss to the Realm of One

With this descent into the Abyss began the search for a personal god I could claim as my own within the Realm of One. This was not a conscious choice until many years later, but from the beginning it became instinctual for me to relate to breath as a sacred gift that could not be taken for granted. Thus within the Realm of One, my natural choice would be a god whose primary medium was Air – a god of wind. Of course, within the global pantheon there are many – including the Sumerian Adad, the Assyrian Anzu, the Egyptian Shu, the Greek Aeolus, the Celtic Danu, the Slavic Stribog, the Ugarit Lillitu, the Mayan Chac-Xib-Chac, the Incan Kon, the Aztec Ehecatl, the Navajo Ahsonnutli, the Iroquois Dajoji, the Hawaiian Pak-a-a, the Polynesian Afa, the Phillipino Anitun Tabu, the Maori Tawhiri-matea, the Aboriginal Bellin-Bellin, the Hindu Parjanya, the Shinto Shina-To-Be (goddess) and Shi-na-Tsu-Hiko (god), Chinese Fei-Lian, the Korumba Danfe, the Yoruba Oya – to name a few (Godchecker, Wikipedia "List of Deities"). I could have chosen any of them, but gravitated toward Odin, probably because he was likely the god of my Germanic ancestors, and because in the late 1980s, I became interested in the runes, which were channeled by Odin.

The runes were of interest to me, as one in a series of oracular systems that had previously captured my imagination – including the I Ching, the Tarot, and astrology. Dr. Martin

CHAPTER SIX

◇◇

D. Rayner, Professor of Physiology at University of Hawaii School of Medicine, describes the function of an Oracle in the Preface to Ralph Blum's *The Book of Runes*:

> An Oracle points your attention towards those hidden fears and motivations that will shape your future by their unfelt presence within each moment. Once seen and recognized, these elements become absorbed into the realm of choice.

Our "hidden fears and motivations" are often the unconscious pivot point around which our core issues are formed. Bringing what is hidden into the light of consciousness then makes these issues amenable to choice. This, in essence, is the soul's work we have contracted on some deep level of our being to address within the Realm of Six. From this perspective, Oracles such as the runes, the I Ching, the Tarot, or astrology[1] become a point of entry to the Realm of Six and to the soul work it requires of us.

As I began to explore the runes in earnest, I naturally became aware of Odin's role in "channeling" the runes, as he hung upside down on the World Tree Yggdrasil. According to *The Poetic Edda* (qtd in Blum 11), Odin describes his initiation by saying:

> I know I hung on that windswept tree,
> Swung there for nine long nights,
> Wounded by my own blade,
> Bloodied for Odin,
> Myself an offering to myself:
> Bound to the tree
> That no man knows
> Whither the roots of it run.
>
> None gave me bread,
> None gave me drink.
> Down to the deepest depths I peered
> Until I spied the Runes.
> With a roaring cry I seized them up,
> Then dizzy and fainting, I fell.
>
> Well-being I won
> And wisdom too.
> I grew and took joy in my growth:
> From a word to a word
> I was led to a word,
> From a deed to another deed.

◇◇

It should be understood here that when Odin speaks of being led "from a word to a word" and "from a deed to another deed," he is speaking of words and deeds in a magical sense. The runes are no ordinary alphabet, but rather a series of shamanic portals to other dimensions parallel to and co-existent with this one that we call ordinary consensus reality. To evoke a rune is to face a portal that compels passage through it. To evoke a rune as an oracular response to some pressing life issue is to face a portal through which it is possible to enter into a more conscious relationship to that issue.

Though runes and other oracles are routinely consulted by contemporary Westerners as a casual exercise in curiosity, the original sense in which these oracles were conceived requires a more deliberate response. In the true spirit of runic tradition, one does not evoke a portal without also taking on a certain responsibility for passage through it. Thus, as Odin is led "from a word to a word," he is also led "from a deed to another deed." These deeds were his passage through the portals posed by the runes, each of which resolved a particular spiritual dilemma, and the sum total of which constituted his shamanic initiation. These deeds can also be understood as stepping stones of passage through the Realm of Six – the dimension in which consciousness is applied through action. As we evoke a rune, therefore, we are tacitly agreeing to act in response to the hidden knowledge of fear and motivation that is brought forth through our contemplation of it. It is this action that constitutes our soul work in the Realm of Six.

In *Tracking the Soul* I described one such "deed" of soul work, undertaken in response to the rune Othila, evoked at the beginning of a Vision Quest in 1987 (Landwehr 54-55). This rune is a right of passage through a portal that requires a new relationship to an inheritance (physical, cultural, or psychic). To pass through this portal, it became clear to me that I needed to release the projections my grandfather had lovingly imposed upon me from birth, and I devised a ceremony to make this hidden source of motivation within me more conscious, and then release it as a matter of intentional choice.

The rest of the story is to be found in the rune I drew to mark the end of this Vision Quest – the exit portal, which marked my next step, the next "deed" on my path of soul work, the "deed" to which I was progressing as a consequence of my quest, and more specifically as a consequence of the soul work I had done in releasing my grandfather's projections. This was the Blank Rune, not originally part of the runes as discovered by Odin, but added by Ralph Blum to represent Odin himself. It was in this moment that I began to suspect that I might be on an Odinic path, although I had no conscious thought to call it that at the time. According to Blum's interpretation:

> Blank is the end, blank the beginning.... In that blankness is held undiluted poten-
> tial. At the same time pregnant and empty, it comprehends the totality of being, all

◇◇◇

that is to be actualized.... Willingness and permitting are what this Rune requires, for how can you exercise control over what is not yet in form? This Rune often calls for no less an act of courage than the empty-handed leap into the void. Drawing it is a direct test of Faith.... Nothing is predestined; there is nothing that cannot be avoided. And if, indeed, there are "matters hidden by the gods," you need only remember: What beckons is the creative power of the unknown.

This rune bears an obvious relationship to the Abyss – blank the end, blank the beginning, pregnant and empty, the totality of being out of which manifestation emerges. Although I did not have this understanding then, drawing this rune nonetheless evoked all the terror of being plunged headlong into the Abyss, which I thought I had undertaken this Vision Quest to leave behind. Instead, I was apparently being told that for all the good soul work I had done, I was essentially back where I started. In my initial response to this rune, I felt as though I had somehow failed, that the gods were laughing at me, discounting the clarity of my intention and the diligence of my effort. I was livid. I felt cheated. As I wrote in my journal at the time, the Blank Rune at first seemed only to confirm my suspicion that this Vision Quest had been a waste of time:

I became angry because I had not received the vision I had come for. I expected some kind of divine revelation, a genuine "aha" experience. The thought occurred to me that life has no meaning or purpose aside from that which I bring to it; that there was nothing out there to come to me. I thought that maybe that was my vision, but felt bitter because I wanted more. I did not want the responsibility for creating the meaning of life; I wanted life to have its own intrinsic meaning. I felt that my vision quest was a failure, and just wanted it to end.

Fortunately I had a wise guide to this quest, Ross Bishop, who helped me to process my experience, and to understand this rune from a different perspective. In retrospect, it seemed obvious that if I were now free from the projections my grandfather had imposed upon me, I was also free to create my life any way I chose. What I wanted was someone or something, some higher power, some god to show me the way out of the Abyss, to tell me what to do. Step One, Step Two, and so on. Instead, what I got was both the freedom and the responsibility to find my own way out of the Abyss. There was no ready-made code of conduct or strategy or well-marked path to guide me. I was being asked instead to hang upside down from my own personal Yggdrasil and channel my own runes, my own secret knowledge, my own wisdom. With the Blank Rune, I was essentially being given my own blank slate on which to write my own truth, extracted from the meaning I ascribed to my own experiences. This was another Odinic initiation, if I could learn to see it that way. This then was the deed set before me, the portal through which I was being asked to pass.

What does this initiation have to do with my ability to breathe, you might ask? In most languages, the etymology of the word "spirit" is identical to or derived from the verb "to breathe," and is intimately associated with the act of creation by which this manifest embodied world comes into being. To breathe is to become aware of the presence of Spirit within the body or the soul's journey, and to participate as a co-creator with Spirit in the evolution of the embodied world. To breathe freely – as I longed to do – was the metaphorical and literal equivalent of becoming an unencumbered agent of the divine, of following in the footsteps of the Creator god. In drawing the Blank Rune I was essentially being told that the power to breathe freely depended upon my willingness to take responsibility for the meaning and purpose of my own life. What could be more fundamental to any soul's journey – but particularly to mine – than that?

Following Odin Into the Realm of Two

Within the runic tradition, this responsibility then becomes my portal through the Realm of One into the Realm of Two, and my rite of passage into becoming. On one hand, taking responsibility for the meaning and purpose of life would seem to be a generic requirement for all souls, part of the process of living that simply comes with the territory of being human. On the other hand, because of the primal wounding I experienced at the beginning of my Odinic journey, it is a task that is likely to pose a major challenge for me personally, the meeting of which will constitute a pivotal rite of passage. In discussing my primal wounding in *Tracking the Soul* (Landwehr 234-235), I said this:

> As a consequence of what I have come to refer to as "the ash pile incident," I began to believe that winning was dangerous to my health and well-being, that I was powerless against unpredictable forces much stronger than I, that other people could not be trusted, and that the world, in general, was not a safe place to be. Something shut down inside of me, and I became much more guarded, less willing to risk fully revealing myself, and less available to fully be myself in the presence of others. To the extent that these early decisions have influenced my choices and my behavior throughout my life, I have struggled to find a satisfactory place for myself within the world, attained only limited success in my creative endeavors, found it difficult to be close to other people, to be myself in group situations, or to feel as though I belonged anywhere, and could claim only limited freedom in situations where freedom was clearly mine for the taking.

Thus, in the throes of this wounding, when it came time to embrace the liberation posed by the possibility of creating my own meaning and purpose, at the end of my Vision Quest,

151

CHAPTER SIX

◇◇

I was unable to do so. Instead, I felt victimized by a cosmology in which the gods had no plan for me, or if they did, were not willing to discuss it. Without a plan, I felt myself simply cast adrift in a world where a plan was necessary for navigation and success in life. No one was going to help me out here. If they did I would reject their help any way. Instead, I was doomed to be an outcast whose dirty little secret was that I did not know where I was going and would be afraid to go there, even if I did. It is in the contemplation of these obstacles to a satisfying life of meaning and purpose that I entered my personal Realm of Two, where ambivalence reigns.

Within the Realm of Two, there are always two prevailing deities - one like Odin who shows the way forward, and one like Loki (the trickster god in Teutonic mythology) who thwarts the process. My personal Loki was the scourge of self-doubt born of an absurd sense of being the only one on the planet whose journey did not come with a map or set of instructions. As an astrologer, I know this is not true. As a wounded soul, however, I must remind myself over and over again that I have the Odinic power within me to progress "from a word to a word" and "from a deed to another deed," whenever I need a portal to the next phase of my life, or even just in response to some circumstance that needs my attention. Meaning and a sense of purpose invariably arise out of my willingness to do that, and each time I do, I experience a bit of healing in relation to my primal wound.

No small part of the process comes through discovering and rediscovering a sense of equality between knowing - being on a path of meaning and purpose - and not knowing - losing track of what gives my life that sense of meaning and purpose. It is always good to be on the path. But when I am not, I have learned - am learning - to recognize this as my opportunity to create something new. In the days following my Vision Quest, for example, as I struggled to understand the Blank Rune, and its significance as a portal of initiation, I published my first book, a gushing New Age manifesto called *The Birth of the Shining One: Moving Beyond Apocalypse Into Godbeing*. In that book I wrote (54):

> *Though the outward flow of events during this time of transition may be marked by unpredictability and in some cases confusion, stepping stones of safe passage will reveal themselves to those who are able to shed their illusions and preconceived notions of how things ought to be, and listen to the truth that emerges in times of altered pace. Certain perspectives, normally concealed innocently within the background context of everyday existence, will emerge from the periphery of consciousness, and reveal themselves to be vital keys to our next step,*

In the blankness posed by a state of not-knowing, in other words, is the opportunity to shed my "illusions and preconceived notions," settle into an "altered pace," and listen to see what wants to emerge "within the background context of everyday existence." This mindset is a radical departure from the conditioning we receive in Western culture, which is to keep it moving at any cost. Instead, it would appear that honoring the Blank Rune – Odin's rune – required me to stand this piece of conditioning on its head in order to channel the secret knowledge hidden at the heart of the blankness.

On the path of inequality through the Realm of Two, conventional wisdom says that knowing is good, not knowing is bad. Not knowing slows us down, disrupts the relentless momentum of progress, reduces productivity, and creates an idle breeding ground for trouble. As the old Calvinist aphorism puts it, "Idleness is the devil's playground." To be good, according to this mandate on the path of inequality, we must ignore whatever inner voice tells us that we really don't have a clue who we are, what comes next, where we are going, or what any of it means, so that we can just keep on keeping on pretending that we do. For most of my life, I have tried to do this, suffering greatly whenever I found myself caught in the dichotomy between my not knowing and the mandate on the path of inequality that says only knowing is ok.

Moving more consciously onto a path of equality – the spiritual challenge within the Realm of Two – means learning to recognize and honor the spiritual opportunity at the heart of not knowing. As I contemplate doing this, I feel a certain relaxation take hold, and in that state of relaxation, my breathing also relaxes. When I struggle up the mountain, against the resistance of my own uncertainty about where I am going or why, I quickly become breathless. Clinging to what I think I know about who I am and where I am going in moments of blankness produces only confusion, self-doubt, self-recrimination, and shortness of breath. Allowing a space for not knowing, on the other hand, returns me to that pregnant place of pure void, where becoming once again is possible.

Following Odin Into the Realm of Three

In the Realm of Three, we begin to gain some perspective on our core issues and experiment with possibilities for mediation – through resolution of conflict on the path of inequality, and creative synthesis and counter-flow on the path of equality. This is essentially where, through trial and error, we begin to learn about the nature of our core issues, and address them. Since $6 = 2 \times 3$, the process begins as we recognize the inequality at the root of our suffering in the Realm of Two, and address this inequality through the tools of rebalancing available to us in the Realm of Three. As we apply these tools consciously and intentionally to the issues that arise through an indulgence of inequality, we undertake the soul work required of us in the Realm of Six.

One of the first tools I discovered on my Odinic path was travel – not for business or pleasure, but for the purpose of seeking answers to questions with no logical solution. Odin

◇◇◇

was a wanderer, and in moments of unknowing I, too, have gravitated to the proverbial road trip in order to gain clarity and perspective, and to simply air out my brain. One of my first major moments of unknowing came after graduation from college. In the Introduction, I described the epiphany I had in the middle of my sophomore titration experiment (see pages 7–9), a revelation that dramatically changed the course of my life. Before this turning point, I had been a chemistry major intent on finding a cure for cancer; after that turning point, I entered a period of unknowing, which throughout the remainder of my undergraduate career, was filled with intriguing bits of knowledge culled from world literature, philosophy, and psychology. My busy pursuit of this knowledge during my junior and senior years in college did not preclude the fact that I was still in a state of unknowing. This became clear to me as graduation drew near, as I didn't have a clue what I wanted to do with this knowledge, much less how to translate it into some kind of viable career path.

While my friends at school were cutting their hippy hair, trading their tie dye shirts for business suits, and interviewing for lucrative jobs at the very corporations they were protesting a few short months before, all I could see as I looked out into my future was blankness. I didn't have a plan, and I didn't have a clue how to develop a plan. Instead, with a couple of close friends, I planned a cross-country road trip. After zig-zaging through many states and many memorable adventures, I landed in a commune called Funny Farm, outside Ashland, Oregon, still not knowing what I wanted to do with my life. In Ashland, I happened upon an astrologer named Sunny Blue Boy, who spent five hours reading my chart, instilling in me a profound fascination for the subject that has not let up since. Although I did not really know what I had discovered at the time, it appears to me now, in retrospect, that this was Odin's gift to me for being willing to hang upside down from the tree of my unknowing.

At various other times in my life, similar expeditions into the unknown have produced additional gifts. Returning to Connecticut after my infatuation with commune life wore off, I rediscovered kundalini yoga, which became the central focus of my life for the next several years and an abiding source of perspective about spiritual psychology (elaborated in my book *Tracking the Soul*). Moving to Florida, in a state of unknowing after my life at the ashram had run its course, I began working with a friend and colleague, Michael Heleus, who had discovered a way to translate astrological cycles into audible sound. Through this work, in turn, I discovered the true imagistic nature of astrological symbolism and our ability to draw from a bottomless internal well of images in response to it. Traveling from my home in Missouri to study with medical astrologer Ingrid Naiman in New Mexico, I became deeply impressed with the planetary cycle as the primary generator of images relevant to the symbolism depicting those cycles. Each time I have taken to the road, either in the midst of some great unknowing, or in response to some inexplicable urge, Odin has met me with some gift, some piece of information or body of information that has somehow furthered my quest. So in the Realm of Three, I learned that one of my resources in learning to embrace the equality of my unknowing was wandering – with or without clear intention.

I discovered a second Odinic resource in a period of unknowing at the end of a relationship in New Mexico. This was the Vision Quest described at the beginning of this chapter and in *Tracking the Soul* (Landwehr 53-55) in which I pulled the Blank Rune. Since then, I have undertaken three additional quests, each one transforming my life in some profound way, often deepening my relationship to the unknown. My last quest, undertaken in 2005, led to the gradual ending of the 13-year relationship mentioned earlier in this book, which in turn led to the profound sense of unknowing that spawned my recent trip to the Southwest, as well as to a new relationship which promised its own hidden knowledge.

A third Odinic resource arises as a reflection of the Teutonic myth about Mimir's Well. Mimir's Well is a sacred spring or well, sometimes depicted as a cauldron at the foot of Yggdrasil, which like the runes, afforded access to hidden knowledge. As the myth has it, Odin traded an eye with the giant, Mimir, who guarded this well, for a drink from it. Ralph Metzner speculates that the well was filled with an hallucinogenic brew, but also notes that "in some German translations, the term used to describe Mimir's well is *märchenreich*, "filled with stories" (220), suggesting that the knowledge to be obtained from the well lies in gaining a deeper perspective on the stories that compose a life. Since my early twenties, I have kept a journal, recording the ongoing story of my life, and it is through this journal (now composing several thousand pages) that I sift for shards of meaning whenever I find myself in a place of unknowing, holding forth the hope and the expectation that these shards will coalesce in a new way that allows a new sense of momentum. I have also drunk my share of hallucinogenic brew, although it is the stories seen in kaleidoscopic fashion – with or without the aid of outside agents to help turn the kaleidoscope – that are the real object of this quest.

All three of these Odinic resources – wandering, vision quests, and sifting of stories – provide the necessary perspective in the Realm of Three to transmute inner conflict, creatively synthesize knowing and unknowing, and establish useful counter-flow. The Vision Quest is by definition, a way of honoring that place of unknowing and declaring its equality with the ordinary knowing of everyday life. It requires us to stop the relentless momentum of routine existence, sit in stillness, pay attention, and open to the hidden knowledge of the moment. Wandering allows me to have a destination, yet at the same time, not know what I will find there, thus setting the stage for a synthesis of knowing and not knowing on the path of equality. Keeping and reviewing a daily journal creates counter-flow between the known events of the past and the unknown future, which converge in the present moment. As I engage these practices, sometimes all three together, I am simultaneously following in the footsteps of Odin – moving beyond the inherent dilemma such a path poses to me in the Realm of Two, where my unknowing becomes a source of stress impacting my ability to breathe.

◇◇

Following Odin Into the Realm of Four

In the Realm of Four, the resources we cultivate set us on a path of simultaneous Ascent toward reunion with Spirit, and Descent into a more consciously embodied life. At the same time, using these tools, we learn to navigate through soul space with a clarified sense of direction. In exploring how my Odinic path plays itself out within the Realm of Four, it is helpful to first translate that path in terms of its elemental associations and the directions of the Medicine Wheel with which those elements are associated.

On the most basic level possible, the inability to breathe is a restriction in the flow of the element Air. In my case, this was caused by ashes being stuffed into my mouth – symbolized as a condition of excess Earth. Earth and Air form a disparate combination, the reconciliation of which poses a special alchemical challenge that is ultimately a major catalyst to spiritual growth. This cross current between Earth and Air then constitutes the most basic level within the Realm of Four on which my core issues can be understood. If Air, for example, (associated with mental acuity) constitutes a state of knowing, then we might postulate that unknowing is represented by Earth.

Although any deity by definition will encompass all four elements in his or her embodiment of Wholeness, the particular ways in which our chosen deity might possibly ameliorate the condition posed by our core issues (plunge into the Abyss) are of greatest importance to us personally. At the most basic level, Odin's Air and my Air represent a natural resonance by affinity that is part of my attraction to this god. Where my own access to Air is restricted, because of some obstruction by Earth (or any other element), Odin's Air will supplement and sustain my own, to the extent that I am able to identify with him and draw upon his strength as a resource. This simple act of identification with Odin increases my access to Air, and in so doing, helps lift me out of the Abyss into which I plunge when I have trouble breathing.

This is true on the literal level, and it is also true in every other arena in which Air has metaphoric meaning – thoughts, ideas and beliefs; communication; social interactions; the conduct of business; decision-making; problem-solving; the pursuit of knowledge; writing; teaching; and the making of a creative contribution to the culture in which I live. When I observe how Odin lived his life and attempt to emulate him as an exemplar of cunning on the field of battle, for example, my own struggle to overcome the belief (acquired during the ashpile incident) *"that winning is dangerous to my health and well-being, that I am powerless against unpredictable forces much stronger than I, that other people cannot be trusted, and that the world, in general, is not a safe place to be,"* I gain the strength to transcend that limiting belief. From an alchemical perspective in the Realm of Four, this is an effect of Odin's Air supplementing my own.

As a god of wind, Odin also naturally brings together Fire and Air, Fire being the force and momentum behind the movement of Air that creates wind. When Fire rises to a certain level of intensity, wind escalates to become a storm. As an antidote to my core issues, within the Realm of Four, Odin's Fire will naturally support my Air, with which it is compatible, by giving it energy and momentum, thus allowing me to breathe easier and at a deeper level of expiration. When I travel, for example, for the purpose of seeking knowledge or answers to intractable questions, I call upon Odin's Fire to move me along my path in life, not unlike a raven (Odin's bird) riding the thermals.

At the same time, Odin's Fire will create a counterbalance to the Earth that is obstructing my Air, in the same way that a controlled burn can clear debris from a forest floor, and help dormant seeds to germinate. I experienced this Fire, for example, in the ashram where an intense practice of kundalini yoga generated a kind of internal heat (called *tapas* in yogic terminology) that burns through blockages in the *chakras*.

The Water drawn from Mimir's Well also creates a counterbalance to my tendency toward excess Earth by placing it in a larger context that often has symbolic implications. The earthy facts encompassed by various life experiences become less literal and more amenable to deeper understanding through the addition of Water. Collecting and sifting through stories for threads of meaning and purpose and deeper understanding is in addition a synthesis of Water and Air – with Water being the repository of our personal and collective memories, and Air being the understanding and hidden knowledge that arises when we share our stories. This alchemical alteration of Earth and synthesis of Air and Water allows a more fluid movement of ideas (Air) that were previously blocked by rigid perceptions (Earth). The alchemical action of Water on Earth can also help ameliorate conditions such as a tendency toward excess tightness – discussed in Chapters Three and Four – which can be understood as a consequence of the ashpile incident.

Lastly, Odin's Earth can serve as a template for bringing my own Earth back into balance. Odin's Earth can be witnessed in his ability to hang upside down in stillness for nine days on the World Tree Yggdrasil, and in the fierce commitment to his own quest that would inspire him to want to trade an eye for knowledge. I access this more Odinic Earth, for example, when I undertake a Vision Quest, enduring stillness of body and mind in nature for the purpose of gaining clarity. On a more everyday level, I access this strength when I push away my doubts so that I might show up at my keyboard and channel some new piece of writing on yet another blank page.

Each of these alchemical consequences of my identification with Odin have their correspondences upon the Medicine Wheel, which in turn, allow me to access a sense of direction, even in the midst of my unknowing. Air upon Air, for example, is a clear movement to the East, where the task at hand becomes an attitude adjustment, or at a deeper level, a closer

◇◇

examination and slow transformation of beliefs. Fire on Air is a prevailing wind from the West acting upon an Eastern disposition. This is where the challenge becomes letting go of preconceived ideas, agendas, and assumptions and opening up to something new – such as I might be more clearly motivated to do, knowing that Odinic travel is an alchemical matter of Fire on Air. Similarly, Water on Air, such as that which occurs during a session of sifting through stories, is – from the perspective of the Medicine Wheel – a matter of learning to see through the heart rather than the mind, and of reconciling South with East. As we contemplate these associations within the Realm of Four when faced with the portal of any shamanic deed, we bring greater consciousness to bear upon the matter at hand, and our issues become more malleable in the Realm of Five.

Following Odin Into the Realm of Five

In the Realm of Five, we first become aware that the core issues at the heart of our embodied life are the propelling force behind a spiritual journey, with implications far beyond the mere facts of our dilemma. To this point, the embodied life appears to be a set of circumstances perched on a slippery slope between the ephemeral experience of happiness and the Abyss immediately below. Our efforts to negotiate this life are largely unconscious, and tend to revolve around an instinctual desire to maximize pleasure and minimize pain. Once we bring consciousness into the equation – which is the essential evolutionary development that takes place within the Realm of Five – we gain the power, and the responsibility to work toward balance, healing, and resolution of core issues, not just so we can feel better, but as our soul contribution to our collective wellbeing.

It is within the Realm of Five that the four elements coalesce into an integrated whole, and we discern our place within it. To use David Bohm's language, we cannot bring consciousness to an inert universe without making the implicate order explicit, nor can we contemplate our experience without gaining insight into the implicate order that underlies it (*Wholeness and the Implicate Order* 259-260). Perceiving the implicate order is, in our language, tantamount to intuiting the divine plan discussed in Chapter Four or natural law as discussed in Chapter Five – an understanding of the embodied world that reflects the intent of the Creator god in whose footsteps we follow. This is the purpose for which we apply consciousness within the Realm of Five to what would otherwise be a dead world.

Perceiving the implicate order or the divine plan, however, is not something we can do just by thinking about it. Instead, we must allow the subject-object barrier to dissolve and engage the world through a process of participation mystique, or what Bohm calls participatory thought: "Participatory thought sees that everything partakes of everything. It sees that its own being partakes of the Earth - it does not have an independent being" (*Dialogue* 99).

It is toward this experience of participation mystique that we will gravitate in the Realm of Five. My epiphany in sophomore chemistry lab was one such experience, since the transformation of inert substance in my titration experiment directly paralleled my own transformation. I was no longer a separate observer watching an external event; I was participating in a revelation of the implicate order with the power to alter the explicate order of my life.

I had an even more seminal experience in September 1986, almost a year before I drew the Blank Rune in my Vision Quest. I describe this experience in *The Seven Gates of Soul* (Landwehr 364-365) as a dream in which I am being followed by a wolf with red eyes. I manage to elude him and get inside my house, but in trying to secure the house against the entry of the wolf, I inadvertently knock down the entire wall, and open myself to an inevitable encounter with the wolf. In retrospect, this encounter with the wolf (one of Odin's animals) can be understood to be an initiation onto an Odinic path. The collapse of the wall can be understood as the collapse of the subject-object barrier, after which it becomes increasingly difficult to maintain separation from the wolf or from the god it embodies.

In the wake of this experience, I chose the medicine name Redwolf, which signaled the beginning of an unconscious identification with Odin. It also signaled my entry into a more collective mode of being, in which my own fate was no longer separate from the fate of the whole in which I was a participant. As I wrote in *The Seven Gates of Soul* (Landwehr 365):

> *The morning after I had this dream and assumed this name, I felt great sadness. I went off into the desert and cried. My sadness stemmed in part from a lingering sense of vulnerability from the dream, but my vulnerability also had a deeper source that seemed to belong to the embodied world and everything in it. In that moment, I cried not just for myself, but also for all the suffering that filled the world.*

In *The Seven Gates of Soul*, I interpret this experience to be a reflection of my image of the world. It is that. But it is also an expression of my entrance to the Realm of Five, following unconsciously in the footsteps of Odin, where my path leads through a resonant field that encompasses the entire web of life, and where each step ripples throughout the web. The collapse of the walls of my home, the subject-object barrier between me and my red wolf, essentially makes me a wanderer just as Odin is a wanderer, and at the end of the trail, I find myself in a place where the whole world becomes my home. Although I have since constructed a physical home capable of keeping out the wolf, I have also learned to function in a world where wolves abound – finding my place in environments as diverse as southern California (where drive-by shootings routinely took place next door to the *ashram* where I studied yoga); Santa Fe, NM (where the glitter of the cognoscente good-life butts up against the harsh desert reality of everyday survival); and Ecuador (where unbelievable beauty struggles to maintain itself against the onslaught of corporate greed and the pressures of overpopulation).

CHAPTER SIX

◇◇

Though I seek, ever more consciously to find and walk the path of equality through this world, I also become aware in Odin's paradoxical presence, that equality must be sought and cultivated amidst severe inequality. As I seek more room to breathe, I cannot help but become aware of how the world is shrinking and in many ways becoming more claustrophobic. As I relax into a place of balance on the looseness-tightness scale, I am confronted with every contributing influence within our collective environment that reinforces an institutionalized preference for tightness. As I follow in Odin's footsteps through the Realm of Five, there is no longer any separation between my own spiritual journey and the spiritual evolution of the entire web of life in which I participate. I am not always able to rise to this level of awareness, but when I am, the meaning and purpose I seek in moments of blankness is scarcely more than the next breath away.

Following Odin Into the Realm of Six

To achieve this state of participation mystique within the Realm of Five means that every word and every deed within the Realm of Six has repercussions that affect the whole, and that alters the arrangement of elements within it. In this Realm, either we function as agents of the divine, or we serve to undermine the natural order through our willful or unconscious resistance to it. Though I follow in the footsteps of Odin, with each step is always the possibility that I will align instead with Loki, and bring chaos to the explicate order. I will, in fact, undermine the natural order within the Realm of Six to the extent that I fail to apply consciousness in the Realm of Five to the mundane reality of my experience in the Realm of Four. Knowing this means that I must listen carefully to each word spoken by my god, and take it seriously as a divine injunction leading to my next deed.

I was given such an injunction when I asked, "What would Odin do?" in the wake of a new episode of difficult breathing. The unexpected answer, which came in the form of an invitation, suggested it was time again to wander – this time to Havasu Falls and Chaco Canyon. Now, that journey will provide a poignant illustration of what it means to enter the Realm of Six, as I examine its details in light of my identification with Odin. It is this alternation of word (listening for divine injunction) and deed (acting) and word (contemplating experience for additional divine instruction) that constitutes the basic rhythm native to the Realm of Six on the Odinic path that I follow.

Wandering Southwest in the Footsteps of Odin

My trip to the Southwest was on a symbolic level an integration of a prevailing wind from the South into a state of being rooted in the West - and simultaneously – an integration of a prevailing wind from the West into a state of being rooted in the South. In Chapter Four, I described the latter dynamic (integrating West into South) as letting go of that which is dearest to us, as each chapter of our lives comes to an end, and entering a place of uncertainty and not-knowing in relation to the future; and the former dynamic (integrating South into West) as a matter of finding compassion for self and others within the midst of suffering, including those who seem to be responsible for that suffering. Thus as I wandered in Odin's footsteps to the Southwest, I was on some level, attempting to affect this integration.

In discussing these directional dynamics in Chapter Four, I mentioned that this current round of unknowing (always an experience in the West) in which I seek greater room to breathe (prompted by difficulty in physically breathing) was taking place after ending a 13-year relationship with a partner with whom I wished to remain friends. At the same time, a prevailing counter wind from the South appeared to be blowing through my life in the form of a new relationship in which love was a potent healing force. This wind also required my new friend to separate emotionally from her former partner and integrate her experiences with him. This trip to the Southwest then became an extension of the same alchemical process in which these relationship dynamics seemed key.

On an elemental level, both the relationship dynamic and my trip to the Southwest became an additional matter of integrating Fire and Water. Through Odin's stature as a god of wind and storms, Fire allows my Air to circulate more freely, energizes me, and burns through whatever obstacles stand in the way of that. Through Odin's trade of an eye for a drink from Mimir's Well, Water opens me to a more heart-centered understanding of my experience, while allowing me to move through that experience with a greater sense of flow. We might surmise, given this preliminary analysis in the Realm of Four, that this trip to the Southwest would facilitate these alchemical shifts.

Odin's Welcome

What made this trip an Odinic trip, as opposed to an ordinary garden-variety vacation? Aside from my expectations, which established a certain intent and mindset about the trip, Odin usually announces his presence in some way that lets me know he is on board for the ride. This trip was no different. Our first night, on the way to Havasu Falls, we camped at Red Rock Canyon, near Hinton, OK, just in time to experience Hurricane Erin rip through with 50 mph winds, gusting up to 80 mph. The storm had been downgraded to a tropical depression shortly after making landfall at Lamar, Texas, but then unexpectedly re-intensified as it moved into Oklahoma in time for our arrival. We were graciously given time to pitch our tent, take a stroll around the campground, and then settle in for the night, before we were pounded for 4 hours straight by unrelenting rain and wind. If we had not actually been

✧✧

in the tent, holding it down, it would have surely blown away. In the morning there was a creek running 3 feet away, where there had been no water at all the night before.

Later, we learned that the storm had caused massive flooding, with over 9 inches of rainfall at nearby Fort Cobb, killed at least 2 people, and forced the evacuation of at least 100 more (NewsOk.com). For some local people at least, it was a plunge into the Abyss. For me, it was a visit by Odin, declaring his presence through his medium of choice. The next day I felt energized (Odin's Fire supplementing my Air), absorbing the impact of the storm on a visceral level – a shift that was recognizable in my voice, when I called my new friend that night. This, of course, makes no rational sense, but from a place of participation mystique, where the barriers between subject and object are erased, in that 4-hour deluge, the storm and I and Odin were one.

Bringing the West Into the South

My first task on this trip, as described alchemically above, was to integrate a prevailing westerly wind into the South - a process of letting go, as another chapter of my life came to an end; and entering a place of uncertainty and not-knowing in relation to the future. In Chapter Four, I suggested that because of this new love that had entered my life, I had essentially left the West behind to enter the South. Yet as our process unfolded in the months since we declared our love, it became clear that before we were free to actually enter this new relationship in the South, there was much letting go yet to be done. For her, there was the painful legal machinery of divorce, a necessary emotional separation, and the ending of a shared dream with her soon to be ex-husband. For me, there was the obvious task of letting go of my former partner - a task I had been working on for 9 months - but as this trip began to reveal, this was merely the first layer of a process that went much deeper than that. Although I wanted to believe that the future was about to unfold, the truth was I was still in that liminal place between the known but fading past, and the emerging but not yet coherent future. My first task on this trip became the necessity for honoring that.

On the way to Havasu Falls and back, we stayed with old friends in Santa Fe. Although these people had been important catalysts to my growth and true soul companions at an earlier stage of my journey, there seemed to be much less of a connection than I had anticipated this time around. Our lives, our interests, our paths through life had diverged since the days of our closeness, and this became painfully apparent to me now. As I remarked on several occasions to my new lover, it felt like time for me to develop a new set of friends.

The friend with whom I was traveling - John - and I had a great time, all in all, and survived the trip without any damage to our friendship. Yet the contrast between his outgoing nature and my characteristic reticence in meeting new people left me feeling oddly alone. If he had not been on the trip with me, I would have been essentially and more utterly alone. On the 10-mile hike into Havasu Falls and on the return trip, John was in much better physical shape than I, and soon left me in his dust to hike by myself – an experience that intensified my aloneness. Though I wrestled momentarily with a sense of abandonment, I soon settled into a more solitary space with gratitude for the silence that allowed me to observe and assimilate my own process.

This was not a new experience for me – following in lone wanderer Odin's footsteps, I have learned how to be alone, and how to use my own strength as a primary resource of self-reliance. But within the context of this trip – where my first task was integrating the West into the South – I began to feel that part of what I was letting go of, before moving into the new, was this old set of friends that I had outgrown, or who had outgrown me. Not only was I leaving an old relationship behind, but apparently much of the past in which it was rooted. This was borne out in the days since returning from the trip, through the experience of additional doors closing. The overall sense I had about this on the trip was that I was passing through a portal into something entirely new – a process that felt quite exciting to me, but that also entailed letting go of much that was familiar and formerly central to my identity.

Bringing the South Into the West

On the second leg of our journey, I became aware of a counter-flow to this sense of ending, when I learned something new about Chaco Culture, and then had another personal epiphany. I had been to Chaco Canyon several times before, and each time has been a sort of homecoming. I once had a vision at Casa Rinconada (the main kiva), in which I saw myself as an Anasazi ceremonial dancer, performing a ritual for the seasons.

The Anasazi, like the Maya, and many indigenous tribes through the world, lived according to natural law, and as part of that way of life, created ceremonies that revolved around the solstices and equinoxes. They had in fact, created a rather sophisticated technology for tracking the solstices and equinoxes, as well as major and minor standstills of the Moon[2]. At Fajada Butte, an isolated central feature of the Chacoan landscape, sits a sophisticated astronomical observatory consisting of two spirals carved into a rock behind three horizontal sandstone slabs. At the summer solstice, a sun dagger (shaft of light) bisects the larger spiral; at winter solstice, two daggers frame it. At the equinoxes, the smaller spiral is bisected by one dagger, while a second passes to the right of the center of the larger spiral. At the major lunar standstill, the Moon's shadow just touches the left edge of the larger spiral, while at the minor lunar standstill, the Moon's shadow bisects it. By observing these moments in their petroglyph light show, the Anasazi were able to develop an agricultural calendar that allowed them to maximize the short growing season of the harsh desert climate in which they lived.

◇◇◇

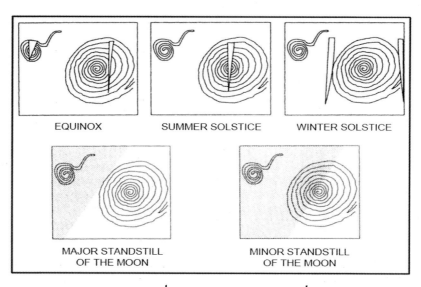

Figure 7: Spiral Patterns at Fajada Butte

Part of my attraction to Chaco Canyon revolves around a fascination with this ancient astronomical knowledge, which also flourished in other areas of the Southwest, as well as among other cultures and at other sites around the world – including Copan (a Mayan site in what is now West Honduras), Palenque (another Mayan site near Chiapas, Mexico), Tenoch-titlan (the Aztec capital where Mexico City is now), Cuzco (an Incan site in southeastern Peru), Glastonbury and Stonehenge in Great Britain, and many temples throughout India, to name a few. More than that, however, I feel drawn by the mystery surrounding the Chaco culture – a mystery in which I somehow feel I have played a part. My sense is that there is a vast body of hidden knowledge rooted in this area, to which I was once privy, but that is hidden to me now. Why else would Odin, god of hidden knowledge, lead me here, if not to remind of what I once knew, and to challenge me to retrieve it from the Mimir's Well?

All of this defies rational explanation, but then again, much of the mystery surrounding Chaco culture also defies rational explanation. Though several generations of archeologists have poured over every shard of evidence, and pieced together a plausible story connecting some of the pieces of this puzzle, many more unanswered questions remain. How could the Anasazi grow corn in the desert, a feat modern agriculturalists have been unable to duplicate? How did the Anasazi transport the 200,000 roofing timbers necessary to complete their great houses, when the nearest source of timber was 27 miles away, and they had no horses, no pack animals, no wheeled carts, no known method of moving materials other than human labor? Why did they build their dwellings with the only entrance on the roof? Why did they build an elaborate system of straight roads, aligned with the cardinal points, and connecting various sacred sites throughout the West, when they had no means of traversing them other than by foot? Why did they suddenly and mysteriously disappear and burn everything behind them?

These questions and others that draw me back for periodic visits are the same that engage a dedicated core of scientists who devote their lives to the pursuit of rational explanations. Thankfully, I am not bound by the same strictures, nor do I feel a need to prove my theories to anyone else. I do thank the archeologists, however, for uncovering certain clues that on this trip seemed to coalesce into a personal epiphany that will perhaps in retrospect, prove every bit as seminal to my Odinic journey as that titration experiment in college or that dream about the red wolf in New Mexico. In a fascinating video documentary about the Sun Dagger, modern Pueblo Indian elders, descendants of the Anasazi will say only that there were some incredible powerful beings among them that had a special knowledge that is lost to us today. After this recent trip to Chaco, I think I know what this special knowledge was.

At a lecture given by one of the dedicated archeologists mentioned above, I learned that one of the most prized possessions of the Anasazi were parrot and macaw feathers that they obtained in trade with the Mayans, and that in the various archeological digs that have been undertaken since the Anasazi disappeared, thousands of these feathers or their remains have been recovered. There are paintings depicting the omnipresence of parrots and macaws throughout the Anasazi great houses, suggesting perhaps that they traded with the Maya not only for these feathers, but for live birds, which they then perhaps bred. Blue feathers were apparently valued for their ability to bring rain, and other feathers served other ceremonial purposes, which were also undoubtedly shamanic. This is where the rational explanation ends, which is not to say that the trade in feathers and birds that seemed to obsess the Anasazi was rational, at least by our standards. Be that as it may, the epiphany I had on this trip to Chaco takes this tantalizing factoid where no rational archeologist would dare to go.

When the Pueblo elders hint that the Anasazi were powerful beings, my sense is that this power came from a shamanic practice, in which the ability to shapeshift into birds was central. Rain was brought to the Anasazi corn patches not by blue feathers used in ceremonies, but by blue-feathered shamans drawing rain clouds through the power of their intention. Timbers were carried from the mountains to the Anasazi rooftops not by human labor, but by bird shamans working together to fly giant twigs back to their nests. Entrances were built in the roofs of dwellings, because as birds, that is how they entered them. An elaborate system of roads was built not for ground traffic, but for aerial guidance of flight patterns. Why did they mysteriously disappear? This remains a mystery, but my guess is that they may have contracted some kind of avian flu – possibly as a shamanic curse – and then they burned their dwellings, so that the epidemic would not spread.

CHAPTER SIX

$\diamond\!\!\!\diamond\!\!\!\diamond\!\!\!\diamond\!\!\!\diamond\!\!\!\diamond\!\!\!\diamond\!\!\!\diamond\!\!\!\diamond\!\!\!\diamond\!\!\!\diamond\!\!\!\diamond\!\!\!\diamond\!\!\!\diamond\!\!\!\diamond\!\!\!\diamond\!\!\!\diamond\!\!\!\diamond\!\!\!\diamond\!\!\!\diamond$

I can, of course, prove none of this, nor do I have a desire to prove anything. I do believe this is a story that I have extracted from Mimir's Well on this trip, which has great power on the symbolic level to shift my understanding of the possibilities of this life, and somehow give me more room to breathe. I also believe that this vision that I had in Chaco is a portal to a new reservoir of hidden knowledge on my Odinic path. Lastly, I am reminded that my new friend was first attracted to me in a dance during which I saw her as a bright feather dancer. Later, in a session of watsu therapy (of which she is a master practitioner), I saw myself as a fallen angel, whose wings were broken and in need of healing.

I am now convinced, after this trip, that aside from letting go of my past, the recovery of these wings is the work I need to be doing in the Realm of Six, the next step on the path to healing my lungs and becoming whole again. I cannot help but feel that pursuit of this ancient bird shaman knowledge and cultivation of this new relationship with my bright feather dancer are the vehicles for this healing that Odin has put before me on this trip, as new pieces of the puzzle begin to fall into place.

And the rune that marks this portal, drawn just now as I close this chapter? The Blank Rune, once again, of course. This lets me know that as I march off into the unknown yet once again, I am on the right track.

Endnotes

1. Astrology can serve an oracular function when it is used to interpret a chart drawn for the moment a query is posed. This use of astrology is called horary, and entails a methodology first perfected by William Lilly in his two-volume series, *Christian Astrology*, published in 1647. Horary astrology has been updated for contemporary astrologers by Olivia Barclay (*Horary Astrology Rediscovered*) and John Frawley (*The Horary Textbook*). Astrologer Geoffrey Cornelius (*The Moment of Astrology*) has further suggested that astrology itself originated as a divinatory art – an assertion that has sparked lively debate within the astrological community.

2. The orbit of the Moon is tilted about 5° from the Earth's orbit around the Sun (the ecliptic). Meanwhile, the ecliptic is tilted 23.5° from the plane of the equator. Over the course of 18.6 years, the Moon shifts from a maximum declination of 28.5° above the equator to a minimum of 18.5° above the equator, and back again. At maximum and minimum declinations (9.3 years apart), the Moon appears to stand still in its position relative to the Earth's orbit, before shifting back in the opposite direction. The reversal at maximum declination is called the major standstill; the one at minimum declination is called the minor standstill.

Chapter Seven: Creativity, Initiation, and the Pivot Point of Change

Of all the numbers from One to Ten, Seven gets the most detailed coverage in *The Theology of Arithmetic*, suggesting a certain complexity that is not easily explained or understood. This complexity is attributed to the fact that 7 is prime - divisible only by itself and 1 - although Iamblichus does not use the word, "prime". Technically, it is not the first prime, since by this definition 1, 2, 3, and 5 are also primes. But One, Two, and Three were understood to be essential building blocks in the generation of a manifest creation, while Five - being half of Ten, and midway through the entire range of numbers from One to Nine - seemed understandable as a conceptual extension and higher, more conscious octave of Two. Seven, by contrast, stands alone. Being as far along in the series as it is, it does not qualify for building block status. Nor does it neatly bifurcate the series, although the arithmologists did note that it was midway between the Tetrad (the inert physical universe) and the Decad (the universe become fully conscious as the embodiment of the Monad). Because it does not neatly fit the pattern developed so far by the arithmological scheme, seven was considered a bit of an anomaly, a sentiment declared in the very first sentence of Iamblichus' chapter on the Heptad: "Seven is not born of any mother and is a virgin" (Waterfield 87).

A motherless virgin has no past, and a future that is full of promise and possibility. I like to think of Seven in this role as a sense of psychic freedom in which something new and unexpected can emerge. This freedom, however, is not free; it must be earned as a consequence of the choices that we make in a state of relative bondage. In Chapter Six, we talked about the runes as portals, alchemical words that lead to deeds. Walking through a portal in the Realm of Six essentially means taking seriously the spiritual injunction from our deity of choice - however it comes to us - by honoring it with action. When we do this, healing occurs, and we take a step toward reunion with our god, that is to say, toward wholeness. The reward for having taken this step in the Realm of Six is a sense of freedom in the Realm of Seven. At the very least, the healing accomplished in the Realm of Six opens a whole new playing field of opportunity, since whatever was healed is no longer an impediment to a full realization of our spiritual potential.

Seven and the Emergence of True Creativity

In a sense, Seven is like the Abyss, but without the trauma and suffering. It is a lifting of the veils imposed by whatever wounds we carry, a parting of the clouds, a glimpse of the Sun

◇◇◇

that illuminates all things, an act of grace within which we experience – for a moment – what it is like to be the Creator in all of His/Her unlimited potential. Of course, healing does not occur all at once, and we will move in and out of the Realm of Six to peel back additional layers of our core issues. But when the clouds do part, we enter the Realm of Seven – born of no mother, and as a virgin – free from the past, and open to an extraordinary future. As we stand before this future, what becomes available to us in our healed state of grace is creativity. We get to make a relatively unencumbered choice about how we want our future to be. About this extraordinary opportunity, Iamblichus says this (Waterfield 89):

> The reason for the Seventh number being an object of reverence is as follows: the providence of the Creator God wrought all things by basing on the first-born One the source and root of the creation of the universe, which comes to be an impression and representation of the highest good, and he located the perfection and fulfillment of completion in the decad itself, and Creator God necessarily considered that the hebdomad was an instrument and his most authoritative limb and has gained the power of creativity,

Iamblichus attributes this property of the Hebdomad (another name for the number Seven) to the fact that it is midway between Four and Ten, where it mediates between an inert material world and a perfected, fully conscious universe. We saw the emergence of divine intelligence within the Realm of Five – itself a prime number – introducing something new into the equation. With Seven comes the possibility for using divine intelligence to create a more perfect world – more humane, more balanced by equality and integration, more supportive of growth and enjoyment of abundance – which in fact is why we are here. What makes this possible is the work on ourselves – the soul work – we do in the Realm of Six. As this work nears completion, our creativity in the Realm of Seven begins to mirror and supplement the work of the Creator, as S/he has envisioned the perfected universe of the Decad. In the Realm of Seven, we become agents of the Divine, that is to say, co-creators.

A form of creativity emerges in every Realm, not just the Realm of Seven. Every choice we make – from the unconscious choice of deity in the Realm of One; to a path of equality or inequality in the Realm of Two; to mediate or not mediate the opposites, and if so, how to mediate in the Realm of Three; to move up or down the *Tetraktys*, and in what horizontal direction in the Realm of Four; to exercise consciousness and live a life in harmony with natural law or not in the Realm of Five; to work toward healing or not in the Realm of Six – all of this creates a world that we subsequently inhabit. Eastern religions call this karma; Christians are more prone to say, "What you reap is what you sow;" New Agers speak about creating our own reality. All these are ways of saying essentially the same thing – that we live in a reflexive universe, where choices and actions bear consequences that must subsequently

◇◇◇

be addressed. Along this path, one thing leads to another, and life progresses one way or another, from one choice to the next.

Until we reach the Realm of Seven, however, this progression seems to have a life of its own that is not entirely within our control. Each choice sets in motion a chain reaction, which proceeds of its own accord, often beyond the influence of our conscious volition. This is especially true when our wounds are triggered, and our core issues reverberate beneath our choices and our actions. Because our wounds and our issues are largely unconscious, we are relatively blind to our part in creating them. Instead we tend to feel as though we are victims of them, which is essentially true, up to the point where we exercise consciousness intention-ally toward their healing – where we walk through the portal of the word into the deed that heals. We are victims of our own unconsciousness and inability to take responsibility for that which remains unconscious. To enter the Realm of Seven at all, we must have broken this pattern by looking our issues squarely in the face and doing something about them. This frees us to create in a different way.

The Creative Power of the Complex Personality

This is not to say that Realm of Seven creativity is entirely conscious, as creativity on any level is always a foray into the unknown. But in the Realm of Seven, we are consciously and intentionally following in the footsteps of our chosen gods or goddesses, having proven ourselves worthy by taking their injunctions to task, and this exertion of spiritual effort on our part evokes a showering of grace that empowers us to become conscious co-creators, moving with the flow of a river much wider and broader than we could possibly encompass or comprehend. We gain access to this river in the Realm of Seven because we are no longer conflicted inside. Instead, all parts of the self come together in a synergistic way to form what former University of Chicago psychology professor Mihaly Csikszentmihalyi calls a complex personality (57):

> *Having a complex personality means being able to express the full range of traits that are potentially present in the human repertoire but usually atrophy because we think that one or the other pole is "good" whereas the other extreme is "bad."*
>
> *This kind of person has many traits in common with what the Swiss analytic psychol-ogist Carl Jung considered a mature personality. He also thought that every one of our strong points has a repressed shadow side that most of us refuse to acknowledge. . . . As long as we disown these shadows, we can never be whole or satisfied. Yet that is what we usually do, and so we keep on struggling against ourselves, trying to live up to an image that distorts our true being.*

169

◇◇

> *A complex personality does not imply neutrality, or the average. It is not some position at the midpoint between two poles. . . . Rather it involves the ability to move from one extreme to the other as the occasion requires. . . . Creative persons definitely know both extremes and experience both with equal intensity and without inner conflict.*

According to Csikszentmihalyi's definition, it is easy to see that what makes the difference between creativity in the Realm of Seven and every other level below it is the fact that there is no longer any ambivalence reverberating in the Realm of Two. The shadow out of which this ambivalence emerges has been dissolved in an alchemical bath of healing acceptance, out of which all of our wild and unruly quirks emerge as resources, cleansed of judgment. It is these resources, previously buried in the shadow that now makes it possible for us to create from a deeper, more integrated place.

The Creative Power of Unmoderated Extremes

Furthermore, once this healing has been accomplished in the Realm of Six, and we come to peace with the opposites in the Realm of Two, our options in each of the other Realms also begin to expand. In the Realm of Three, our task had previously been to somehow mediate the opposites – either the conflict between them, or through creative synthesis and counter-flow. Whether conflicted or not, these solutions all imply that the opposites can only exist in the same space if their relationship is somehow controlled by intervention. Once the ambivalence toward the opposites has been healed, however, such restraint is no longer necessary, and this opens up a fourth possibility: that the opposites simply be allowed to exist side by side, each at times front and center in the full intensity of its extreme, neither censored or repressed, each available when the situation at hand seems to call for it.

Once I achieve a certain level of healing with relation to the tightness-looseness scale discussed in Chapter Three, for example, I will no longer be compelled to find a happy medium between tight and loose, or to try to set up a counter-flow between them. Instead, I will have the freedom to be outrageously tight in one moment, and shockingly loose in the next. I will be able to save my money religiously, living like an ascetic monk for months on end, then spend like a drunken sailor on a lavish rampage, where money is no object. I might seem silent, introverted, and observant on one day, and gush forth unrelentingly on the next with the social equivalent of Tourette's Syndrome. I will be free to study the great masters of my craft with diligence and unrelenting focus, then equally free to throw out everything I have learned from them and break all the rules in creating something of my own. In fact, I will generally become quite unpredictable – a true wildcard to others and even to myself – which is not to say that whatever I do will not have intention. As unnerving as this can be to those who are less liberated, I will nonetheless be claiming the full complexity of my being, and in so doing, gaining access to a whole new world of creativity, possible only in the Realm of Seven to the extent that the healing of ambivalence toward the opposites is completed in the Realm of Six.

As Csikszentmihalyi suggests, the complex, creative person is essentially a walking contradiction. In particular, such a person will exhibit the following paradoxical freedoms (58-76):

> Creative individuals have a great deal of physical energy, but they are also quiet and at rest.

> Creative individuals tend to be smart, yet also naïve at the same time.

> A third paradoxical trait refers to the related combination of playfulness and discipline, or responsibility and irresponsibility.

> Creative individuals alternate between imagination and fantasy at one end, and a rooted sense of reality at the other.

> Creative people seem to harbor opposite tendencies on the continuum between extroversion and introversion.

> Creative individuals are . . . remarkably humble and proud at the same time.

> In all cultures, men are brought up to be "masculine" and to disregard and repress those aspects of their temperament that the culture regards as "feminine," whereas women are expected to do the opposite. Creative individuals to a certain extent escape this rigid gender role stereotyping.

> Generally, creative people are thought to be rebellious and independent. Yet it is impossible to be creative without having first internalized a domain of culture. And a person must believe in the importance of such a domain in order to learn its rules; hence, he or she must be to a certain extent a traditionalist.

> Most creative personas are very passionate about their work, yet they can be extremely objective about it as well.

> Finally, the openness and sensitivity of creative individuals often exposes them to suffering and pain yet also a great deal of enjoyment.

Creativity and the Channeling of Divinity

In the Realm of Four, these internal contradictions lead to an even more fundamental paradox. No longer burdened by the weight of core issues, the complex, creative individual who has entered and learned to dwell in the Realm of Seven will naturally begin to ascend

◇◇

up the *Tetraktys* toward reunion with the One. At the same time, the creative soul will have acquired a capacity to act as an agent of the divine, and within this capacity will naturally bring the full numinous power of Spirit down into the embodied world. At this level, creativity will essentially be a channeling of the divine, and whatever such a person creates will naturally bear the power of its Source. Thus, once a soul has entered the Realm of Seven, there will be a strong counter-flow in the Realm of Four, simultaneously moving up and down the *Tetraktys*.

Moving down the *Tetraktys*, the creative soul begins the creative process with an immersion in Spirit, or what Pythagorean scholar John Opsopaus describes as an Inebriation of Love. On the mundane level, this space of inebriation can be understood as a passionate obsession and utter absorption in the creative process – a state in which time and space disappear along with the subject-object barrier between the artist[1] and her emerging creative product. This is the state of being that Csikszentmihalyi refers to as "flow" (111-113). When flowing, the creative soul essentially disappears into the creative process, which emanates from a place beyond the artist's conscious volition. We could endlessly debate the nature of the mysterious source of this creativity produced in a state of flow, and to a large extent it will depend upon the state of consciousness of the artist. Yet, from a Pythagorean perspective, the source within the Realm of Seven is the divine.

What happens in this state of immersion is that the four elements of the manifest world become far more malleable. The creative person essentially sees the four elements, not as discrete conditions, but as vortices of energy that interpenetrate and perpetually modify each other, then steps into the midst of these vortices and alters the flow. A lesser artist will intentionally alter the flow through some conscious idea that he wishes to impose upon it, but an artist functioning from within the Realm of Seven will alter the flow simply through his willingness to participate in it, as a fully conscious observer who draws forth from the elements that which they have revealed to him through their dynamic, ever-changing interpenetration of each other.

As Lao Tsu says in Stanza 48 of the *Tao Te Ching*:

> *In the pursuit of learning, every day something is acquired.*
> *In the pursuit of the Tao, every day something is dropped.*
>
> *Less and less is done*
> *Until non-action is achieved.*
> *When nothing is done, nothing is left undone.*
>
> *The world is ruled by letting things take their course.*
> *It cannot be ruled by interfering.*

◇◇◇

Creativity and Non-Intervention

On the second tier of the *Tetraktys*, experienced from within the Realm of Seven, the creative soul changes the world by letting things take their course – functioning as what the Taoists might call a *wu-wei* artist. *Wu-wei* means doing by not doing. This is of course a great paradox that bears explanation, yet no explanation is possible that would satisfy the mind in pursuit of learning. So instead, I will attempt to illustrate, again by referring to the now familiar tightness-looseness scale. In Chapter Four I suggested that adjusting the tightness-looseness scale was, on the elemental level, a matter of playing with the relationship between Earth and Water, making mud at the earthy end of the continuum and slurry at the watery end. Working with the Earth-Water continuum in this way is a form of creativity, but one exercised by what I referred to above as a lesser artist – that is to say, one creating somewhere below the Realm of Seven. By contrast, an artist functioning from within the Realm of Seven would approach the same task somewhat differently.

Such an artist would begin by noting that Earth and Water were already in relationship to each other. Earth naturally absorbs moisture from the air, a great deal when it rains, and a lesser amount when it is dry. Conversely, Water naturally contains Earth in the form of microorganisms, plankton, algae, fish, and other forms of aquatic life. In reality, Earth and Water are not separate, but constantly interpenetrating each other. If this is so, then one not need combine Earth and Water in order to adjust the tightness-looseness scale. One need only observe how tightness and looseness interpenetrate each other, and – here is the essential intervention of the *wu-wei* artist – give conscious consent to this interpenetration.

The week I wrote this chapter, for example, I was fiercely committed to my writing. The energy and momentum was there, and I was focused upon it. Although my pattern the previous few months had been to spend my mornings responding to emails from my new friend (our communication had been quite prolific), while writing this chapter I put my writing first. I intentionally moved, in other words, to the earthy, tight end of the tightness-looseness scale. A few weeks later, she came to visit and I devoted all my attention to her, letting my writing go entirely. When she was here, I moved to the watery, loose end of the tightness-looseness scale. But all I really did, during both periods, was simply give my consent to the interpenetration of Earth and Water, more specifically to the natural rhythm between them, riding the waves of shifting balance as they rose and fell in each moment. This is essentially how the *wu-wei* artist works.

On the second tier of the *Tetraktys*, experienced from within the Realm of Seven, our awareness is focused in the moment, but also within a larger context that understands the waxing and waning of the elemental dance in terms of cycles. Because the complex personality is able to contain both extremes of any polarity within her psyche at the same time,

◇◇◇

without the need to affirm one at the expense of the other, she can simply allow both to be present. But this is not a passive allowing, for the complex personality is also poised to shift at a moment's notice, when the situation at hand, or her own internal rhythms, require it.

Creativity and Body Wisdom

Astrology allows us to track these rhythms, and then map them to our own experience. But the creative individual, functioning in the Realm of Seven, has no need for astrology. Instead, she attunes to the rhythm intuitively and through a certain level of body wisdom that is in touch with all the elements simply because she is composed of them. When I find it difficult to breathe, for example, I know the momentum of my life has shifted me out of balance toward too much Earth, and this awareness will simply allow me – in the Realm of Seven – to shift back toward increasing Air. It is this level of awareness of the second tier of the *Tetraktys* that makes possible the kind of creativity that does not require doing. The truth is that our bodies are self-regulating mechanisms, if we do not allow our minds to get in the way, and our minds do not get in the way if there is no wounding to create resistance to the flow of life. So, to the extent that we have done the soul work required of us in the Realm of Six, entering the Realm of Seven means simply letting the embodied life live itself through us. This embodied life has its own intelligence and its own sense of direction, and we need not attempt to direct it, or shape it, nearly as much as we think we do.

In the Realm of Seven, we can step back, as it were, and enjoy life as an expression of the divine plan unfolding through us. To see life this way is to marvel at its ingenious construction, which on the level of Bohm's implicate order, is an expression of orchestrated intelligence permeating every level of creation, even down to the so-called inanimate world. When the Lakota Sioux speak of rocks as the memory-keepers of the Earth, or the Ayahuascan shamans of South America speak of the plants of the rainforest as teachers, or an unorthodox scientist like Rupert Sheldrake speaks of the intuitive capacities of dogs and cats, they are not speaking metaphorically. On the first tier of the *Tetraktys*, as perceived from within the Realm of Seven, even the elements have their own intelligence, which they graciously impart to us to the extent that we are open to meeting them without preconceived ideas. Fire, Water, Earth, and Air are each an ancient repository of wisdom that transcends our mind's attempt to comprehend what the body – composed of these elements – already knows.

Anyone who has ever gazed into a campfire for hours on end, or sat by the ocean, or hiked up into a mountain, or really listened to the wind, has sensed this intelligence, even if they could not put it into words. The complex individual who has learned to exercise his creativity in the Realm of Seven not only senses it, but is able to draw upon it, to absorb it into his being, and then recycle in a way that gives human expression to it. Moving downward through all four tiers of the *Tetraktys*, the creative individual in the Realm of Seven brings the divine down to Earth by reminding us that everything – the Earth especially – is sacred ground. As eco-philosopher David Abram observes (ix):

> *For the largest part of our species' existence, humans have negotiated relationships with every aspect of the sensuous surroundings, exchanging possibilities with every flapping form, with each textured surface and shivering entity that we happened to focus upon. All could speak, articulating in gesture and whistle and sigh a shifting web of meanings that we felt on our skin or inhaled through our nostrils or focused with our listening ears, and to which we replied – whether with sounds, or rush of waves – every aspect of the earthly sensuous could draw us into a relationship fed with curiousity and spiced with danger. Every sound was a voice, every scrape or blunder was a meeting – with Thunder, with Oak, with Dragonfly. And from all of these relationships our collective sensibilities were nourished.*

The Cost of Failure to Enter the Realm of Seven

The world needs this sensibility, born of movement down the *Tetraktys* into a heightened awareness of the intelligence built into the very fabric of the manifest world. Without it, we suffer greatly from the illusion that human beings rule the world. We don't. We are merely guests here – as were the dinosaurs before us – and some would say, we have overstayed our welcome. We are not invincible, as countless plagues and natural disasters throughout history that have plunged thousands, even millions headlong into the Abyss, can attest. To the extent that we ignore the overriding intelligence of the Earth or pretend to be superior to it, it is likely that intelligence will see to it that we are sloughed off like the disease that we have become, in order to restore balance and health in the Realm of Six – on a planetary scale.

Already human fertility rates are beginning to decline, while human immune systems are becoming more vulnerable to ever more potent diseases. New strains of tuberculosis, for example – such as XDR TB – are resistant to all known antibiotics, including fluoroquinonolones, extremely potent drugs with serious side-effects such as depression and musculoskeletal problems. The death rate for XDR TB is about 50% in the general population; among those with lowered immunity, 85% will die (Goetz 96). An estimated total of 33.4 million people were living with AIDS in 2008, and this figure has increased 11.9% since 2001 – about 50% higher in East Asia, 230% higher in Eastern Europe and Central Asia, and nearly 300% higher in Oceania (Australia, New Zealand, Fiji and Papua New Guinea) (World Health Organization). Each year 10.9 million people worldwide are diagnosed with cancer and there are 6.7 million deaths from the disease (Cancer Research, UK). Although billions of dollars are spent on cancer research every year, cancer rates in some parts of the

◇◇

world have doubled since 1940, while with each passing decade the median age of those con-
tracting cancer goes down (Ryan). Why is this happening? Because we have not yet learned
that fouling our nest, year after year with toxic chemicals, radioactive waste, and reckless
consumption of non-renewable resources will eventually catch up with us.

Nor are human beings the only species being affected by our reckless pride. According
to a disturbing report in *Mother Jones* magazine (Whitty 38):

> *When we hear of extinction, most of us think of the plight of the rhino, tiger, panda,
> or blue whale. But these sad sagas are only small pieces of the extinction puzzle. The
> overall number is terrifying. Of the 40,168 species that the 10,000 scientists in the
> World Conservation Union have assessed, 1 in 4 mammals, 1 in 8 birds, 1 in 3 am-
> phibians, 1 in 3 conifers and other gymnosperms are at risk of extinction. The peril
> faced by other classes of organisms is less thoroughly analyzed, but fully 40 percent
> of the examined species of planet Earth are in danger, including up to 51 percent of
> reptiles, 52 percent of insects, and 73 percent of flowering plants.*

According to Harvard biologist Edward O. Wilson, by the most conservative estimate,
the rate of species extinction on this planet is currently 100 times the background rate, and
could be as high as 10,000 times the background rate. "Bracketed between best- and worst-
case scenarios, then, somewhere between 2.7 and 270 species are erased from existence every
day, including today" (Whitty 38). At this rate, Wilson predicts that half of all plant and
animal species on the planet will be extinct by 2100.

It may be naïve to assume that if enough of us enter the Realm of Seven, and learn to exercise the kind of creativity that is essentially divine grace in action, we will save ourselves or
the planet. But what choice do we have? It is clear that our present course is suicidal. What
have we got to lose by pretending that this garden planet is our teacher, and the intelligence
that permeates it on every level is a potential antidote to the fatal hubris that has guided us so
far? For thousands of years, Western civilization has been driven by the Platonic injunction
to attempt to escape the prison of the body, which by extension includes this material planet
on which we live, eventually casting it aside with our triumphant release. Eastern civilizations
have sought a similar escape from the endless wheel of earthly reincarnation. With such an
attitude as our cultural conditioning, is it any wonder that we have failed to fully value this
embodied world from which we are attempting to escape? What if our true purpose here was
not to escape, but to listen and learn from the intelligence of natural law and our own body
wisdom, and then bring as much of this awareness as possible to bear upon our collective
human endeavors? Could this possibly slow our agonizing slide toward the Abyss?

I once had a vision in which I saw the Earth as a radiant jewel. It was exquisitely beautiful, and as I was told by the voices in my head narrating this vision, it was inviolable in some

dimension beyond human comprehension. Then I was shown the same planet, but this time as a cinder, charred beyond recognition. The voices told me that this was the Earth in the third dimension (the physical realm of time and space), as it would exist in the foreseeable future, were we to continue our present course. The Earth itself would remain jewel-like and inviolable in this other dimension, and eventually regenerate itself in the third dimension. But meanwhile, it was entirely possible that Earth would become temporarily uninhabitable in the third dimension, and human beings would not survive to see the garden paradise restored. Let us hope this fate – terminally sad for us, just a bad case of the flu for the Earth – does not come to pass. Personally, I believe that what happens to us as a species depends upon how willing we are, collectively and individually, to heal our wounds in the Realm of Six, and enter the Realm of Seven as co-creators with Spirit and channelers of divinity.

The Realm of Seven and Spiritual Initiation

In esoteric lore, the number Seven is often associated with ritual, ceremony, and the process of spiritual initiation. Although it takes different forms in different cultures, the pattern is essentially the same. The candidate for initiation must endure a series of tasks or trials in order to prove himself worthy of secret knowledge, which in the end, allows him to transcend the limitations of the embodied life. The initiation process – of which the Pythagorean Ascent up the *Tetraktys* is one – is ultimately the same promise of escape that conditions our collective mindset, but typically reserved for an elite core of superior beings who must first prove their superiority.

Archetypally, the initiation process is associated with the hero's journey[2], which is undertaken as a pathway toward reunion with Spirit, although significantly, most heroes are charged with the final task of returning from their journeys to share what they have learned with society. The most enlightened of these returns can be understood as the Descent back into embodied life for the purpose of infusing that life with divine consciousness, much as we have been speaking of it in this chapter. It is the Ascent, in particular, however, that has been traditionally associated with the number Seven, and it is to this part of the hero's journey that we will turn our attention now. About the Ascent, Joseph Campbell says this (*Hero With a Thousand Faces* 147):

> The paradox of creation, the coming of the forms of time out of eternity, is the germinal secret of the father. It can never be quite explained. Therefore, in every system of theology there is an umbilical point, an Achilles tendon which the finger of mother life has touched, and where the possibility of perfect knowledge has been impaired. The problem of the hero is to pierce himself (and therewith his world) precisely through that point; to shatter and annihilate that key knot of his limited existence.

CHAPTER SEVEN

◇◇

The problem of the hero going to meet the father is to open the soul beyond terror to such a degree that he will be ripe to understand how the sickening and insane tragedies of this vast and ruthless cosmos are completely validated in the majesty of Being. The hero transcends life with its peculiar blind spot and for a moment rises to a glimpse of the source. He beholds the face of the father, understands – and the two are atoned.

As is suggested by this passage, the hero's journey typically involves a separation from the Mother (and all things associated with the archetypal Feminine) and after a series of trials, privileged entry into the secret knowledge held by the Father (and all things associated with the archetypal Masculine). The realm of the Feminine, which includes the hero's Achilles tendon, and the "key knot of his limited existence" is the test the hero must endure, before he can rise to glimpse the source of his being, and claim his identity with the Father. The problem with this model of initiation is that through it the hero's journey becomes a path of inequality between Masculine and Feminine. Following such a path provides a momentary individual escape from the Abyss – "the sickening and insane tragedies of this vast and ruthless cosmos" – but does nothing to transform the Abyss itself.

While the hero attains his personal enlightenment, back on planet Earth, millions are still dying daily from diseases wrought by environmental contaminants, species are disappearing at an incomprehensible rate, wars are still raging, and our third dimensional home is gradually becoming inhospitable to life. Can we rest easy knowing that all of this will be "completely validated in the majesty of Being" once we have undergone the necessary shock therapy to blast us through the "knot of limited existence?" Or is that knot – which from our perspective represents the work to be done in the Realm of Six – the legitimate focus of our god-given capacity for attention? Is the passage through this knot a matter of "piercing" – that is to say, a full-bore Masculine frontal assault? Or is it more a matter of relaxing into the knot, working it like the kneading of bread, and gradually undoing it – that is to say, undergoing a Feminine process of entering into a more conscious and more liberated relationship to it?

As Jungian scholar Joseph Henderson points out (151-152):

. . . often the great initiation stories begin only after the Hero Quest has initially ended in failure. . . . The theme of failure of initiation seems to imply some tendency for the initiate to forget to honor (or even to notice) significant vestiges of the old feminine religion of the Earth. . . . The Grail – like cauldrons, stones, and other magical, food-producing objects – is a symbol for the ancient wisdom of the Earth Mother and her sibylline connection with the unknown powers. Where these are unknown to the initiate-as-quester he is tempted to revert to the security of the hero myth and its con-

soling religion of the Father. Consequently the true initiate cannot remain a father's son only, any more than he remained a mother's son in the beginning. And so our modern Quest Hero, in the next episode of this dream, leaves his father and pursues his journey alone into still stranger archetypal country at a deeper level than before.

The true process of initiation, in other words, must begin on a path of equality, not culminating with the *hieros gamos*, but embarkng from it as a point of departure. This is made possible, not by piercing the knot of our earthly limitations, but by settling into it, allowing it to untangle of its own accord through our acceptance of it, and our willingness to care for it, in Thomas Moore's sense of the word (xiv-xix). We must learn to encompass it in the womb of our own gestation, slowly ingesting its potency, rather than attempting to cure it, or transcend it, or be done with it. This only occurs when all ambivalence is dissolved. It is at this point, that the initiate – the complex, creative individual we met on the Descent – begins to ascend the *Tetraktys* in such a way that the journey no longer leads to escape from the manifest creation, but rather to its restoration to garden paradise quality through loving care and creative attention.

Initiation as Participation Mystique

The process begins on the bottom rung of the *Tetraktys* with a deep appreciation for the beauty of the Earth, and great compassion for the suffering of all its sentient beings, including the Earth itself. This appreciation and compassion must arise, not from a place of magnanimous superiority, but rather from a state of participation mystique in which the initiate recognizes herself to be at one with the elements, and at one with the Earth. The fire in her eyes and the light from the Sun are one and the same; the fire in her belly rumbles with the same primal force as churns up from the core of the Earth in volcanic eruptions. The water that predominates in her body – as blood, hormonal secretions, and spinal fluid, to mention a few of its more obvious forms – is the same water that flows through rivers, lakes and oceans. The earth that forms her bones and her skin is the same earth that blankets her larger body as mountains, valleys, forests, and plains. The air she breathes is the same air the Earth breathes – each inhale a gift from the Earth, each exhale a return of the favor to it. Only from this place of participation mystique is the initiate in a position to begin the Ascent - not as some member of a spiritual elite, but as a representative, an embodiment of the sacred Earth and of everything we hold in common.

On the second tier of the *Tetraktys*, the Pythagoreans understood our threefold task to be calming the non-reasoning mind, quieting the inner discourse of the reasoning mind, and a non-willful surrender to the Divine Nous, the spark of the Divine in each of us. From the perspective of the initiate climbing the *Tetraktys* from within the Realm of Seven, however,

◇◇◇

this is where we claim our body wisdom, create a space for Spirit to dwell within us, as us, and allow its light to shine through us so that the Earth we inhabit and that inhabits us might be illuminated. As we realize our identity with the Earth, our senses come alive as channels of communication through which Spirit speaks to us most directly. As we learn to listen to these messages that come to us through our bodies, we become inhabited by Spirit – which *is* this awareness focused by our presence upon the Earth. From this place of presence, our interactions upon and with the Earth become informed by a broader, deeper, more all-inclusive perspective. In this place, we gain the memory of the rocks, we hear the plants and animals speaking to us and learn what they have to teach, and we develop an intuitive understanding of the natural cycles that flow through us and through the living cosmos.

On the third tier, we begin to understand the ancient polytheistic religions that saw gods and goddesses in every tree, every waterfall, and every shooting star – not as theory or concept, but as a living truth. Continuing our Ascent, we experience the Earth as a sacred playground, where all the deities dance, make love, and sometimes cavort with mischievous delight, all so that we can one day awaken to and share in their antics. Having followed this far in the footsteps of our chosen deity, we enter a sacred grove populated by gods and goddesses embodied in Earth and Water, Fire, and Air, stone and flower, wolf and raven, Sun and Moon. And we become one of them, understanding why we were born into this particular form at this particular time, and then serving this holy function as sacred ceremony performed for the benefit of the entire web of life in which we are privileged to participate.

Lastly, to the fourth tier, we enter that place where words cannot go, melting into an understanding of how the Earth and its Creator are one. Having reached this place, we have not escaped anything, but rather lifted ourselves – now one with the Earth – up to the place where we are revealed to be Spirit incarnate. From this place within the Realm of Seven, untold parallel universes are born, extending like petals from the stamen of the flower that is our own, and where we are co-creators on a level and a scale not imaginable from within this one.

The Realm of Seven and the Necessity for Collective Initiation

The initiation that takes place in the Realm of Seven on the Ascent up the *Tetraktys* is not the prerogative of the few. It is the mandate of the many, for unless we can somehow enter the Realm of Seven as a species and create a culture that encourages a conscious Descent into participation mystique, we will not possess the creative capacity to address the burning issues that threaten to overwhelm us in the Realm of Six – overpopulation, rampant consumption of limited non-renewable resources, the impending end of the global oil economy, proliferation of toxic chemicals into the environment, loss of habitat and mass extinction of non-human species, increasing displacement of human refugee populations through genocide and environmental catastrophe, the emergence of new pandemic diseases for which there is no cure, political instability caused by senseless wars, the spread of nuclear and biochemical weapons technology to rogue states, and global warming, to name a few of the challenges we bring with us into the 21st century.

Without a conscious collective entry into the Realm of Seven, we will lack the creative capacity to address these issues because the fundamental split that exists within the human psyche in the Realm of Two remains unhealed. This split will continue to drive us forward on a path of inequality, where a devolutionary force leading to conflict, intolerance, disrespect, irreverence, and competition will undermine everything we might hope to achieve as a progressive, humane, creative culture of life. It is no exaggeration to say that our survival as a species, and as a civilization depends upon our ability to do this – to claim our complexity as creative agents of the divine, and to simultaneously ascend and descend the *Tetraktys* together as a cultural initiation process.

Obviously, individuals will continue to enter the Realm of Seven sporadically as they heal their personal wounds and discover the heightened creativity that becomes available to them in the ensuing state of grace. Sociologist Paul H. Ray and psychologist Sherry Ruth Anderson have coined the term "cultural creative" to describe such people. Cultural creatives are passionately involved in their own personal growth (conscious participation in the Realm of Six) *and* in creating a socio-economic infrastructure that supports the cultural ideals of the 1960s – civil rights, social justice, gender equality, tolerance for all people, non-violent conflict resolution, ecological sustainability, a more egalitarian business model, participatory democracy, increased government and corporate accountability, and foreign policies based on cooperation, diplomacy, and interdependence – all contributing factors toward a collective path of equality.

Ray and Anderson claim that 50,000,000 Americans are cultural creatives (FAQ), an estimate I feel is probably a little high, but that such people exist - and I know more than a few - is encouraging. It suggests that as a society, at least some of us are entering the Realm of Seven, where - as the Pythagoreans suggest - true change becomes possible. The question is, "Are enough of us entering this Realm fast enough to turn the tide before it swamps us?" This remains to be seen, and I would argue that the fate of our species and our civilization depends upon it.

The Realm of Seven and the Pivot Point of Change

The Pythagoreans recognized the Realm of Seven to be a kind of crisis point in the affairs of human beings, which they referred to as "critical time" (Waterfield 99):

CHAPTER SEVEN

<><><><><><><><><><><><><><><><><><><><><><><><><><><><><><><><><><><><><><><><><>

> *They called it 'critical time' because it encompasses, in a short span of time, activities when they are in crisis and are tending to health or sickness, or to generation or destruction. They called it 'Chance' because, just like Chance in myth, it controls mortal affairs.*

There is much discussion in *The Theology of Arithmetic* about the progression of disease, particularly fever, and the relationship between Seven and the point at which "it reaches one of the two possibilities – attack or remission" (Waterfield 97). Likewise, according to Iamblichus (94):

> *All seeds appear above ground, during growth, in the course of the seventh day or thereabouts, and the majority of them are seven-stemmed for the most part. Just as fetuses were sown and ordered in the womb by the hebdomad, so also after birth in seven hours they reach the crisis of whether or not they will live. For all those which are born complete and not dead come out of the womb breathing, but as regards the acceptance of the air which is being breathed and by which soul in general acquires tension, they are confirmed at the critical seventh hour one way or the other – either towards life or towards death.*

As with many assignments of number to specific natural processes, one might debate the literal role of the number Seven in the turning point of fever, sprouting of seeds, and viability of fetuses and living babies. But it is clear that conceptually, the Pythagoreans associated the number Seven with crisis and critical moments in any process that mark the difference between success and failure, life and death. Within the context of our previous discussion, we might postulate that our fate – individually and collectively – depends upon whether or not we attain the level of complex creativity associated with a healing of ambivalence, and can or cannot ascend to a level of participation mystique with the creation, while channeling the divine back into it.

Thus, in our efforts to move beyond the Abyss through the numbers into a viable realm of becoming, the Realm of Seven is critical in determining whether we actually will or not. If we do, our becoming is ensured as a graceful reunion with the One, out of which a fully conscious Creation will gradually evolve as we reach the Decad. If we don't, the Abyss will likely reclaim us. Thus the Realm of Seven is a turning point, determining our fate. The Pythagoreans called it Chance, but what happens in the Realm of Seven really depends upon how consciously and conscientiously we approach the tasks of soul work required of us in the Realm of Six, and through that soul work how completely we heal the primal split in the Realm of Two that lies at the root of all ambivalence. Depending upon our choices, something shifts in the Realm of Seven toward breakthrough or breakdown.

The Implications of the Realm of Seven Within the Chakra System

Working through our core issues in the Realm of Six is the rite of passage into the Realm of Seven. This work can take place in any *chakra*, or between any two *chakras*, and will be quite unique to each individual. In *Tracking the Soul*, I elaborated in some detail how core issues related to each *chakra* can be mapped astrologically, and it is to this map that one must refer before knowing where the soul work in the Realm of Six will be focused.

In the Realm of Seven, however, the work returns to a more generic form. Here the task for everyone is the integration of the second and fifth *chakras*, a combination I referred to in *The Seven Gates of Soul* as the creative *chakras*. As I wrote in *Tracking the Soul* (Landwehr 113):

> I call these two chakras together, creative, because working in tandem they produce the sensibility, if not necessarily the capability of the artist. I use the word "artist" here in the broadest possible sense of the word as one who, aligned with an intention-al sense of self that is endowed with purpose and a receptive awareness of connection that seems to want to invite further exploration, is compelled to give expression to something that transcends the self and enriches the embodied world.

This "artist in the broadest possible sense of the word" – who is capable of "giving ex-pression to something that transcends the self and enriches the embodied world" – is the complex personality, who has healed the fundamental split at the root of ambivalence in the Realm of Two. This split is first encountered in the second *chakra*, where we develop distinct preferences for one side of the split or the other, and create a life that revolves around what I have called resonance by affinity. Our core issues introduce a second kind of resonance – by wounding – into the equation, when our affinities lead to excess, deficiency, and imbalance. This is where my affinity for tightness, for example, leads to a deficiency of looseness and an imbalance on the tightness-looseness scale. Healing our core issues, wherever they may lie, requires us to explore the deficient polarity through a third kind of resonance – by contrast. By learning to become relatively comfortable with looseness and gradually shifting toward the loose end of the spectrum, I begin to develop an expanded repertoire in my response to life. This expansion broadens in the Realm of Seven until the entire spectrum, including extreme tightness and extreme looseness are options that no longer throw me off balance.

What permits me to maintain my balance, even when operating at the extreme of the tightness-looseness scale, or any scale, in the Realm of Seven, is an awakened and clear fifth *chakra*. The fifth *chakra* is where we discover and begin to live the unique individuated truth of our being and find our place within the whole. In order to individuate, we must have

◇◇◇

danced intimately with all three types of resonance in order to learn where our tendencies toward imbalance lie, and how to shift them when we need or want to, in order to more effectively address whatever challenge is at hand. We still have our preferences and our idiosyncrasies, but we are no longer enslaved by them. At times we will express ourselves through them, but this will increasingly be a matter of choice. At times, we will act in ways that utterly defy all expectations.

Writing about the fifth *chakra* in *Tracking the Soul*, I suggested that taking our place within the world was largely a matter of being at home within one's own skin and one's own psyche, which in turn allows one to be at home anywhere one happens to be. When we are functioning through a clear fifth *chakra* in the Realm of Seven, this sense of being at home ripens into a more full-bodied experience of participation mystique. On this level, being at home means adapting to the circumstances in which one finds oneself and contributing in that way, to the wellbeing of the whole. When I drive in heavy traffic, for example, I must necessarily become a tighter driver in order to avoid having an accident. When traffic thins out, I can relax into a looser space. A developed capacity for participation mystique allows me to flow through traffic – and through life – with a minimal amount of strain.

At the same time, a developed sense of participation mystique means I am more sensitive to larger issues that affect the whole. With an affinity for Earth and tightness, I also feel an affinity for the forest, where a thick tangle of trees provides a satisfying sense of place. At various times in my life, I have also been moved to fight to protect forests that were in danger of being clearcut, including a patch of forest adjacent to my home, which through my efforts has subsequently become a part of this protected land cooperative where I live. With a deficiency of Water and a more awkward relationship to looseness, I have likewise been sensitive where available water has been compromised or the threat of compromise existed – fighting to stop a city where I once lived from fluoridating their Water supply and refusing to buy Exxon-Mobil gas ever since the Valdez oil spill in Prince William Sound, Alaska in 1989.

I would not claim that these small acts were expressive of complex creativity in the Realm of Seven, but when an expanded repertoire along any polar scale meets a broadened sense of participation mystique, one's own self-interest becomes seamlessly integrated with the greater good of the larger whole. It is this kind of integration that takes place within the Realm of Seven, as the creativity *chakras* (second and fifth) begin to work in tandem as a vehicle through which channeled divinity can address our collective wounding.

Endnotes

1. The word "artist" is used here in a very broad sense to depict anyone creating from within the Realm of Seven, where life itself is the medium.

2. The hero's journey and the process of initiation, as we are speaking of them here, can apply to both men and women, although traditionally, especially within most patriarchal societies, these were opportunities largely reserved for men, and denied to women. For both men and women, the hero's journey is essentially an integration of the archetypal Masculine, achieved as we each face our personal challenges on the road to self-actualization.

Figure 8: The Rune Inguz

Fertility, Ing - the hero/god, the quest for balance & harmony

Chapter Eight: The Wisdom of the Serpent

The arithmologists were obsessed with exploring the many ways in which numbers could be formed – by addition or multiplication, squaring or cubing – as well as the ways in which numbers shared various properties and series of numbers were related to each other. With Eight, the arithmologists had a field day, as the following passage from *The Theology of Arithmetic* will attest (Waterfield 101-102):

All the ways in which it is put together are excellent and equilibrated tunings. First, it results from the only two numbers within the decad which are neither engenderers nor engendered[1] (I mean, from 1 and 7); then, it results from the two which are even-odd, one potentially, the other actually – i.e. from 2 and 6; then it results from the first two odd numbers – i.e. from 3 and 5 (and this is the combination which is elementary for the generation of cubes, and is the first such sum, since the cube before it, 1, comes about without combination, while the one after it results from the next three odd numbers – 7, 9 and 11 – and the one after that from 4 continuous odd numbers – 13, 15, 17 and 19; and fourthly, it results from 4 twice taken, and 4 is the only number which both engenders and is engendered. The consequence is that 8 is completed by means of the first two engendered numbers, and from their opposites (numbers which engender) and from the number which contains both characteristics. . . . Hence they used to call the ogdoad 'embracer of all harmonies' because of this marvelous attunement, or because it is the first to have been attuned and multiplied so as to be equal-times-equal-times-equal, which is a most lawful generation.

The Realm of Eight and Harmony

The words harmony – *armonia* in Greek – and arithmology derive from the same root, *ar*, which means "to fit together" (Online Etymology Dictionary), suggesting that the study of harmony (arithmology) reaches a certain climax in the number Eight. Pythagorean scholar Alister Cameron notes that the Latin word *ritus* derives from this same Indo-European root, pointing toward a religious dimension to this study (26).

While the origin of the word religion is somewhat open to debate, "popular etymology among the later ancients (and many modern writers) connects it with *religare* 'to bind fast,' via notion of 'place an obligation on,' or 'bond between humans and gods' " (Online Etymology Dictionary). Beyond their interest in number, the Pythagorean arithmologists were seeking to recreate the harmony inherent in the One within the world of human affairs. The promise of harmony was both the nature of the covenant between God and human, and the obligation faced by humans, who were created by God and thus bound to Him. Numbers were the vehicle for the fulfillment of this covenant, and the number Eight in particular was the peak of its expression.

Although the Decad symbolized the fully conscious creation – the full embodiment of the One within His creation – the Octad was where human beings (potentially) reached the pinnacle of their development. Within the Realm of Eight, humans reaped the harvest of every previous effort to align with a Creator god and address the core issues that contributed to their sense of separation from that god. In the Realm of Eight, the illusion of separation was dispelled, and everything within creation suddenly fit together harmoniously. Beyond the Octad was an unfolding of grace that did not depend upon human effort.

As the embracer of harmonies, the number Eight brings balance to the entire spectrum of relationships first encountered in the Realm of Two – the polarities, the possibility of consciousness of those polarities, and the sense of ambivalence that the very notion of polarity seemed to introduce into the creation from its inception. In the last chapter, we saw how the healing of core issues in the Realm of Six led to a higher-level complex creativity in the Realm of Seven. Through the exercise of this creativity – which is a channeling of divine intent – all things come into harmony in the Realm of Eight.

To the arithmologists, this property of Eight derives from the fact that $8 = 2^3$. This cubing is a redemption of everything that was problematic for the arithmologists in relation to the number Two. In various ways, the numbers Three and Five helped alleviate the arithmologists' anxiety, but it was only really as the troublesome Two was cubed that the problem was solved. The harmony encompassed by Eight is implied in the Realm of Four (2^2), since the four elements and the four directions reveal a mandala-like order beneath the apparently chaotic complexity of the manifest Creation. Yet until conscious beings (emerging in the Realm of Five) actually apply an awareness of this order to the disorder of core issues in need of healing (in the Realm of Six) and learn to serve as creative agents of the divine (in the Realm of Seven), the harmony inherent in the divine cannot be fully experienced. In the Realm of Eight, after all prerequisites of the intervening Realms between Four and Eight have been met, it can be.

The Myth of Cadmus and Harmonia

The redemption attained in the Realm of Eight, however, implies an important paradox. For within the arithmological scheme of things, Eight is essentially engendered by Two – the very factor that caused all this trouble in the first place. How could something "evil" – that is to say, contrary to the One – generate its own redemption? The answer, I believe, lies in a clue placed by Iamblichus at the end of the paragraph extolling the virtues of the Octad as the "embracer of harmonies" (Waterfield 102):

> So when they called it (the Octad) 'Cadmean,' they should be understood to be refer-
> ring to the fact that, as all historians tell us, Harmonia was the wife of Cadmus.

In Greek mythology, Cadmus was the brother of Europa, a maiden who was abducted by Zeus in the form of a bull and then raped on the isle of Crete. When Cadmus set out to find his sister in the wake of this rape, he stopped at the oracle at Delphi for advice. The oracle told him to forget about his sister, but to follow a certain cow instead, and when the cow became tired and sat down to rest, upon that spot he should found a city. Against all reason, Cadmus did what he was told – as we have already seen, sometimes the injunctions of the gods make no sense – and followed the cow to what became the city of Thebes.

Upon reaching his destination, Cadmus sent his men to fetch some water at a nearby spring, so that he could ritually sacrifice the cow, as he was instructed to do by the oracle. At the spring, a Serpent guardian who was sacred to the god Ares killed most of Cadmus' men, and was in turn killed by Cadmus. Responding to another divine injunction in the wake of these deaths – this time from Athena (Zeus' daughter) – Cadmus sowed the teeth of the Serpent in the ground where the cow had stopped. From the ground sprang up armed men, who proceeded to kill each other until only 5 were left. These 5 then became the heads of the founding families of Thebes.

Before he could get on with the business of establishing his city, however, Cadmus had to appease Ares for the death of his guardian Serpent. To do this, Cadmus spent 8 years as an indentured servant to Ares, at which point Ares gave Cadmus his daughter Harmonia to be his wife. Cadmus and Harmonia married, had 4 daughters, and together ruled Thebes for a very long time. At the end of their lives, Zeus – with whom this whole tale began – turned them both into great Serpents to signal their transition from mortal to divine status.

The Realm of Eight and the Reclaimed Wisdom of the Serpent

This myth holds several keys to the power of the number Eight to embrace all harmonies. At the beginning of the myth, Cadmus' conscious intention – to follow the bull who had raped his sister – was countermanded by the oracle at Delphi, who admonished him instead to follow a cow, the feminine counterpart of the bull. If the bull can be understood to be a representation of the Masculine One, then the cow must be understood to be a guide to the Realm of Feminine Two. Surely Cadmus must have wondered, "How could following a cow be more important than rescuing my sister?" But he did it anyway, suggesting that the first step in following the footsteps of any god will be moving in what appears to be the opposite

◇◇

direction dictated by reason. Before harmony can prevail, in other words, we must first pass through the irrational Realm of Two, and face our ambivalence toward our destiny.

On this path, Cadmus naturally encounters conflict. The Serpent who belongs to Ares, god of war, kills most of Cadmus' men and is killed in turn by Cadmus. Out of all this killing eventually springs the foundation of his town - the 5 men left standing. The 5 men reference the Realm of Five in which consciousness emerges as a property of the Creation itself, and as a resource that will eventually become the root of human creativity. Consciousness arises as a consequence of wrestling with the ambivalence that breeds conflict in the Realm of Two. Harmony, in other words, is the product of consciousness applied to disharmony. The goal is reunion with the One, in whom harmony exists as a matter of course. But harmony in human affairs is not something we can merely take for granted as a divine gift. We must earn it through effort in the embodied world, where divinity appears to be in conflict with a counterforce.

In this myth, as in the myth of the Fall, this counterforce is represented by a Serpent. In Eden, a Serpent tempts Eve and through her, Adam, to taste the fruit of the knowledge of good and evil. This knowledge is essentially the duality inherent in the Realm of Two, placed into a moral code marked by judgment. In Cadmus' myth, a Serpent is the guardian of a well that contains the water he needs to perform an important ritual. Although water has many symbolic implications, some of which were discussed in Chapter Four, here it can be understood in an alchemical sense as the universal solvent in which such distinctions as good and evil, right and wrong, light and dark, male and female, and every other polarity in the Realm of Two dissolve. It is the antidote, in other words, to the knowledge of good and evil that wrecked everything after the Fall.

But Cadmus kills the guardian to this antidote, because he is not yet ready to claim the wisdom that it embodies. He has followed his cow into the Realm of Four, but consciousness - the gift of the Realm of Five - does not emerge until he kills the Serpent and sows its teeth. Killing the Serpent is the ultimate act of unconsciousness, since Cadmus has no idea who the Serpent is or its importance, only that it has killed his men, and thus stands in the way of the building of Thebes. Yet as Joseph Campbell reminds us (*Creative Mythology* 154):

> *Wherever nature is revered as self-moving, and so inherently divine, the serpent is revered as symbolic of its divine life. And accordingly, in the Book of Genesis, where the serpent is cursed, all nature is devaluated and its power of life regarded as nothing in itself.*

The true gift in the Realm of Five, which Cadmus is not yet ready to claim, is the realization that nature itself is self-moving, that is to say, infused with divine intelligence and thus divine. As guardian of the sacred well, Ares' Serpent signifies that before one can access the

◇◇◇

antidote to the pain and suffering aroused in the Realm of Two through the knowledge of good and evil, one must experience the Creation (Realm of Four) as intelligent (Realm of Five). When Cadmus kills the Serpent, he fails to recognize its intelligence. Nonetheless, through divine grace, he has an experience of participation mystique in which he witnesses this divine intelligence in nature at work. Sowing the Serpent's teeth (the embodiment of this intelligence), the foundation of his city emerges through an act of effortless creativity.

In this moment, Cadmus has a brief glimpse into what is possible in the Realm of Seven. Before he can enter this Realm, he must first indenture himself to Ares in the Realm of Six, where the task is to release himself from the core issues that prevented him from seeing the Serpent as divine. This task takes 8 years, at which point, he gains Harmonia's hand in marriage. Working through his core issues (rooted in a judgmental attitude toward polarity) is what makes an embrace of harmony possible, and working through his core issues is essentially a matter of realizing the divinity of his own nature and by extension, all of nature. In this way, he comes to embody the archetypal wisdom of the Serpent, a spiritual attainment that is recognized at the end of this life when he actually becomes a serpent. It is the attainment of this wisdom of the Serpent in turn that allows him to govern the city of Thebes, along with his wife Harmonia.

The Realm of Eight and the Mystery of the Caduceus

However, there is more to the wisdom of the Serpent than this. As Occultist Manly Palmer Hall claims (LXXII):

> The ogdoad was a mysterious number associated with the Eleusinian Mysteries of Greece and the Cabiri. It was called the little holy number. It derived its form partly from the twisted snakes on the Caduceus of Hermes and partly from the serpentine motions of the celestial bodies; possibly also from the Moon's nodes.

The Eleusinian mysteries are often associated with the myth of Demeter and her daughter Persephone, who is abducted into the Underworld by Hades and compelled to return there on a cyclical basis. Some scholars have suggested that this myth and the cult that revolved around it are but one variation of a theme originally set in motion by Orpheus. The story of this sublime troubadour who traveled the Underworld in search of his beloved Eurydice, and nearly brought her back, began our exploration of the numbers in the Realm of Zero. As Mark Morford, professor of Classics at the University of Virginia suggests (265):

> Although there must have been differences among the various mystery religions (some
> of them probably quite marked) obvious to the ancient world, we have difficulty

191

◇◇◇

today in distinguishing precisely among them. It seems fairly certain that the major common denominator is a belief in the immortality of the soul and a future life.

The caduceus is best known as Mercury's staff, a magician's wand composed of two snakes, but it also appears in other cultures - including the Indus civilization, circa 2500-1500 BCE in what is now India; ancient Egypt and Sumeria around this same period; and Aztec culture in the worship of the plumed Serpent god Quetzalcoatl (Campbell, *The Mythic Image* 281-292). Aside from its association with wisdom, in its ability to shed its skin and grow a new skin, the snake is also associated in these cultures with immortality. The two intertwining snakes of the caduceus (whom some interpret to be copulating) refer back to the Realm of Two, but in such a way that the original ambivalence associated with that number is gone.

The ambivalence in the Realm of Two is replaced in the caduceus by an intertwining that gradually ascends a central shaft. The suggestion is that through this intertwining of the opposites, and the absorption of the wisdom that this intertwining entails, a certain ability to rise above the downward pull of mortality is attained. In the practices espoused by Patanjali in the *Yoga Sutras* (circa 2nd – 5th centuries CE, exact date uncertain), a solar *nadi* (subtle nerve channel) and a lunar *nadi* intertwine around a central channel through which *kundalini* (associated with a coiled serpent) rises, leading to a liberation of the soul from earthly limitations. The renowned physician Asclepius, also carried a caduceus, and was killed by Zeus when his unparalleled skills in medicine (and perhaps magic) brought the dead back to life.

It would appear then, that through its association with the Eleusinian Mysteries and the caduceus, the Realm of Eight is the repository of a secret knowledge that leads to immortality. The secret knowledge is apparently that which first emergences in the Realm of Seven, as the ambivalence associated with Two is healed. In this Realm, it manifests as complex creativity that allows one to channel the Divine. In the Realm of Eight, as one channels the Divine - only possible when dual snakes are intertwining - the immortality of the Divine is attained. This does not mean that the body will not eventually fall away. It does mean that the apparent duality between life and death that exists in the preceding Realms no longer applies in the Realm of Eight, and that the falling away of the body is of no more consequence within this Realm than the moulting of a snake skin.

The few hints we have from those who knew the secrets of immortality - Hermetic magicians following in the footsteps of Hermes, Asclepian healers pushing the cutting edge of their esoteric arts, yogic practitioners, Taoist and Toltec shamans, and Tibetan masters of the bardo state - suggest that the physical body was not necessary to their attainment (Eliade 30-31). Instead, most immortality practices were aimed at the same goal as the Pythagorean Ascent discussed in Chapters Four and Seven - reunion with the One. In this context, the

◇◇◇

wisdom of the Serpent takes on a different meaning. According to Joseph Campbell (*The Mythic Image* 300):

> As the idea of transcending death can be interpreted in two ways, so too the image of the serpent. The first, more popular way is of reincarnation, eternal return, death, but then rebirth, as in the waning and waxing moon, or the serpent's sloughing its skin. However, one can also think of an ultimate transcendence of this everlasting round through an attainment of unfailing light, like that of the sun; and this is the aim of yoga. The waxing moon, night by night, approaches the fullness of the solar sphere, to which it attains the fifteenth night; after which, night after night, it again falls back into darkness. According to the mystical way of the Wisdom of the Yonder Shore, however, the ultimate aim of a life is to have one's light remain at the full – once having attained through many lifetimes to that apogee, to leap from Moon to Sun in consciousness and let the body go its way, like a waning moon; and it was just such a victory that the Buddha achieved in his final birth. Enthroned on a lotus pedestal, he is shown here being borne up from the watery abyss by a pair of serpent kings. So may we too be raised up by virtue of the serpent power to the status of a Buddha, a sun god, on a lotus throne of undiminishing light.

If the Serpent can be understood to be a symbol of the Ascent, as this passage seems to imply, then this suggests that the way to reunion with the One is through the Two, the path of the Serpent. This is what the Pythagorean arithmologists have been saying all along, although they doubted that piece of illogical wisdom, even as they asserted it. By the time Two has engendered Eight (the cube of 2), however, all doubt is gone, and the Serpent is revealed to be inseparable from the One.

In the early centuries after the birth of Christ, there were Orphic cults that asserted the Serpent in the garden was the first appearance of the Savior, but these were discouraged after the Church began to consolidate its power (Campbell, *The Mythic Image* 296). This suggests that what we are attempting to transcend in our quest for immortality on the path of the Serpent in the Realm of Eight is not embodiment itself, but everything that keeps us from experiencing embodiment as sacred.

The Realm of Eight and the Cyclical Power of Legacy

If the goal of embodiment is not transcendence of the earthly plane, but rather bringing consciousness more deeply into it, then an immortality that results in a mere ascension to the lotus throne of a Buddha is incomplete. The Serpent we are following through the Realm of Eight belongs *in* the Garden, which is where its wisdom must be applied. It is the application of this wisdom to the affairs of this life, then, that will constitute our real bid for immortality.

What matters is not how much light we take with us when we die, but how much we leave behind. Whatever we have managed to contribute in the Realm of Seven through complex creativity will continue to assert an influence within the embodied world after the body is gone. Those whom we remember now, long after they are dead, for their contribution to our collective human culture, are those who managed while they were living to enter the Realm of Eight – where they now enjoy a certain level of immortality within the collective consciousness. Each of us in our own way will live lives that reverberate into the future, whether we are remembered or not. It is this reverberation that will ultimately provide our only real claim to immortality.

For ordinary mortals, not actively engaged in some form of intentional spiritual Ascent, immortality has generally meant the capacity to leave behind a legacy, which is the sum total of the wisdom acquired through a lifetime and captured in a form capable of outlasting the body. For some, this legacy will be embodied by children or grandchildren. For others, it will be by a body of creative work. For those who have managed to attain a more complex form of creativity in the Realm of Seven, the legacy may have lasting consequences for the collective.

The Wisdom to be Harvested Through Awareness of Planetary Cycles

Whatever form our legacy takes, it is a process that takes a lifetime to ripen. In reference to the earlier passage from Manly Palmer Hall, it can be understood to evolve through the "serpentine motions of the celestial bodies." This motion expresses itself in terms of planetary cycles, and may be understood to constitute a third dimension of the wisdom of the Serpent. It is through absorbing the lessons encoded in the various planetary cycles that compose our birthchart, that the first two levels of serpentine wisdom are gradually released into consciousness.

The ways in which planetary cycles provide this opportunity were discussed extensively in my previous book, *Tracking the Soul*, and will be further elaborated in Part Two. In the meantime, it will be helpful to note that each planetary cycle contains eight primary stations – marked by angular relationships between planets called aspects – at which learning is possible. These eight primary aspects are the beginning conjunction (0°), the waxing sextile (60°), the waxing square (90°), the waxing trine (120°), the opposition (180°), the waning trine (120°), the waning square (90°), and the waning sextile (60°). The return conjunction (0°) completes the cycle and begins a new one.

◇◇◇

Alternately, we might consider the critical points of an 8th harmonic cycle (360° divided by 8) – the beginning conjunction (0°), the waxing semisquare (45°), the square (90°), the waxing sesquiquadrate (135°), the opposition (180°), the waning sesquiquadrate (135°), the waning square (90°), the waning semisquare (45°), and the return conjunction (0°).

All planetary cycles are continuous, and there is never really a moment when they are not in play. Often, however, at the critical junctures of the cycle marked by these aspects, the nature of the cycle will become more obvious, and the opportunity to apply consciousness in an intentional way more compelling.

When planetary cycles reference our core issues, our task will ultimately be to transcend the ambivalences that make those issues what they are. This will involve learning to negotiate one or more polarities – at first by mediating the extremes, and then gradually by embracing the extremes in complex creativity. Each cycle, and each planetary trigger to the pattern representing our core issues, will provide an opportunity for us to take another step in this process.

Sifting Through Planetary Cycles for Wisdom in the Realm of Eight

On the day I had my realization about the bird shamans of Chaco Canyon, for example, there were six major planetary cycles moving through an 8th harmonic phase (in an angular relationship that was some multiple of 45°, which is 1/8 of a full 360° circle):

> Sun waning semisquare Neptune
> Moon waning sesquiquadrate Moon
> Mercury waxing square Moon
> Moon waxing semisquare Sun
> Mars conjunct Moon
> Moon conjunct Venus

The fact that three of these 8th harmonic aspects are transits to my natal Moon, including a critical phase of the lunar cycle (Moon in aspect to the Moon), draws attention to the Moon. As I will discuss in more detail in Part Two, the Moon – which is the astrological symbol for the archetype of the Feminine – signals a point of entry into the Realm of Two, where the task is to assimilate some aspect of reality that raises internal ambivalence. What could be more ambivalent than a piece of information – hidden knowledge on an Odinic path – that defies the logic of the rational mind? The other two planets transiting my natal Moon on the day this piece of irrational wisdom was imparted to me are Mars and Mercury,

◇◇

forming a square to each other in the sky. In *The Seven Gates of Soul*, I associate my natal Mars-Mercury square with Odin, a god of the wind (Mercury) whose name means "furious," "mad," or "wild" (Mars). In that book, I suggested that (Landwehr 370):

> *When I follow my Mars-Mercury trail through soul space, I land on Odin's doorstep. . . . This is not a rational process; it does not yield an objective truth that can be applied to anyone else; it does not describe a causal relationship between inner experience and anything in the outer world. It is not a truth that a scientist, a scientifically-oriented psychologist, nor even a traditional astrologer would necessarily recognize as an orthodox interpretation of my experience. In some circles, it might even be considered hallucinatory or delusional. Following these threads of meaning through my Mars-Mercury story does, however, enrich my understanding of the subjective truth of my being.*

This Mars-Mercury trail could be understood as a pathway to Odinic wisdom within the Realm of Eight, perhaps particularly as it relates to my Moon. This trail has many threads – including Mars-Mercury and Mercury-Mars, plus in relation to this bird shaman revelation, Mars-Moon, Moon-Mars, Mercury-Moon, and Moon-Mercury – and not all memories related to all key moments of each of these cycles will be meaningful. Nonetheless, as we track these various cycles, we will find many clues to the subliminal wisdom weaving through them. Through such intentional scrutiny, seemingly mundane reality can be seen to have multiple levels of significance normally missed, just because we are not looking through a lens that reveals them clearly.

If I look back to the key moment in my Mars-Moon cycle (its waning semisquare) immediately preceding my bird shaman revelation, I see that I was contesting a traffic ticket in court, but unable to work up my usual obsession about this ticket, because I was beginning to fall in love. If I allow myself to free associate to this experience, I was able to rise above my drama and just be a neutral witness. This experience, which was part of the cycle related to my bird shaman revelation, can – in some way that defies rational explanation – be understood to be a touchstone moment for me in understanding how bird shamanism works.

During Mercury's previous 8th harmonic station to the Moon (its waxing semisquare), my new friend (who was out of the country) and I talked on the phone for the first time – in retrospect, another example in my mind at least, of bird shamanism at work, which seems to have to do with the ability to transcend space and time. To this day, we marvel at the fact that we were able to communicate like this across the miles, and create a relationship of intimacy, even though our actual time together had been minimal. Even though we take phones and email for granted, these vehicles of technological magic are the contemporary cousins of the transportation of roofing timbers 27 miles through no visible means of transportation, and

◇◇

lively trade with another culture 1000 miles away before the age of cell phones and Internet. Certainly the bird shamans of Chaco, who lived well before phones or email were even conceivable, would look upon these modern conveniences as magic – although I am guessing that they were able to duplicate this magic without the devices that we find necessary now.

Be that as it may, by tracking various threads of my Mars-Mercury-Moon story (previous moments in time when Mars or Mercury was in 8th harmonic aspect to my natal Moon) through past and future, it is not unreasonable to assume that I can learn a great deal about bird shamanism, as it applies specifically to me.

The Nodal Cycle and the Wisdom of the Serpent

A second possible clue to this process of collecting Serpent wisdom within the Realm of Eight was provided by Manly Palmer Hall, when he suggests that Moon's nodes might also have something to do with the number Eight. The Moon's nodes represent the points at which the orbit of the Moon around the Earth intersects the orbit of the Earth around the Sun. This point gradually shifts backwards in time over the course of an 18.6-year cycle, which was identified in Chapter Six as one of the celestial phenomena that the Anasazi were tracking at Fajada Butte (see pages 163-164).

When a New or a Full Moon occurs while the Moon is close to its own nodal axis, an eclipse takes place. In *Tracking the Soul* I suggest that New and Full Moons are especially potent openings to the seventh *chakra* (Landwehr 357-362). In terms of our discussion here, the seventh *chakra* might be understood as the portal through which a bird shaman rises above whatever particular melodrama happens to be taking center stage at the moment and becomes a neutral witness. It is also the place where space and time as we know them dissolve into an omnipresence of being where communication across time and space becomes as simple as thought. There are many ways to elaborate the metaphor of the seventh *chakra*, but a bird shaman would surely appreciate those I have chosen in this paragraph.

Even more interesting is a consideration of what happens at an eclipse. Any syzygy (New or Full Moon) represents a potent alignment between Sun and Moon that allows an exchange of energy between them. This is not necessarily a literal exchange of measurable energy, but more like a dialogue in which all things solar and all things lunar, and all polar opposites, become permeable to each other. This is the point at which the small circle of yin at the heart of yang and the small circle of yang at the heart of yin in the symbol for the Tao begin to pulse with a more noticeable energy of their own. The archetypal Masculine becomes more receptive to the Feminine; within the light, shadows become more animated and accessible to consciousness; a tight disposition becomes more amenable to mediation and/or a counter-flow with looseness; and so on. The actual astrology by which these possibilities

197

are triggered – or not – will be explored in more detail in Part Two. For now, I simply note the heightened possibility of interchange that occurs at these two points of the lunar cycle.

Normally, during a New Moon, the solar agenda (Masculine, light, and tight) is dominant; while during a Full Moon, the lunar agenda (Feminine, dark, and loose) moves to the foreground. During an eclipse, this rule of thumb is reversed: solar eclipses (occurring only during the New Moon) become more lunar, since the Moon eclipses the Sun; and lunar eclipses (occurring only during the Full Moon) become more solar, since the reflected light of the Moon is blocked by the Earth. What this means is that the exchange between polarities that is enhanced during a normal syzygy becomes an outright pole reversal during an eclipse.

In the language I have begun developing in this book, it is during an eclipse that the One becomes the Two and the Two becomes the One. The Serpent rejected in the Garden of Eden becomes recognized as the Redeemer, the antidote to the same ambivalence in the Realm of Two that previously made the Serpent seem dangerous. It is in such moments, then that the wisdom of the Serpent becomes most readily available. Among European astrologers especially, the nodes of the Moon where eclipses occur are known as the dragon's head and tail – a dragon being an especially potent serpent, exponentially more fearsome but also entrusted with the guardianship of a more valuable treasure.

Many indigenous cultures as well as Christianity, Hindu, Scandinavian and early Greek cultures considered an eclipse to be a bad omen, a reversal of the natural order that required some propitiation of the gods (primarily related to whichever heavenly body was eclipsed) so that everything could be put right again (Cashford 324-329)[2]. This attitude of fear arises from the very ambivalence to which the wisdom of the Serpent is antidote. If one is not ready for the antidote, then it exists only as poison. When one is ready, then an eclipse can become a portal through which the antidote is administered.

The Wisdom to be Harvested Through Awareness of Eclipse Cycles

Manly Palmer Hall refers to these eclipses at the dragon's head and tail when he implies that they are somehow related to the same constellation of symbols as the number Eight, the dual snakes of the caduceus, and the serpentine motion of the planets. This is a constellation of symbols we have identified with the wisdom of the Serpent, where the dissolution of ambivalence toward the Two makes possible an awakening of a deeper subjective truth. If a given eclipse seems to be associated with an experience in which such a truth is revealed, the tracking of memories related to an eclipse cycle can be revealing of additional layers of meaning and nuance.

◇◇

Eclipses are associated with other eclipses through a sequence called the Saros series. Eclipses within a given Saros series take place when the geometry[3] between Earth, Moon, and Sun is nearly identical, generally at an interval of 18 years 11 days 8 hours. A given series will begin with a partial eclipse, then shift a bit farther north or south in declination with another eclipse 18+ years later, gradually passing through a total eclipse phase, then out the other end to a final partial eclipse. The total process takes 1226 – 1550 years and can involve 69 – 87 eclipses before passing beyond the range at which an eclipse is possible, thus ending the series. The tracking of these successive eclipses, 18+ years apart, within a given Saros cycle, can result in a deepening of subjective truth.

My bird shaman revelation, for example, took place a couple of days before a total lunar eclipse that is part of Saros Series 128. In addition to the revelation of this eclipse, already described, it is possible to extract serpentine wisdom through a consideration of other eclipses in the same Saros series. Within my life so far this includes previous eclipses on August 17, 1989; August 6, 1971, and July 26, 1953. The first eclipse is – as close as I can tell – within a month of the ashpile incident discussed in Chapter Six that set my Odinic journey in motion. The second eclipse in August 1971 coincides with my graduation from college and my first Odinic journey to southern Oregon, where I met Sunny Blue Boy and fell in love with astrology. During the third eclipse in August 1989, I was on a Vision Quest at Lake Catherine in the Sangre de Christo Mountains outside of Santa Fe, NM. About a seminal moment of serpentine wisdom during this quest, I wrote (Landwehr *The Lightning of Opheukos* 132):

The third day of my vision quest began with a thick wall of fog that enveloped the shores of the mountain lake where I made my camp, and sank my psyche into a debilitating depression. I could not see six feet in front of my face; the trees around the lake were completely obscured; the sky a bleak, unfriendly shade of gray. It was cold and windy: I did not want to go through this day. Having written my brains out the first two days of my vision quest, I could muster only enough energy to stare bleakly off into space. Finally, with a supreme effort, made more melodramatic by the third day of a water fast, I crawled back into my sleeping bag and sought the blessing of escape.

After awhile, however, a voice inside my head suggested that if I wanted to begin my inner work, I should go outside. Though mind and body both dragged their heels in resistance, some larger (deeper? wiser?) part of me dragged both mind and body outside, then down to the lake, then out onto a rock jutting out into the water. "Lift the fog," the voice challenged me.

Feeling somewhat ridiculous, intimidated by the audacity of this impossible task, I nonetheless decided to give it a try. After all, I reasoned in my delirium, if some part of me was bold enough to create this heroic labor, some other part of me must be

◇◇◇

capable of performing it. So, I prayed to that part of myself, and watched in disbelief as the fog lifted just enough to reveal the opposite shore, then settle back down again.

Encouraged by this fluke success, I decided to go for it in earnest. I began repeating over and over to myself, "I am perfectly clear. The fog is lifting," turning to face all four directions in turn, then randomly toward any direction I imagined needed my encouragement. Amazingly enough, within an hour, the fog had lifted entirely around the lake, revealing a clear ring of green trees, dripping with moisture. Patches of blue appeared in the sky, and high on my unbelievable success, I tried to call forth the Sun.

As I undoubtedly overextended myself, perhaps sliding into the old trap of heroic hubris, the Sun failed to materialize, and after awhile, cold and tired, I crawled into my sleeping bag. In what I saw as my failure to attain the comfort of the Sun, I sank back into my depression. The fog, however, never did return. It was only later that I recognized the magnitude of what I had done.

Whether I actually lifted the fog or not is beside the point. The synchronicity between my intention and what was happening around me in the external world gave me a glimpse of another order where rational planning plays a minor, almost comical part. My life, I realize now, is not about manifesting a rational plan, but about entering this other realm, learning to live there, and perhaps showing other people the way inside. If I can cut my moorings of external security, a greater buoyancy, a deeper magic, will carry me through this world.

This other order "where rational planning plays a minor, almost comical part," is the path I have been describing in this book as following in the footsteps of Odin, an order where the injunctions of the gods often take one in a direction opposite to that which would appeal to reason. This is the quest for serpentine wisdom that has propelled me through a process of participation mystique – *I am perfectly clear. The fog is lifting.* – from a nearly fatal wounding on an ashpile in childhood to the discovery of the saxophone, kundalini yoga and astrology in my youth, to the exploration of my own subjective truth through a series of Vision Quests, to this new love and the possibility for entering more deeply into an understanding of bird shamanism capable of helping me grow back my wings. This is not a journey that would necessarily be meaningful were my goal to write a linear autobiography of outer events, but within the context of the Saros series I just tracked, it clearly does demarcate the path.

The Wisdom of the Serpent and the Practice of Recapitulation

This is the kind of whole life perspective that not only becomes possible in the Realm of Eight, but without which we cannot enter the Realm of Eight at all. We enter this Realm only as we engage in what Carlos Castaneda refers to as recapitulation. Recapitulation is a Toltec shamanic practice of remembering our past, but it is not remembering in the ordinary sense. For beneath the outer events of our daily existence is an internal movement of energy that registers on more subliminal levels. We experience it as sensation, and at other times as emotion, but it goes deeper than that.

Within the Toltec tradition, recapitulation involves accessing what Castaneda refers to as our Second Attention. First Attention is that which we normally apply to our experience. Second Attention is rooted in a state of non-ordinary reality, where everything we appear to experience is merely the gateway through which being is transformed. It is our experiences of movement through this gateway that we are trying to remember.

We don't remember these experiences in the ordinary way because they are not linear, and they do not contribute to the continuity of what we think of as our identity. In fact, quite the opposite is true. In moments of Second Attention, the cracks in identity are exploited to reveal something more luminous, and more fluid beneath outer appearances. In such moments, identity is revealed to be but a temporary container for something far more durable that cannot be contained. It is this durable essence, continuously reverberating beneath the fog of our experience in First Attention, that we are trying to remember in our recapitulation practice. We often miss it when it is happening, because we are too focused on outer experiences, and secondarily upon our internal responses to outer experiences. These responses are merely the veneer covering a more fundamental shift in being that may not even begin to register until years later.

This book is partly a personal exercise in recapitulation. Pulling together the various threads of my Odinic journey from past and present allows me to see and feel the numinosity of what outwardly appears to be a very ordinary life. Indeed, from the perspective of First Attention, it is. But from the perspective of Second Attention, no life is ordinary. Each is a journey into wholeness and harmony with a divine implicate order, implied within the Realm of One, and gradually unfolded through the other Realms parallel and co-existent with it. In the Realm of Eight, as we gradually evolve toward the wisdom of the Serpent to be found in a recapitulation of our lives, the outer trappings of those lives fall away. What is left is the numinosity of Second Attention, the radiant luminescence of the One that was there all along, but inaccessible because of our preoccupation with the ego-based self and the melodramas swirling around this self.

It is only as this preoccupation with the ego-based self falls away, that we begin to reap the harvest of wisdom buried within our lives. This is a wisdom that requires genuine dispassion to embrace – a dispassion that comes when all judgments about good and evil, right and wrong, light and dark are released, and when we no longer identify quite so strongly with the one caught up in the struggle between these polarities. The self with which we normally iden-

◇◇

tify – through the exercise of First Attention – has been wounded and redeemed, battered and boasted, exalted and crushed, and takes those ups and downs, ins and outs, seriously. In the Realm of Eight, we begin to see that all that matters within this story is the gradual release of something capable of surviving and transcending the story. Buddhist nun Pema Chodron talks about dropping the story line of our lives to enter a place of true compassion; Carlos Castenada talks about erasing our personal history and dropping the human form; other teachers and traditions refer to this surrender of small ego-based identity in different ways. The end result, however, is the same – in the Realm of Eight, we become transpersonal vessels for the movement of transformative energy, permanent residents in Second Attention.

When we talk about the Realm of Eight as a gateway to immortality, this is really what we mean. Whatever it was we thought or hoped might survive on the near shore of our quest for immortality becomes a fading afterthought as we actually approach it. What lives on after us does not have our name stamped upon it. But to the extent that we are able to recapitulate our lives, and then let the outer story line go, we can experience a state of being that is beyond death, because it is aware that there is nothing left of us capable of dying. It is this mystery to which we are seeking entry in the Realm of Eight.

Astrological Recapitulation

Astrology can facilitate our entry into the Realm of Eight: first because it is an extraordinary tool for recapitulation and secondly, because its symbolism reveals the nature of the transformation underlying the events of ordinary life and makes them more accessible to Second Attention. Astrology approaches the recapitulation process through the technique of the cyclical history. Since the strongest patterns in our natal charts refer to the core issues likely to command the greatest portion of our First Attention, the triggering of these patterns – particularly by the transpersonal planets, Uranus, Neptune, and Pluto – will provide the most potent openings to Second Attention.

As we track our memories related to these cycles, looking to the eight cardinal points of the cycle for emphasis, we are likely to stumble into the life experiences we outwardly recognize as turning points, moments of catharsis or crisis, breakdowns in the fabric of business as usual, and occasionally breakthroughs to another level of being and possibility. Harbored within these experiences are also strong sensory and emotional memories, and to evoke these is to stand on the threshold of Second Attention. With distance and in retrospect, we are in a position to perceive what was actually transformed in these moments, and – most importantly – to identify less with the events themselves, and more with the transformation. To the extent that we can do this, the wisdom to be found in our experience can be harvested, and we begin to dwell more permanently in the Realm of Eight. We develop a dispassionate distance from the events of our lives, and are able to function more freely within that dispassion.

As we observe the astrological symbolism involved in the transformative cycle, we gain a deeper understanding of the nature of what is being transmuted and how. A Saturn-Venus cycle brings a different sort of transformation than a Uranus-Mercury cycle. The language of astrology allows us to make these distinctions. The experiences are personalized, even as we gain transpersonal distance from them. We begin to see that whoever we thought we were and whatever we thought we were doing, all of that was just a vehicle for allowing the energy associated with the planets – which are named after gods and goddesses – to re-create existence in accordance with the movement of the cosmos.

As we move into a state of Second Attention in our practice of astrology, we begin to understand that we ourselves are merely the channels through which these planetary gods and goddesses do their work. Knowing this, and giving our consent to it, their wisdom becomes our wisdom. There is no longer any difference between the heaven that they occupy and our Earth – perhaps the last polarity to collapse on the way to the Realm of Ten, where everything on Earth and elsewhere in the material universe becomes fully conscious.

Entering the Realm of Eight Collectively

It can be argued – as I did in *The Seven Gates of Soul* (Landwehr 174-177) – that this embodied world is by definition a playground for the perpetual interplay of opposites. We can also glimpse the occasional implied collapse of some polarity within the world that leads, at least potentially, to the wisdom of the Realm of Eight on a collective scale. If we step back to consider history cumulatively, we begin to realize that however broad the window, the historical processes that provide a glimpse into the Realm of Eight are ongoing, with ever more subtle shades of completion evolving through cyclical time.

Before the Civil War, for example, blacks in this country were routinely pressed into slavery. Slaves were freed after the Civil War, but then found themselves in circumstances offering extremely limited options. Rapid mistrust, exploitation and segregation of blacks continued for another century, until great waves of civil unrest, rioting, and the insistence of many brave men and women, white and black in the 1950s and early 1960s, ushered in an era of increasing rights, and slow begrudging equality. Even now, nearly 50 years later, the battle for true equality between the races continues, and the polarity between black and white has not yet been completely erased.

Still, the conditions under which slavery existed a mere 150 years ago are unthinkable today to even the most unapologetic racist. Blacks have risen to positions of authority and

◇◇

power in nearly every professional arena, including the highest levels of government. Within the Realm of Eight, we have together been gradually working toward a color-blind society in which race is no longer a pivotal point of division or separation. Even the neo-conservative call for an end to affirmative action can be understood, in principle at least, as a backhanded recognition that race should not be a factor in our collective decision-making process. Should this process of deconstruction of polarity ever reach its conclusion, identity based on race will become inconsequential. The recent election of Barack Obama – son of a black father and a white mother – as US president is both symbolic of this possibility and a significant achievement on the long and arduous path to it.

On a larger scale, we can see the workings of the Realm of Eight within the world in the drawing and redrawing of the world map, through which countries appear and disappear, and empires rise and fall. Take the current war in the Middle East, for example, which will likely eventually redraw the map yet again. In the 8th Century BCE, the dominant player in the Middle East, specifically in what is now southern Iraq, was Chaldea, a country (and later an empire) that no longer exists. The Chaldean Empire lasted until 539, when it gave way to the Archaemenid Persian Empire, which spanned three continents, encompassing the modern states of Afghanistan, Pakistan, Turkey, Iraq, Saudi Arabia, Jordan, Israel, Lebanon, Syria, significant portions of Egypt, and Libya. The Archaemenid Empire fell to the conquests of Alexander the Great circa 330 BCE, which erased its boundaries and again redrew the map of the world in the image of the Greeks, encompassing not only what is now modern Greece and the Archaemenid Empire, but also parts of India, stretching as far east as the Himalayas. Alexander's Empire gave way to the conquest of the Romans in the 1st century CE, which in addition to Greece and most of Europe, also encompassed Egypt, Libya, Tunisia, Algeria, and Morocco. Meanwhile, various desert warlords attempted to regain those portions of the old Persian Empire subjugated by the Greeks, an effort that eventually crystallized as the Sassanid Empire in the 3rd century CE, centered in what is now modern Iran. The Sassanid Empire fell in the 5th and 6th centuries, partially to the Romans, and partially to the White Huns from central Asia, probably originating in what is now modern China. The Arab world was reunited by Mohammed in the 7th century, only to fall again to the Turks in the 11th century. And so it goes, down to the recent wars in Iraq and Afghanistan, which are bound to have reverberating consequences for the entire region, on into the future, until another reshuffling of the ancient deck produces a different set of cards.

The history of this region, however, is less important than the underlying transformations that have occurred through an intermingling of cultures from the various parts of the world that combined, splintered, and then recombined. With each new conquest, old polarities broke down and new ones were formed, so that identity within this world has necessarily become a perpetually shifting amalgamation of worldviews. Despite the efforts of each new conquering nation or tribe to dictate a more dogmatic worldview, layers upon layers of cultural cross-pollination have produced a complex network of subtle differences within each nation state.

The dilemmas posed by the current Israeli-Palestinian conflict, for example, in which two distinct groups of people essentially occupy the same land, are endemic throughout the Middle East, where vestiges of former civilizations intermingle in the very dust beneath the feet of those who live there now. We continue to pursue resolution to these difficulties along a path of inequality. Yet there also exists within this fertile and volatile cauldron of perpetual transformation, the wisdom of the Serpent through which the partial truths shared by warring factions may one day somehow fuse into something human culture can point to as its proudest achievement. The timeframe for this is unfathomable, given the history that has played itself out within the range of our collective First Attention. From the perspective of Second Attention, however, it is likely that some deeper more archetypal process is at work. On this level, it is not hard to envision the endless battle in the Middle East as a modern-day version of the myth of Cadmus, sowing the Serpent's teeth that sprang into the armed men whose fighting eventually resulted in the foundation of Thebes.

The Implications of Realm of Eight Within the Chakra System

The deeper process of Second Attention represents the culmination of human effort to recreate the harmony inherent in the One within the embodied world. It seems fair to assume that within the *chakra* system, this is a possibility realized only as the so-called higher *chakras* are integrated. In *Tracking the Soul*, I discuss the limitations of assuming a hierarchical arrangement of the *chakras* (Landwehr 10-13), so strictly speaking, this integration is not an advanced attainment, as much as it is an ever-present possibility that must be accessed with clear intention. This intention can be understood as a recapitulation process centered in the fifth, sixth, and seventh *chakras*.

In the fifth *chakra*, we align ourselves with a truth that is unique to each of us, and arises as a consequence of our individual quest for a life of meaning or purpose. In the fifth *chakra*, we learn where we naturally sit in relation to each of the polarities we encounter, and consequently where our various tendencies toward resonance – by affinity, wounding, and contrast – lie, thus fulfilling the admonition of the Delphi Oracle to "Know Thyself." If we can do this without judgment, accepting ourselves unconditionally, then our recapitulation process reframes our entire life as an adventure in learning. Out of this comes a sense of self that is free to be itself without censure, apology, or any necessity for redemption.

When the fifth *chakra* is integrated with the *sixth*, we gain the flexibility and mobility to move anywhere on any polar spectrum. This is where the Two becomes the One, and

CHAPTER EIGHT

<><><><><><><><><><><><><><><><><><><><><><><><><><><><><><><><><><><><><><><><>

the Serpent in the garden becomes the harbinger of wisdom. For when the sixth *chakra* is clear, there is nothing within the manifest creation that is not a source of teaching. We shun nothing, and we embrace it all. Out of this all-inclusive embrace comes a deeper sense of self that is inherently wise, because it harbors no ambivalence about the true nature of this manifest Creation, which is – on this level – Spirit incarnate. All the particulars are merely a veil through which Spirit can be seen at work, from the grand cosmic scale to the subatomic. With this integration, we further recognize ourselves to be a thinly disguised carrier of Spirit. Each of us has a job to do – one that will require all the wisdom we can muster, but also arise quite naturally out of who we are, and have become, as a consequence of our unique lifetime of experience.

When the fifth and sixth *chakras* become integrated with the seventh, even this final veil falls away. We give consent to fulfill the cosmic function for which we were born, and whoever it was that gives the consent fades away as an echo upon the wind. Here is where, in the Realm of Eight, we attain immortality because there is no one left to die, and nothing left that death is capable of shifting. To be in this place is to be a pure agent of divine will with no interference from personality, the complications of core issues, or circumstances that would previously have presented an obstacle.

In the Realm of Eight, where the embracer of all harmonies opens up unlimited possibility for the realization of a balanced, fully conscious universe, we are all called to pour our cupful of ocean water into the common stream, and together follow that stream back to the ocean. To do so is to participate selflessly in the unfolding of something exquisite and sublime, beyond the messy outworking of human trials, beyond the wars and genocides and unthinking ecological devastation wrought on the path of inequality in other dimensions, to rise to the occasion for which human beings were created, to at last play our part in the grand scheme of things without attempting to hijack the scheme for our own petty purposes. Whether we as a global culture ever reach such a lofty pinnacle of accomplishment depends upon whether or not we can put aside our judgments about good and bad, right and wrong, light and dark, and merely participate with full attention in the miracle unfolding every day around us, within us, beyond our knowledge or capacity to know.

Like a silent, scintillating Serpent coiled in the darkness, just beyond our sight, the Realm of Eight and the promise that it holds reverberates with both numinous power and deadly potential. To embrace it, as the harbinger of wisdom that it is – because we have erased all sense of ambivalence about the Two, in all of its many guises, from our psyches – is to rise above the human predicament and participate in the divine. To stumble into it blindly, and react with fear, is to fall – like Eurydice into the Underworld Abyss, like Adam and Eve out of paradise, like Cadmus into the necessity for servitude to another round of apprenticeship. Depending on how we respond to the challenge of the Serpent, the Realm of Eight becomes a final chance to see the truth and erase the false barriers between One and Two, or a last dark round of souls vying for their place within a hierarchy destined only to enslave them.

Only the wisdom of the Serpent can set us free from these dark rounds, can complete the journey of becoming began in the Realm of Zero. But the wisdom of the Serpent is not available just for the asking. We must earn it. We must consciously and earnestly move through our illusions about the true nature of this embodied world by realizing the ultimate futility of inequality, at every other level, in every other Realm, by embracing the equality of One and Two. To do that is no small accomplishment, and there is no guarantee that the human species will meet the challenge successfully. If we don't, whatever new experiment survives our fatal follies, will try again. For this is the nature of the Serpent, and the various cycles, small and large, that it sets in motion.

Endnotes

1. A number that engenders is one that becomes a factor of another number (i.e. since 2 x 2 = 4, 2 can be said to engender 4. A number that is engendered can be factored into other numbers (i.e. 4 is engendered by 2). A number that is not engendered is a prime (i.e. 5 or 7).

2. There are notable exceptions to this view. The Bangala of Africa, for example, believed that during an eclipse the Sun and the Moon came together in a secret tryst of intense mutual passion that was concealed in darkness (Cashford 327).

3. The geometry between Earth, Moon, and Sun takes into account the necessary alignment between: Sun and Moon at New or Full Moon; the convergence of the Moon and its own node (where the lunar orbit and the ecliptic, or plane of the Earth's orbit, intersect); and the distance between the Earth and the Moon. When all three of these conditions are identical for two given eclipses, these eclipses are said to be part of the same Saros cycle.

Figure 9: The Rune Isa

Ice, the frost giant Ymir, standstill, the rune of Fate

Chapter Nine: Fate and the Portal Between Dimensions

In the preceding chapter, I suggested that in the Realm of Eight human beings reach the pinnacle of their evolutionary development. As a consequence of choices set in motion at other levels, we either attain a certain wisdom that releases us from the struggle of learning from our mistakes, or we insist on a certain willful ignorance that compounds them in a dwindling spiral downward toward devastation in the Abyss. Within the sequence of numbers presented so far, there have been ample opportunities at each level to correct previous errors along the path of inequality, and choose differently. Seen in its entirety, the numbers between the Monad and the cube of Two represent an evolutionary school for learning how to live in harmony with the One, how to align ourselves with the undying truth that will set us free from the limitations of ordinary mortal life, while at the same time, drawing Spirit more deeply into it. In the preceding chapter, I spoke of this liberation as immortality, recognizing that it had less to do with the mere preservation of bodily existence, and everything to do with passing on a legacy capable of fostering life and a more refined quality of life.

The Realm of Nine and Fate

Once we reach the Realm of Nine, the opportunity for learning ends, and we receive a report card of sorts, in the form of what the Greeks called Fate. Fate can seem an antiquated concept in a post-modern world conditioned to believe in unlimited possibilities. To call something fated today is to essentially give up on it – to relinquish one's power to influence it, and to passively allow it to take its own course, since no amount of intervention in its process will make a difference. Regardless of how despairing we are about our lives, few of us are ready to admit such terminal defeat, and will keep trying to better our lot, one way or another, until the day we die. But this attitude implies that Fate is something imposed on us externally, and that with sufficient will or power of intention, perseverance and positive thinking, we can eventually shake our fate as though it were just a temporary head cold. What the Greeks meant by Fate was much more durable than that.

At every other level, the more malleable post-modern definition of Fate might well apply. Indeed, I have implied this throughout my discussion of number to this point. In choosing a deity to follow in the Realm of One, in choosing equality over inequality in the Realm of Two, in choosing to mediate the opposites in the Realm of Three, in choosing a more conscious relationship to the elements and four directions in the Realm of Four, in choosing to live in harmony with natural law in the Realm of Five, in choosing to address our core issues through the soul work required of us in the Realm of Six, in choosing to exercise complex creativity in the Realm of Seven, in choosing to use our hard-earned wisdom to create a life of harmony in the Realm of Eight, we are basically exercising our power to modify Fate. At each of these levels, the choices we make imply the possibility of creating a more flexible relationship to the circumstances of our lives, and gradually improving them. Within the Realm of Nine, however, there is a certain relentless inertia to the flow of events born of the cumulative weight of our choices that is extremely difficult, if not impossible to reverse. It is in this sense that we must understand the Realm of Nine as the crystallization of Fate.

The Arithmological Power of the Square

To understand how the number Nine relates to Fate, it will be helpful to consider a basic arithmological principle. Within the system devised by Pythagoras, the squaring of a number essentially makes the essence of that number more tangible, more durable, more entrenched. We saw this first when 2 was squared to become 4, creating a manifest world of substance from the potential encompassed by the polarities within the Realm of Two. At this level, hot and cold are no longer merely concepts. They impact the embodied soul as sensory experiences – triggered by fire and ice, summer and winter, volcano and glacier, stove and refrigerator; emotional experiences – anger and fear, passion and detachment, joy and loneliness; and life circumstances – noisy neighbors and distant in-laws, gambling casinos and church services, wars and sexual escapades, births and deaths. Within the Realm of Four, the principles associated with Two – duality, ambivalence, conflict, and the quest for balance – are all played out on a much more fully realized stage, as an intrinsic part of the embodied world. In a very real sense, the solidification of duality (Realm of Two) within the manifest world (Realm of Four) constitutes the first level of what we are now calling Fate, since to be born into such a world as an embodied soul is by definition to be limited – by mortality, gravity, entropy, space and time, and other forces endemic to a physical reality.

The same principle of entrenchment occurs when 3 is squared. The Realm of Three is where choice first becomes possible in relation to the polarities; the Realm of Nine is where the consequences of choice become tangible, durable, and entrenched in a second level of Fate. In the Realm of Three, we gain the perspective necessary to understand where we are in relation to any given polarity – i.e. toward the hot end or the cold end – at any given time, and by natural preference. In the Realm of Nine, this perspective loses its flexibility and essentially becomes our reality, an actual embodiment of that which we perceive it to be.

In *The Seven Gates of Soul* I speak of this solidified perception of the world as the image, initially formed through our experiences, but then gradually becoming an *a priori* factor dic-

◇◇◇

tating our experiences (Landwehr 261-262). It is within the Realm of Nine that our image of the world gains this power to determine our reality. In the Realm of Three, we can choose between hot and cold. The more often and the more consistently we choose hot instead of cold, the more likely it is that within the Realm of Nine, we will find ourselves living in a hot world – battling rashes, ulcers, insect bites and sunburn; dealing with situations that make us angry; living impulsively with reckless abandon; hotly debating our point of view; and so on.

Aside from being the last square within the single-digit numerical sequence, Nine gained its association with Fate or finality through its unique position as the last number in the sequence. The next number, the Decad, was essentially a repeat of one (1 + 0 = 1), beginning the entire sequence again (11 = 1 + 1 = 2; 12 = 1 + 2 = 3; etc). Thus whatever had evolved through the number Nine had reached its natural limit, and could go no further without beginning again. In this slightly altered understanding of the number Nine as the end of the evolutionary road, it is possible to see a numerical correlate to the Eastern concept of karma, in the sense of returning to reap what we sow. The Realm of Nine is where our actions (and our choices) bear karmic consequences, which precipitate another turning of the wheel of reincarnation – i.e. Nine becoming One again (10 = 1 + 0 = 1). I will say more about these cyclical implications of the number Nine toward the end of this chapter, but first let us get back to the more restrictive notion of Fate.

The Realm of Nine and the Morai

The ancient Greeks envisioned Fate as a triad of crones, called the Morai. Clotho wove the thread of Fate, Lachesis measured it, and Atropos cut it. As they did so, a human life took its characteristic shape. The Greeks envisioned this entire drama as taking place before birth. I prefer to think of it as a cumulative process unfolding continuously throughout a life within the Realm of Nine, an unconscious substratum of consequence humming quietly below the surface of conscious choices made within the other Realms.

That Nine functions unconsciously in the background is suggested arithmologically by the fact that adding 9 to any number produces the same number, when addition continues long enough to produce a single digit number. For example, 9 + 6 = 15; 1 + 5 = 6. And 9 + 8 = 17; 1 + 7 = 8. In this sense, the arithmological function of Nine is similar to that of the number Zero, with which it shares this property. Both the Abyss and Fate are ever-present simultaneously with whatever else might be going on. It is our fate that determines how close the Abyss is in any given moment. Our fate in turn, depends upon how consciously we have chosen a path capable of leading us out of the Abyss. The way this works, as I envision it, can be understood with reference to the functions of the three crones that are together responsible for spinning the web of Fate.

⟡⟡⟡

The Weaving of the Thread

The process begins with Clotho, who weaves the thread. The threads that Clotho weaves can be understood to come from a number of sources that are in some sense determinative of Fate. One thread – perhaps the primary contributing factor – arises through the cumulative consequences of our actions. Every time we make a choice, and act accordingly, Clotho draws forth a bit more thread to weave into the mix. If I choose to move to Cincinnati, or father a child, or study law, then doing so marks various threads of Fate that will play themselves out as the circumstances of my life. My experiences in making these choices will be different than they would be were I to make a different set of choices – say moving to Los Angeles, remaining single, and studying medicine. The cumulative weight of our choices in this life, to some extent, determines the nature of circumstances that we face now. Eastern concepts of karma and reincarnation would additionally encompass choices made in past lives.

A second thread of Fate comes through our family lineage. A soul born to a wealthy family of well-educated Caucasian academics growing up in a pastoral university environment will have a different fate than a soul born to a poor black family in an urban ghetto, where gang violence and drug trafficking are the everyday norm. In addition to these external differences, each family will tend to have its group characteristics, tendencies or issues, which are passed from generation to generation – e.g. an entrepreneurial spirit, a love of music and literature, a sense of compassion toward those less fortunate, a problem with drugs or alcohol, a tendency toward co-dependent relationships, chronic poverty, strong racist biases, religious zeal, a genetic predisposition toward heart disease, or any number of other seemingly self-perpetuating patterns taught or otherwise transferred by parents to their children. While it is possible to break these patterns, and children often do, especially in adolescence, there is also a great deal of inertia associated with them. It is this inertia that gets passed on from one generation to the next through a second thread of Fate woven by Clotho in the Realm of Nine.

A third thread of Fate is dictated by the collective mindset of the culture into which we are born. To some extent, the family thread will reflect this mindset. But the collective thread will also encompass a broader worldview that is a product of the time and place we are born, regardless of family background. A black person growing up in southern Alabama during the 1950s, for example, will have an entirely different fate than one growing up in New York City during the 1980s, or in Dafur, Sudan today. An intellectual male growing up in an aristocratic family in 17th century England would have had a different life experience than one growing up in the macho barrio culture of Hispanic Los Angeles in the 20th century. A woman giving birth to a child out of wedlock in Amsterdam today will have an entirely different set of reproductive rights than one having the same experience in Plymouth, Massachusetts, during the era of Nathaniel Hawthorne's *The Scarlet Letter*. All of these cultural determinants pose a potent thread within the weave of Fate drawn by Clotho in the Realm of Nine.

Perhaps the most potent thread of all, however, is posed by the simple fact that we are born into physical bodies that will, over time, gradually succumb to the forces of entropy and die. If life were unlimited by mortality, we would have endless time in which to fiddle with Fate – to change our personal habits, to rise above patterns of family and cultural conditioning, to outlast the forces of inertia through the sheer vitality of an unbridled spirit. But this is not the case. Every life is a race against time, a race to become conscious of our path and walk it to its natural conclusion before the Earth reclaims us. The fact that none of us knows when exactly that will happen renders this thread of Fate a bit more precarious than the others. The others will in some sense outlast us, but this one will break, typically at the most inopportune time. While it is possible to imagine the family Fate being passed on through our children to our grandchildren, or the Fate of a culture working itself out over the course of generations, the thread of mortality is an absolute limit imposed on the Fate of the individual human soul.

Those who believe in reincarnation will argue that this limit is illusory, since we will return in another body and another life to pick up where we left off in our last life. There is great appeal to this idea, and I don't necessarily discount it, although I did argue in *The Seven Gates of Soul* that belief in reincarnation was essentially a distraction for the embodied soul wanting to make maximum use of the opportunity to become fully conscious in this life (Landwehr 23-24). From the arithmological perspective, the possibility of reincarnation is certainly implied in the recycling of numbers after Nine through addition of their composite digits. Yet, even if this is so, the Fate of a soul is altered upon death. None of the threads contributing to Fate in this life – neither the personal habits or preferences, the family influence, or the culture – would remain intact, even if the soul did pass into another body. There might be some karmic thread that would make it possible, on the level of soul, for you to still be recognized as you, but the circumstances of your new life – and thus your Fate – would be entirely different.

Fate as the Natural Condition of the Embodied Soul

Fate, then, is necessarily the condition of a soul born into a body. Embodiment seals Fate as a set of conditions and limiting factors that will impact the soul in this life, and these are not that easily altered once set in motion. It is this set of limiting factors that Clotho weaves into our fate within the Realm of Nine.

CHAPTER NINE

◇◇◇

A natural corollary to this idea of embodiment as a primary determinant of Fate extends beyond the body itself to encompass the fact that we live on a physical planet with natural limits. These natural limits in turn condition not just our personal lives, but also the life of the collective, and of the entire web of life in which humans participate. As I suggested in Chapter Five, the exercise of consciousness that becomes possible in the Realm of Five is still bound by the fact that it is within a physical universe that we must do so. Thus in order to exercise consciousness responsibly, and in such a way that we do not return to the chaos of pre-creation, we must do so within the confines of natural law. Within the Realm of Nine, the extent to which we have done this or not done this will come home to roost as the Fate of the planet on which we live and die. Some, such as the scientists at Earthwatch Institute, have argued that in many areas we are pushing the Earth past its natural limits, in some cases, past the point where rebound is possible.

A case in point is Russia's Aral Sea, a 25,000 square mile saline lake, once home to 24 native species of fish, and a thriving fishing culture scattered throughout numerous villages on its shores. Today, because of an aggressive agricultural program, driven by state-mandated production quotas of fruit, vegetables, rice and cotton, massive use of pesticides, and insatiable appetite for fresh water, the Aral Sea is an ecological disaster, suffering probably irreversible damage (Wolff 213-214). Irrigation requirements of this agricultural program have dried up the primary fresh water tributaries of the Aral Sea – the Amu Darya and Syr Sarya rivers – reducing the depth of the Aral Sea by about 46 feet, cutting its surface area nearly in half, and tripling its salinity, all within the span of a single generation. Now what is left of the sea has become too salty to support any fish at all, and once-thriving fishing villages find themselves landlocked, 20 miles from their former fishing grounds. What is more, the agricultural fertility of the land around the Aral Sea has also greatly diminished as the soil is now "so encrusted with salts and the residue of massive amounts of fertilizers, pesticides, and herbicides dumped on the adjacent fields that the wind picks them up in swirling clouds of toxic dust and scatters them across the surrounding region" (Wolff 214). Groundwater in the area is also hopelessly contaminated and, by some estimates, even the climate has changed to reduce the growing season by 10 days a year.

From the perspective I am presenting in this book, what has happened at the Aral Sea, and is happening in many places around the planet, is happening in the Realm of Nine, where cumulative consequences become virtually irreversible. This is a huge factor impinging upon our collective Fate. As we envision the Realm of Nine in its cyclical implications, we can weasel our way out of the thought that what we are doing is irreversible. Indeed, in Chapter Seven, I told of a vision I had in which I saw the Earth as an inviolable jewel in some other dimension, where despite what we did to it in the third dimension, it retained the power to regenerate itself on some time scale perhaps unfathomable to us. In Hindu mythol-

◇◇

ogy, the *Mahabharata* speaks of a cosmic cycle of four *yugas* or ages, encompassing a period of 4,320,000 years, envisioned as the in-breath and out-breath of the Creator God, Brahma (Campbell, *Oriental Mythology* 116). Within this large cosmic cycle, manifest creation comes into being and is ultimately destroyed, then resurrected perhaps in a different form during the next cycle.

From this perspective, the number Nine is not irreversible in an absolute sense, although the time scale on which it becomes reversible is such that it is likely that the Creation we recognize now would be utterly different than the next one, and endowed with an entirely different Fate. Although the numbers do repeat themselves as they cycle through a single human life and the life of the cosmos, within the Realm of Nine, that which passes to the next cycle is only loosely related by a thin thread of essence to that which is left behind. The Fate that Clotho weaves is contained and in some sense limited by the cycle to which it refers.

Measuring the Thread

After Clotho weaves the thread, Lachesis measures it. Although Greek mythologists are sketchy about the mechanism for this measurement, it seems plausible that the out-working of Fate proceeds cyclically, and is measured by the movement of these cycles. This, of course, is a specifically astrological perspective. It would be quite natural to Greek sensibilities to think in terms of astrological cycles, since astrology was considered by the Greeks to be the basis by which the manifest universe could be understood to embody rational order (Tester 18), at least since the Greeks adopted it during the Hellenistic period. Astrology was often associated with Fate within the Greek mindset, particularly through the influence of the Stoics (Tester 52), whose philosophy could largely be summed up by the poet Manilius' famous line: *Fata regunt orbem, certa stant emnia lege* (The fates rule the world, all things are established by a settled law). This settled law was seen to operate through planetary agencies, whose cyclical motions measured the Fate established by the law.

The crone Lachesis can be understood to measure Fate in accordance with various astrological cycles. First among these would be the day, which represents one revolution of the Earth about its axis. In astrology, this daily cycle generates the twelve houses, which represent the gamut of circumstances that define a life – everything from the type of body into which the soul is born (1st house), its material circumstances (2nd house), the educational opportunities available to it (3rd house), and on to its calling and career path (10th house), the cultural atmosphere conditioning its expression in life (11th house), and the institutional limitations of the society in which it lives (12th house). The circumstances of a life are measured astrologically by the condition (planets ruling a house or dwelling within it and their relationships to other planets) of the 12 houses.

CHAPTER NINE

◇◇◇

The second measure of Fate is governed by the month, which is essentially a lunar cycle – the time it takes the Moon to move through all of its phases from New Moon to Full Moon and back to New Moon again. In astrology, the Moon is associated primarily with feeling and memory. As a measure of Fate, the Moon most accurately portrays our internal response to the circumstances of our life, that is to say, how we perceive our lot in life, our attitude toward it, and consequently how functional we are within the bounds of Fate. Those who are born to unlimited wealth and opportunity, for example, may suffer because of an attitude of entitlement or arrogance or lethargy that limits their capacity to take advantage of their Fate. Conversely, those who grow up in limited circumstances, but approach their experience with an attitude of curiosity, self-confidence, and hard-working perseverance, can maximize the opportunities within what might otherwise seem to be a more severely limited range of motion. This attitudinal flexibility in relation to Fate allows Fate to breathe with us, and becomes a measure of who we are in relation to it.

A third measure of Fate is the year, which represents the annual revolution of the Earth around the Sun. Beyond circumstances and our attitude toward circumstances is a deeper aspect of being that propels us into embodiment for a purpose. When we incarnate, each of us must separate from the One, differentiate ourselves as an aspect of the One, and enter the world as a unique soul, distinct and apparently separate from other souls. We undertake this arduous journey, so that the One can express itself through us in a particular way, conveying an intimately personal message, making a one-of-a-kind contribution to the collective that we are uniquely configured to make. In part, the circumstances of our lives, reflected by the measure of daily rotation, embody conditions necessary to the actualization of our purpose. In part, they will provide a seeming impediment to the actualization of our purpose, which at a deeper level is designed to strengthen us, teach us what we need to know, and train us to be more effective in conveying the message we have taken on. This level of Fate is symbolized by the Sun, which embodies ego – the force necessary to propel us through our lives, and grow into the purpose for which we have taken birth. At a more evolved level of possibility, the Sun also represents the spark of Spirit within us that animates us, gives us life, and grows like a seed within us until we are fully capable of mastering our Fate.

Beyond the Earth, the Moon, and the Sun are the other planets, each of which provides an additional measure of Fate. Mars – whose orbit around the Sun is marked by a 2-year cycle – describes our capacity to actualize our purpose; Mercury and Jupiter – encompassing 1- and 12-year cycles respectively – describes our capacity to process information and learn from our experiences. Saturn – with a 29 ½-year cycle – governs our capacity to handle adversity. These various measures of Fate, mapped to planetary placements and the cycles through which these placements are experienced, determine our lot in life. Within the Realm of Nine, Fate becomes more or less deeply entrenched over time, depending on the choices that we make of critical junctures of these cycles. These various measures of Fate also interweave in complex ways – primarily through astrological aspects and rulership patterns[1] – to create a sense of Fate that is fairly unique to each individual.

Planets beyond Saturn also measure Fate, but these cycles are more accurately understood within the Realm of Nine as collective Fate, since their durations – 84 years for Uranus, 168 for Neptune, and 248 for Pluto – transcend the length of the average human life. When these transpersonal planets form a major aspect each other, these larger cycles can provide an opening through which collective Fate impacts the individual, or generations of individuals. Historical trends, and major changes within the global political or cultural climate, the background atmosphere within which we must all live out our personal fate, can be mapped to Uranus-Neptune, Uranus-Pluto and Neptune-Pluto cycles, and on a more detailed scale to Jupiter or Saturn cycles with these same three outer planets. When the transpersonal planets form aspects to one or more personal planets[2] within the context of an individual birthchart, they provide an opening through which an individual may potentially contribute to the mitigation or the further entrenchment of our collective Fate.

We will observe these relationships in more detail in Part Two when we explore the more specific astrological ramifications of the Realm of Nine. For now, it is enough to know that when Lachesis measures our fate, she does so – or can be conceptualized to do so – within the context of planetary cycles.

Cutting the Thread

After Lachesis measures Fate, Atropos cuts it. Whereas the weaving of Fate and its measurement encompass a kind of logic that can be mapped to the movement of a life and the astrological cycles that weave through it, the cutting of Fate – which we may ultimately equate with death – would appear to be a random, capricious force beyond the reach of reason or logic. None of us knows when death will come, or where the measured fabric of Fate will be cut. It is this inability to know when or how our Fate will be cut that gives Fate the terrible power to trump any choice we might make in any of the other Realms, and forces us to consider each choice carefully. For if we do not, the cutting of our Fate may catch us by surprise, and leave too many frayed threads dangling. Too many frayed threads dangling in turn may translate into an unraveling of everything we have worked to achieve and a catastrophic backslide into the Abyss out of which we have been struggling to emerge. Entire civilizations have fallen to the hand of Atropos, and what might appear to be her arbitrary power of discretion.

But is this power entirely arbitrary? To weave Fate carefully out of the fabric of our choices, then measure it against an ingenious backdrop of exquisitely calculated planetary cycles, only to have it cut capriciously at the end of life seems meaningless, if not utterly cruel.

If we can indeed ascribe a meaning to the Fate that accrues as a consequence of our choices, then Atropos' function within the determination of Fate must also somehow be a reflection of consequences.

Religions have tried to interpret Atropos' cutting of Fate as a matter of adherence to a moral code, an absolute authority that renders the end of life a moment of judgment, rather than a random imposition of finality. This strategy is at work, for example, when fundamentalist Christians declare AIDS to be a punishment for what they judge to be deviant lifestyles; or when fundamentalist Islamic terrorists justify a jihad against Westerners with reference to our cultural decadence. As we have seen throughout this book, however, such judgment propels the soul along a path of inequality that ultimately ends in the Abyss. When religious judgment continues into the Realm of Nine, it assumes the mantle of apocalyptic vision, where the final schism between One and Two produces rapture for a chosen few and utter annihilation for the rest of us; or spiritual glory for a martyred jihadist, and the bloody aftermath of a terrorist attack upon his victims. Even if one lands on the preferred side of such a scenario, it is not likely that what remains of the Creation will be distinguishable from the Abyss out of which it previously emerged.

Given that the moral code results in no net gain, as measured on the path of becoming through the numbers, there must be some other more fundamental force at work behind Atropos' cutting of the cloth of Fate. For if the number Nine is to mean anything at all as the place where the consequences of our choices register, then the cutting of Fate – essentially rendered at the moment of death – must somehow represent a matter of ultimate consequence. If this ultimate consequence is not a moral imperative, then what is it?

The Mythological Origins of Fate

According to Hesiod and Homer (the two primary sources of our knowledge of Greek mythology), the three crones representing Fate are the daughters of Zeus and Themis. Themis is a fertility goddess, whose name means "the right" or "established custom." Zeus, of course, is the chief Olympian god, ostensibly also a primary keeper of the established order. Given the official seal of propriety that envelops this couple, at first it does appear that the Morai, in their role as dispensers of Fate, are governed by the measure to which those who receive their Fate have adhered to a culturally sanctioned moral code. From this perspective, Atropos' function can be understood as a final dispensation of reward or punishment, directing the saintly to the Elysian Fields and sinners to Hades. But for the most part, the Greeks did not envision their afterlife as a merit system (Morford & Lenardon 277).

Nor does this moral code make sense in terms of the mythology. Anyone who knows anything at all about Greek mythology will recognize the assumption that Zeus' fatherhood of the Morai lends a moral context to the notion of Fate as a supreme piece of irony. For all his official power, Zeus was one of the least moral of all the gods. Zeus was prototypically amoral, especially when it came to the expression of his prodigious sexual appetites, which he exercised freely with anyone – mortal or immortal, male or female – who struck his fancy. If the fact that Zeus is the father of the Morai suggests that Fate is a karmic allotment for moral rectitude, then either Zeus himself must be beyond the reach of Fate, or there must be some other explanation. Indeed, many who wrote about the gods in later Greek literature suggested that Zeus was above the moral code. Perhaps, however, it is the assumption that Fate – spawned by Zeus – bows to a moral code that must be re-examined.

Joseph Campbell reminds us that ". . . at no time in the history of proper Greek thought does the idea appear of a book of moral standards revealed by a personal god from a sphere of being antecedent to and beyond the laws of nature" (*Occidental Mythology* 180). Instead, he suggests that Zeus' power to flaunt the law existed only within the domain of strictly human affairs, and that beyond this domain Zeus was bound like the rest of us by something higher than a moral code, which Campbell identifies as natural law. Campbell goes so far as to consider Zeus "merely a masculine name for what is otherwise known as Moira" (*Oriental Mythology* 180). If this is so, then the Morai, daughters of Zeus representing Moira, are ultimately agents of natural law.

What becomes solidified as one's Fate in the Realm of Nine, depends upon how closely the life one has chosen in the other eight Realms conforms not to manmade laws, but to natural law. As a fertility goddess bound to the Earth out of which fertility arises, Themis – the mother of the Morai – is also a harbinger of natural law. The union of Zeus and Themis is essentially a marriage of Spirit (sky) and Matter (earth), which produces the conditions of natural law under which all embodied souls must function. The fact that each soul is bound by natural law, and that its choices and actions will be measured against natural law, is the determining factor giving rise to Fate.

Natural law implies natural limits, which is where Atropos cuts the cloth of Fate. Adherence to natural law produces health, well-being, abundance, fulfillment, and in general, longevity; life in willful ignorance or denial of natural law produces disease, malaise, poverty of spirit, disillusionment and in general, a shortening of life. In the Realm of Nine, what prompts Atropos to cut the cord is not mere whimsy, but the cumulative pressure of choices made in keeping with or in opposition to natural law. There are, of course, exceptions to these generalizations, but in the end the gift of life is conditioned by how respectful we are of Life – that is to say, of the entire web of Life that weaves through this creation as a conscious force tending toward harmony and synergistic cooperation. If we try to create a life outside of this larger web of Life, in denial of it, or as an attempt to claim special privilege within it, we push Atropos to the final cut that seals our Fate.

CHAPTER NINE

◇◇

Death is a natural phenomenon, and death is inevitable. If the cumulative weight of our choices in defiance of natural law does not precipitate an early death, our bodies will eventually return to dust as a consequence of natural law. Some will die because of their choices. Others will become the casualty of collective choices and be killed by wars, plagues, toxic environments, genocides, and other misguided adventures back into the Abyss. Regardless of our choices – individual or collective – we all will die. The hope is always that we will die having learned from our mistakes, and pass on what we have learned to future generations, but this is by no means guaranteed. Whatever we fail to learn will come back to haunt us within the Realm of Nine as the Fate of our collective future.

Fate as an Invitation to Conscious Embodiment

Whatever we do learn can potentially alter the fate of future generations, and of the embodied world in general. Taken together the three Fates represent not just a set of limitations imposed upon the human experience, but also a strategy for making the most of it. As Clotho weaves Fate from the consequences of personal choices, familial and cultural inheritance, human mortality, and the natural limits of the Earth in which our experience is embodied, we find ourselves ensconced in a unique set of circumstances that essentially defines who we are. As we enter into these circumstances, fully conscious, and with a clear intention to learn from our experiences, then the fate that is woven by Clotho becomes our classroom, and our springboard to higher, more fully integrated possibilities for human evolution.

If I accept my familial inheritance as a third generation German-American, I enter a rich and complex world, where it is possible to learn a great deal. Immersion in Teutonic mythology allows me to identify with the god Odin, explored at various places in previous chapters, and understood in the context of Fate within the Realm of Nine as a predisposition to follow the path I am following. When I contemplate all that I have gained in doing so, I must conclude that Clotho's weaving of Fate has limited me only to the extent that I have resisted it.

I had my first inkling of this in my teens when contemplating Hitler's prominent place within German history. This meditation occurred at a time in my life when I was also questioning my relationship to God and the Catholic Church, discovering my sexuality, and facing the prospect of being drafted in an unjustifiable war in Vietnam. In reading *The Rise and Fall of the Third Reich*, I had to wonder how could human beings, supposedly created in the image of a just and merciful god, sink to such levels of depravity in their treatment of each other? How could an embodied soul, capable of such intense orgasmic pleasure, also be capable of choosing to inflict such incomprehensible pain on others? How could a government, supposedly of the people, by the people, for the people, fail to acknowledge the horrendous lessons of history, and demonize other people as subhuman and expendable?

These were questions my adolescent brain could not comprehend without a sense that there must be some deeper, more inexorable force rumbling unconsciously below the level on which human beings appeared to be making choices. No one in their right mind, I reasoned, would make the kind of choices that would lead to the willful slaughter of 6 million Jews or the napalming of innocent women and children half a world away. Perhaps this force – which I now understand as a collective Fate we have somehow contracted with each other to resolve – is genetically encoded, or hard-wired into our psyches. Perhaps it is inherent in the nature of this world of duality, where One splits into Two, and then seeks to heal the split.

Whatever the mechanism behind it, it appeared that my personal choices would be forever conditioned by this collective, perhaps more cosmic rumbling of Fate. In some way, the necessity for dealing with Fate in this larger sense would somehow render my personal choices more poignant in their role as catalysts to my growth, and perhaps to the collective growth of the entire generation of souls into which I was born.

A Personal Encounter With Fate

The Vietnam War required my generation of males to define who they were in relation to the larger Fate that drove the US to genocidal madness. Some rationalized the war through a propagandized sense of its noble purpose, and signed up for an adventure in surrealism they could not have anticipated. Others – driven solely by Fate – went because they felt they had no choice. Others resisted in anger, some through a selfish desire to escape civic responsibility. A significant number made hard choices to leave the country or otherwise evade the draft on principle, refusing to participate in a collective Fate that undermined their humanity and denied the humanity of strangers. All were forced to choose, one way or another, within a larger context in which our collective Fate appeared to hold the dominant hand.

I chose to become a conscientious objector, writing out an impassioned 17-page statement of my opposition to war and violence to submit to my draft board. My argument was based partly on the injunction of my inherited Christian religious tradition, in which Yahweh had said, "Thou Shalt Not Kill," and Christ had advised us to turn the other cheek and "Love thy enemy as thyself." I also invoked the principle of *ahimsa*, or non-violence to any living thing that lay at the heart of my then current practice of yoga. To supplement my own statements, I procured references from a priest, a minister, and a rabbi, as well as my neighbor who had been a Major General in the army in World War II, all attesting to the sincerity of my beliefs. Whether a testimony to the persuasiveness of my case, or my obvious unsuitability for military service, my draft board sent my conscientious objector status by mail, without even requiring a face-to-face interview – an unprecedented easy approval in an era when all claims to conscientious objection were eyed with hard suspicion.

CHAPTER NINE

Though grateful for my reprieve, and vindicated in the knowledge that my personal sense of integrity was powerful enough to mitigate what would have otherwise been my fate, uncomfortable questions remained. To be honest with myself, I had to acknowledge that despite the sincerity of my beliefs, I did not know how I would actually respond were I to be placed in a situation where I either had to kill or be killed. Was I capable of killing? In giving this question serious consideration, I concluded that I was, but that in procuring my conscientious objector status, I was choosing not to place myself in a situation where I would be forced to make that choice. Was I being brave in resisting the pull of Fate in order to assert a deeper, more life-affirming truth? Or was my choice not to place myself in the line of fire, in order to avoid a harder testing of that truth, merely a coward's cop-out? I could not unequivocally answer this question.

A deeper stratum to my dilemma was revealed to me on a trip to Germany in 1999. Upon entering the city of Rothenburg, a walled city preserved in its medieval ambience, I learned from the hotel clerk that the entire town had once been an enclave of my ancestors. The name Landwehr literally means "keeper of the land," and was used as an epithet to describe the common foot soldiers of the army of the Austrian Empire, of which Germany was then part. During the Middle Ages, Landwehrs surrounded the town of Rothenburg to protect the wealthy Germans inside the walled perimeter from roving bands of marauders.

Given this bit of history, it would appear that a distinctly military thread of Fate was woven through my family fate. My father had fought in the Philippines in World War II; my grandfather had emigrated from Germany to this country during World War I. It could be that by refusing military service during the Vietnam War, I was breaking the chain of family fate that had previously bound the Landwehr clan to the collective military misadventures of our culture.

Yet within the Realm of Nine, where Fate continues to reverberate, perhaps I have merely taken the ancestral call to war to a less obvious level. In Chapter Four, I spoke of the constant pull to battle I have experienced in this life, standing up for principle in relation to various environmental and political issues, dealing with local politics within the land cooperative in which I live, attempting to come to terms with the anger that is evoked in me as I contemplate various modern injustices large and small that seem to relentlessly plague our world. Some battles, I have learned, I can safely sit out with little compromise to my integrity; others I cannot. Some I have no power to influence; others perhaps I do. Hopefully over time, I am becoming more skilled in discerning the difference between battles I can or must win and those I can but need not. In any case, within the Realm of Nine, I am constantly being called by Fate to ask the question, "to fight, or not to fight?" In some respects, in an ongoing echo of my decision to become a conscientious objector, I have created a life for myself that serves as buffer against the necessity for fighting unnecessarily. But to the extent that I live within the Realm of Nine – as we all do, regardless of our personal choices – the questions at the heart of Fate never really go away.

This is because Fate arises out of the cumulative weight of the choices we have made in Realms Zero through Eight. These choices – and the existential questions that give rise to the necessity for them – can be understood differently within each Realm, because like Zen *koans*, these questions must be approached from many different perspectives before they can be addressed at a deeper level of consciousness than can be accessed by the rational mind. With each of the previous Realms, I have discussed the dynamics of number in terms of the *chakra* system, because the *chakra* system provides a useful model for confronting the multi-dimensional complexity of these questions. As these questions are addressed in each of the Realms, all seven *chakras* are engaged in a way that requires us to weave together various threads of consciousness into a comprehensive response that reflects the potency of divine intelligence at work. When we are successful in doing this, individually and collectively, Fate yields. Until then, we are subject to Fate.

Fate and the Chakra System

Within each Realm, we are challenged to make a leap in consciousness that can be understood as an integration of two or more *chakras*; a shift of energy, awareness, and emphasis from one *chakra* to another; or the shift of a previous integration of two or more *chakras* into a third. The Realm of Six is unique in that it requires each of us to address our individual core issues, which can involve work in any of the *chakras* or in some combination of them. To recapitulate briefly, the following challenges are required of us in Realms One through Eight:

In the Realm of Zero: a shift from the 1st *chakra* to the 4th and 6th

In the Realm of One: a shift from the 4th and 6th *chakras* to the 5th

In the Realm of Two: a shift from the 3rd *chakra* to the 2nd and 4th

In the Realm of Three: an integration of 3rd and 6th *chakras*

In the Realm of Four: an integration of 1st and 7th *chakras*

In the Realm of Five: a shift of the Realm of Four integration (1st and 7th *chakras*) into the 4th

In the Realm of Six: the cultivation of a more conscious relationship to one's personal core issues, centered in one or more *chakras*

In the Realm of Seven: an integration of the 2nd and 5th *chakras*

In the Realm of Eight: an integration of the 5th, 6th, and 7th *chakras*

CHAPTER NINE

◇◇

In the Realm of Nine, our task is to shift the integration of the Realm of Eight back to the first *chakra*. Since we began our journey through the numbers in the first *chakra*, it is fitting that we should end there, and bring the sequence of numbers full circle. In so doing, we encompass the entire sequence of numbers in which various shifts and integrations have potentially equipped us to approach the challenges of the first *chakra* with a higher level of consciousness, a deeper creative capacity, and a more seasoned ability to respond. To the extent that we have successfully met the challenges posed to us in each of the other number Realms, we face the Abyss that first took our breath away in the Realm of Zero, now infused with the divine intelligence of a Creator god fully capable of transmuting Fate.

In the Realm of Zero, we learn to respond to the horror of the Abyss – as it registers in the first *chakra*, where our survival is at stake – from a place of compassion (in the fourth *chakra*) and nonattachment (in the sixth *chakra*). In the Realm of One, we follow a path with heart (fourth *chakra*), and discover the deeper truth of our being (sixth *chakra*) as it is aligned with a deity in whose footsteps we are following ever more consciously and intentionally (in the fifth *chakra*). In the Realm of Two, we make a shift from a Masculine orientation of domination in a field of competition to a Feminine way of being in which enjoyment (second *chakra*) and interconnectedness and cooperation (fourth *chakra*) are paramount. In the Realm of Three, we learn to exercise our personal power (third *chakra*) with increasing perspective (sixth *chakra*) about how we fit as an integral part of the greater whole. In the Realm of Four, we ascend to an aerial perspective where navigation becomes possible (seventh *chakra*), and descend into the material plane (first *chakra*) with consciousness of its elemental construction and an increased capacity to initiate intentional alchemy. In the Realm of Five, we channel these capacities for navigation and elemental alchemy through the heart (fourth *chakra*), where they become the basis for living in harmony with natural law and participating in the world as increasingly conscious agents of divine intelligence. In the Realm of Six, we turn that intelligence inward and do the work of healing whatever issues or blockages stand in the way of a full embrace of our own divinity. In the Realm of Seven, we discover a fully liberated sense of complex creativity, as we embrace (in the second *chakra*) the extremes of our nature without judgment and stand in a more fully realized truth (in the fifth *chakra*). In the Realm of Eight, we attain the wisdom of the Serpent in the embrace of that truth it all of its wild and fertile authenticity (in the fifth *chakra*); the clear sight of the prophet in a full recapitulation of all that we know in Second Attention (in the sixth *chakra*); and the ultimate freedom of the soul who has dropped the story line of her life (in the seventh *chakra*) to become the deity in whose footsteps she has been following. To the extent that we are able to do this, the Abyss reveals itself to be Ain Soph and a portal opens into a higher realm of possibility than can be glimpsed in this one, encompassed by the single-digit numbers. As enough of us move through this portal, the world that we share is gradually transformed into a more humane, more egalitarian, more habitable and hospitable place to be.

To the extent that we do not successfully meet the challenges of the number Realms – either individually or collectively, the Realm of Nine becomes the measure of our failure. Few, if any, will complete the circuit of the numbers without courting the Abyss through everything left undone. What we view historically as our evolutionary progress as a species will be the byproduct of bringing the consciousness of a Creator god into creation, successively and simultaneously by meeting the challenges of the number Realms. What we view as intractable problems become more deeply entrenched within the Realm of Nine as the cumulative Fate of future generations – the endless war in the Middle East raging in one form or another since Biblical times; the use of war itself as a solution to human problems; the glaring inequalities between rich and poor, which seemingly widen with each passing generation; and the increasingly rapid deterioration of ecological health around the planet – to name a few of the more seriously troubling and more seemingly perennial issues facing us all. Fate in this sense is the inheritance we pass on to our children and our children's children – an inheritance that includes both our successes and our failures; everything we have achieved and everything that remains unhealed, broken, and complicated by a willful refusal of the divine gift of consciousness.

Fate as a Cyclical Opportunity for Growth

The good news is that the unresolved questions at the heart of Fate tend to reassert themselves astrologically as cyclical phenomena. As the pivot point of the next cycle, the Realm of Nine offers seemingly endless opportunities to revisit those thorny conundrums that have not been adequately addressed in previous rounds, always with the hope that in this round, they will be. These opportunities are endless at least until Atropos makes the final cut. Though we can't anticipate when the final cut will come, we can rest assured that it will come unexpectedly, while most of us are in mid stride, on our way to what we hope will be a better future. Just as our mortality puts an end to the illusion of unlimited time in which to live out our individual lives, so too is it possible that the extinction of our species will one day cut short our collective striving to finally get it right. Until that day, we have the opportunity to track and more consciously mitigate our Fate through the increasingly more skillful navigation of interpenetrating astrological cycles – so that future generations need not suffer needlessly as we, and every previous generation, have.

In Part Two, I will explore in more detail some of the mechanisms by which the threads of number weave themselves astrologically into a collective Fate, as well as the birthcharts of extraordinary individuals who have succeeded in altering Fate within each of the number Realms. In concluding this chapter, I will provide the briefest glimpse into the workings of our current Fate, as it is playing itself out astrologically.

225

◇◇◇

The Threads of Fate Weaving Through the Vietnam Era

As suggested earlier, our collective Fate tends to reveal itself astrologically through the cycles of the transpersonal planets: Uranus, Neptune, and Pluto. In discussing my personal encounter with Fate earlier in this chapter, I spoke of the Vietnam War as an opportunity for my generation of males to define who they were in relation to the collective Fate. Like other moments of collective Fate, the Vietnam War can be understood with reference to these larger transpersonal cycles.

As the Vietnam War was moving to the foreground of collective consciousness, more or less[3] from August 1964 to September 1967, Uranus and Pluto were conjunct each other. Pre-eminent astrological historian Richard Tarnas associates the planet Uranus with the Greek god Prometheus, who stole fire from heaven and championed the liberation of the creative human spirit. When Uranus and Pluto work together, this Promethean agenda becomes intensified, accelerated, and often exaggerated to titanic proportions. According to Tarnas, key moments in the Uranus-Pluto cycle are typically marked by "widespread radical social and political change and often destructive upheaval, massive empowerment of revolutionary and rebellious impulses" (143-144). My response to the Vietnam War, as well as the collective response of my generation coming to age in the volatile, radical, wildly rebellious and recklessly experimental era of the Sixties (marked astrologically by a key station in the Uranus-Pluto cycle), fits the archetypal tone and the zeitgeist of the time quite well.

From March 1965 to January 1967, Saturn was also opposed (180°) to Pluto, marking a sharply contrasting archetypal keynote. According to Tarnas, Saturn-Pluto cycles are marked by "increased calls for moral vigor and social restraints, censorship and repression, puritanical standards of conduct, severe punitive judgments . . . and wars against enemies perceived and described as evil" (230). When Saturn-Pluto cycles overlap with Uranus-Pluto cycles, we have a polarized situation tending sharply toward a path of inequality in the Realm of Two. Such periods are characterized by "the exacerbation of tensions between authority and rebellion, order and freedom, structure and change" plus at its most radical and intense, "the rapid ascendancy of authoritarian and totalitarian forces" galvanized to meet the "revolutionary and rebellious" spirit of Uranus-Pluto (Tarnas 222-223).

It was during this brief window in which both Uranus-Pluto and Saturn-Pluto cycles were in critical stations, that then President Lyndon Johnson escalated and intensified the Vietnam War to its most feverish pitch. In March 1965, the US began Operation Rolling Thunder – the prolonged peak of aerial campaign that dropped nearly 8 million tons of bombs on Vietnam, "four times the tonnage dropped during all of World War II, in the largest display of firepower in the history of warfare" (The History Place). Also in March, President Johnson authorized the use of napalm, a highly toxic petroleum product causing severe burns, asphyxiation, unconsciousness and violent death in firestorms with self-perpetuating winds up to 70 miles an hour. In May 1965, the first 3,500 combat troops arrived in Vietnam; by the end of this Saturn-Pluto period in January 1967, troop levels had increased over a hundred-fold to 389,000. The number of Americans killed in action similarly rose from a monthly average of 172 during 1965 to an average of 770 in 1967 (Simon).

While the war effort (Saturn-Pluto) intensified during this period, so did the protests (Uranus-Pluto). In March 1965, the SDS (Students for a Democratic Society) organized the first teach-in against the war at the University of Michigan, sparking a strategy that would be repeated at 35 campuses across the country. In April 1965, 15,000 students gathered in Washington to protest Operation Rolling Thunder. In May 1965, the first draft card burnings took place at a student protest rally at UC – Berkeley, where Lyndon Johnson was burned in effigy. In August, the destruction of civilian villages in Da Nang, suspected of harboring Viet Cong, was shown on television, further infuriating the antiwar movement. By October, anti-war rallies in 80 cities around the US, London, Paris and Rome attracted 100,000 protestors. In November, 31-year old pacifist Norman Morrison lit himself on fire below the third-story window of Defense Secretary Robert NcNamara at the Pentagon, while later that month, 40,000 protestors surrounded the White House, calling for an end to the war. In January 1967, Martin Luther King's support showed that "the antiwar movement had reached maturity as the entire nation was now aware that the foundations of administration foreign policy were being widely questioned" (Barringer).

Mainstream public support for the war would begin to deteriorate after that, but from this point forward, so would the violence and the escalation of police brutality against war protesters. By this confluence of Saturn-Pluto and Uranus-Pluto cycles, the battle lines were sharply drawn – between the US military (Saturn) and the Viet Cong (Uranus); and between the US government (Saturn) and the antiwar protesters (Uranus) at home. Coincident with the war in Vietnam and its protests were widespread urban racial riots in over 120 cities across America, in which the Civil Rights movement pitting black against white reached its bloodiest pitch. Looking back, this would appear to be a period of history in which the dark consequences of our cumulative choices came home to roost in the Realm of Nine.

The Historical Consequences of Our Collective Insistence on Inequality

Across these battle lines, we can see the path of inequality drawn in rather dramatic fashion in the book of collective Fate. The question is, what have we learned from our collective experiences during this critical moment of history – and from many others like it? Are we learning anything at all?

CHAPTER NINE

◇◇

As Tarnas points out, Saturn-Pluto and Uranus-Pluto cycles also converged during the years 1929-1933 (using his broader orbs) when Uranus was square to Pluto and Saturn was opposed. These were the years of the Great Depression, the rise of fascism in Germany, Italy and Japan, and the rise of the equally brutal Stalinist regime in Communist Russia – all rather dramatic and violent outworkings of our collective Fate around an entrenched and institutionalized insistence on inequality.

Prior to that, there was a similar alignments during the years 1845-1856, coincident with a wave of revolutionary spasms, creating widespread upheaval throughout Europe – in Paris, Berlin, Vienna, Budapest, Dresden, Baden, Prague, Rome, and Milan – including the overthrow of governments, and massive revolts against the forces of repression. In the US, this was the period of tension preceding the Civil War, which erupted in 1861, when Saturn formed an exact square with Uranus.

Prior to that, the only other confluence of Saturn-Pluto and Uranus-Pluto in relatively recent history was the Reign of Terror in the midst of the French Revolution, from 1793-1796, during which 25,000 suspected enemies of the revolution were beheaded by guillotine, including those who had started the revolution a few short years before. The guillotine itself is a symbol for the sharp divide that separated the two rival factions – the Girondins and the Jacobins – across a devastating divide of inequality.

Looking back at this history, a reasonable person might ask what we are doing to ourselves in our collective insistence upon a path of inequality. What are we learning from the horrendous mistakes we make pursuing this path? Is it possible, given our collective refusal to learn the lessons of history, that we can avoid repeating these mistakes again, in increasingly costly fashion? In an age when our blind use of fatal technologies makes it possible to annihilate ourselves, and seriously compromise the life-sustaining capacities of the planet on which we live, can we continue along this path without tempting Atropos to make the final cut of Fate that will end the human experiment?

These might all seem to be rhetorical questions, except for the fact that we are headed once again for another confluence of Saturn-Pluto and Uranus-Pluto cycles. Using tight orbs, Saturn will square Pluto from October 2009 to June 2011, while Uranus will square Pluto from June 2010 to March 2017. This Cardinal Grand Cross – as the alignment is called – is an event long anticipated by astrologers, who are predicting everything from catastrophe to catharsis. If history is any indication, we are at the very least beginning another period of intensified polarity between the Promethean forces of liberation and the Saturnian forces of repression that will last approximately seven years, with the battle lines being drawn right now – from June 2010 to June 2011 – just as this book comes to press. If we allow a more liberal orb, as Tarnas does, this Saturn-Pluto period extends from November 2008 to August 2011.

◇◇◇

The Hand of Fate in the Early Obama Era

The beginning of this extended period saw the historic election of Barack Obama, which was hailed nearly universally as a development of immense Promethean promise. Obama was welcomed with great relief as a champion of progressive reform by cultural creatives around the world, who in general felt battered and bewildered after eight years of despair during the arch conservative, secretive, destructive Bush administration that undid much of what they had worked for since the Sixties. As feminist writer Judith Butler put it, we took to Obama with "uncritical exuberance." Some of this may have been warranted; some of it may not have been. As of this moment in the fall of 2010, it may be too early to tell.

Yet as we enter the Saturnian phase of this Uranus-Saturn-Pluto triumvirate, recent events begin to give even the most hopeful among us, pause. While large corporations and banks received unprecedented government bailouts and handed themselves hefty bonuses at the beginning of this period; unemployment continued to rise to 25-year highs. Nearly 3,000,000 people were handed a mortgage foreclosure notice in 2010 alone (as of May 2010), with "a million properties . . . somewhere in the pipeline still, [and] a further 5 million serious delinquencies stacking up" (Smith). Economists remain cautiously optimistic about the prospects of an economy that has received its worst jolt since the Great Depression (during the last confluence of Saturn-Pluto and Uranus-Pluto cycles). The inequalities that fuel this current economic crisis are a profound and deeply entrenched system of inequality that does not promise to resolve any time soon. If anything, current trends seem to indicate that we are moving more deeply into these inequalities.

In December 2009, President Barack Obama decided to increase the US military presence in Afghanistan to 100,000 troops – a decision that parallels that of Lyndon Johnson to raise troop levels in Vietnam during the Saturn-Pluto opposition of 1965-1967. In February 2010, NATO-led forces launched a major offensive, Operation Moshtarak (perhaps in an echo of the Vietnam Era Operation Rolling Thunder), to secure government control of Taliban-controlled Helmand province, thus escalating the war to a troubling pitch. Obama has promised a withdrawal from Afghanistan beginning in July 2011 (the end of the Saturn-Pluto period), while Defense Secretary Robert Gates warns us that ". . . it is very early yet and people still need to understand there is some very hard fighting and very hard days ahead" (qtd in Jaffe). His pronouncement is eerily reminiscent of those made in Vietnam and Iraq and in every other war fought by the US since its inception.

War is not an unusual event in US or world history, but given the confluence of astrological cycles related to an especially intense outworking of Fate, our collective refusal to learn the lessons of previous wars may come back to haunt us in ways we might prefer not to anticipate now. By any measure, the complex, perennial, hydra-headed conflict in the Middle

East is one of the most enduring realities in the Realm of Nine. During this convergence of Saturn-Pluto and Uranus-Pluto cycles, we face the real danger of dropping more deeply into this particular Abyss. In the Realm of Nine, the more deeply into the Abyss we go, the more difficult it becomes to get back out.

On April 20 2010, there was an explosion and fire on the British Petroleum Transocean drilling rig Deepwater Horizon in the Gulf of Mexico. Two days later, the rig sank, and a 5-mile oil slick spread across the Gulf. On April 25, the US Coast Guard reported that 1,000 barrels of crude oil per day were spilling into the Gulf, a figure that was raised to 5,000 barrels a day 3 days later. Within a month, estimates were ranging from 20,000 to 100,000 barrels a day. In the ensuing clean-up effort, over 2,000,000 gallons of toxic dispersant Corexit were sprayed onto the oil slick and into the leak site 5,000 feet below the surface of the ocean, creating what Texas Tech University toxicologist Ron Kendall called "an eco-toxicological experiment" (qtd in Goldenberg).

At this juncture, the EPA remains concerned about the long-term effects of the spill and the "clean-up" on the ecological health and viability of the Gulf. According to Assistant Professor Michael Blum of Tulane University's Department of Ecology and Evolutionary Biology, "There's immediate shock to system, immediate toxicity and immediate mortality - birds, dolphins, marine mammals oiled. The mortality is relatively small in comparison to the potential effect that may accumulate over time. Things are not as bad now as they likely will become" (qtd in Cardinale).

Meanwhile, according to an investigation by Associated Press, there are over 27,000 abandoned oil wells in the Gulf of Mexico from a host of companies including BP, dating back to the 1940s, many of them badly sealed. The report describes the Gulf as "an environmental minefield that has been ignored for decades" (Wray). While BP and other corporate executives fall all over themselves, pointing the finger of blame, it is clear, a long range deeply entrenched attitude of casual denial, irresponsibility, and willful ignorance of natural law has plummeted the Gulf into an abysmal state that may become semi-permanent in the Realm of Nine.

Our Difficult Passage Through the Realm of Nine in the Years Immediately Ahead

Assuming we continue along our present path of inequality, this period of Uranus' current square to Pluto will be a difficult period for humankind and for the planet, perhaps during which several long-term and increasingly entrenched patterns reach a critical point of very unlikely return. Unless the Promethean spirit of Uranus-Pluto can galvanize a collective awakening to match the intensity and ferocity of the Saturn-Pluto forces of Orwellian repression and entrenched inequality, the Abyss looms more closely and more menacingly than anyone of us – on either side of this divide – would wish upon each other or the world that we share.

This is not to say that Saturn is bad and Uranus is good. Saturn at its best can help to guide and shape the wild, capricious, reckless energy of Uranus in responsible, productive, constructive ways. Together, Saturn and Uranus can create a world in which equality, respect for natural law, and divine intelligence work hand in hand for the greater good of our collective evolution toward a healthy embrace of our compassion, wisdom, and fully functional complex creativity. To the extent, however, that we have become polarized in our relationship to these two very different archetypal forces, then Pluto's entry into the mix will intensify and exacerbate that schism in the collective psyche. When this happens, whatever lessons we have failed to learn in the preceding Realms will become a more deeply entrenched sense of Fate in the Realm of Nine.

During this critical period, divisive economic disparities between rich and poor, semi-permanent states of war, environmental catastrophes, racial divides, ethnic hatreds, and every other manner of atrocity at the bottom of the Abyss can potentially solidify as a set of repercussions we no longer have the power to mitigate through any act of choice. Within the Realm of Nine, this would be the intrusion of Fate upon a long-standing experiment in choice gone awry – the sad culmination of a history of misuse of power and the divine gift of consciousness, pursued in willful ignorance of natural law along a path of inequality.

The Cutting of the Thread of Fate and the End of the Mayan Calendar

This same period of crisis in relation to the current Uranus-Pluto cycle coincides with the end of the Mayan calendar in 2012. Much has been written about this cosmological event, associated by the Mayans themselves with a supreme moment of Fate – the ending of the manifest creation, as we currently know it, and the beginning of another, entirely new reality – an event that can be understood as a final cutting of the thread of our collective fate by Atropos.

Astronomically, this date – December 21, 2012 – represents a close conjunction of the winter solstice Sun with the crossing point of the Galactic Equator and the ecliptic. Mayan calendar enthusiasts have long puzzled over the meaning of this crossing, and inconclusive speculations abound. As reported by Mayan scholar John Major Jenkins, a breakthrough in understanding came in the 1994 book, *Mayan Cosmos* by epigrapher Linda Schele. Schele

◇◇◇

notes that the Galactic Equator crosses the ecliptic in late Sagittarius, a region of space that corresponds to a dark bifurcation of the Milky Way by interstellar dust clouds. The Mayans called this rift *xibalba*, translated by Dennis Tedlock as the "Black Road," where the Hero Twins Hunahpu and Xbalanque must journey to battle the Lords of Xibalba. (334, 358).

According to Jenkins, the Black Road symbolized the Underworld – a place of death, around which life was believed to revolve. When a planet, the Sun, or the Moon, entered the Black Road, Mayan shamans gained entrance to the Underworld, where it was possible to perform their rituals and ceremonies, designed primarily to mitigate the forces of death, and thus ensure the continuation of life. When the Sun enters the Black Road at the winter solstice – the darkest moment of the Earth's annual revolution around the Sun – the forces of death will be, theoretically at least, at peak capacity. In terms of the arithmological scheme outlined here, it is at this moment that humanity as a whole will enter the Realm of Nine, where our capacity to mitigate our collective fate will conversely be at its lowest ebb.

This exceedingly rare astronomical event ends a 5,120-year cycle, called a *baktun*, and also implies the beginning of a new cycle. This current *baktun* began in 3,114 BCE, a date to which the Mayans ascribe the birth of their deities (Jenkins). Presumably then, the end date of this cycle would correspond to the death of these same deities, and perhaps to the birth of an entirely new pantheon. In 3,114 BCE, Sumerian culture, centered in Babylon, was reaching its peak. Mayan culture had not yet emerged as a civilization to be reckoned with, and a different set of gods predominated. One can look back at this previous date in history and observe with some relief that the world did not end on that day, despite a previous changing of the divine guard. Many *baktuns* have come and gone, since hominids, the precursors to human beings, first appeared on the planet 1,750,000 years ago (Campbell, *Primitive Mythology* vi).

The ending of this current *baktun*, however, marks the first time in 26,000 years that the ending of the Long Count coincides with the Sun's passage through the Black Road at winter solstice. Previously, this event would have occurred around 24,000 BCE, the era of the Cro-Magnon man, an early human hunter-gatherer society with primitive flint tools and weapons, and the rudimentary beginnings of a shamanic religious practice evidenced in cave paintings. Anthropological evidence suggests that Cro-Magnons banded together in small tribes in order to more effectively cope with an extremely difficult life (Wikipedia, Cro-Magnon), although it would be a stretch to call Cro-Magnon culture "civilization" in the same sense in which we would use that word today. This upcoming ending of the Mayan Long Count, in other words, marks the first time that the Sun has entered the Black Road at winter solstice since human civilization existed.

What will happen at this time is anyone's guess. Among predictions circulating, as reported by a skeptical *USA Today* article (McDonald):

◇◇

Journalist Lawrence Joseph forecasts widespread catastrophe in Apocalypse 2012: A Scientific Investigation Into Civilization's End. *Spiritual healer Andrew Smith predicts a restoration of a "true balance between Divine Feminine and Masculine" in* The Revolution of 2012: Vol. 1, The Preparation. *In* 2012, *Daniel Pinchbeck anticipates a "change in the nature of consciousness," assisted by indigenous insights and psychedelic drug use.*

The article assumes that such predictions are "a complete fabrication and a chance for a lot of people to cash in." Yet as one contemplates the accumulating evidence that human beings have collectively abused our privileged position within the web of life, misusing the divine gift of consciousness in willful defiance of natural law, since the age of Cro-Magnon man, it is not unreasonable to assume that there will eventually come a point of reckoning. The movement of the Sun (our source of Light and Life on this planet) through the darkest part of its cycle into the dark region of the Black Road ought to at least give us pause to consider the possibility that this moment of reckoning may be at hand.

If there is such a thing as Fate woven from the fabric of our collective choices, measured by various cosmological cycles, and eventually terminated by Atropos – working within the bounds of natural law, then the ending of the Mayan calendar is a choice candidate for timing the cutting of the thread. The lights will probably not go out precisely at midnight on December 21, 2012, but unless something critical in our collective consciousness has shifted by then, the ensuing *baktun* may well witness the end of the human experiment – an end that we have little power to forestall in our waning millennia.

Nine as a Portal to Other Dimensions

Whatever else this "final" cutting of the thread of Fate might mean, it is useful to speculate from the perspective of the arithmological tradition. The number Nine is not merely the end of the sequence of single-digit numbers, but also the pivot point at which the cycle repeats itself. Beyond the number Nine is the Decad, which ideally represents the full embodiment of consciousness, the One made manifest through the creation in such a way that there can no longer be any distinction between Spirit and Matter. At this level, the final polarity collapses, and that which we identify as an earthbound reality conditioned by polarity ceases to exist. On this side of that final collapse, it is hard to imagine what would replace the familiar experience we call reality, since the very language we use to describe this reality is conditioned upon a separation between subject and object that would no longer exist.

This is not to say that a small cadre of intrepid scouts have not attempted to imagine this outcome, for there is a vast bounty of mystical literature, poetry, and folklore that hints

◇◇◇

at what such an experience of unity with Spirit might be like. The mystical 13th century Sufi
poet Rumi, for example, suggests (Barks 136) that:

> If anyone wants to know what "spirit" is,
> or what "God's fragrance" means,
> lean your head toward him or her.
> Keep your face there close.
> > Like this.
>
> When someone quotes the old poetic image
> about clouds gracefully uncovering the moon,
> slowly loosen knot by knot the strings
> of your robe,
> > Like this.
>
> If anyone wonders how Jesus raised the dead,
> don't try to explain the miracle.
> Kiss me on the lips.
> > Like this. Like this.
>
> When someone asks what it means
> to "die for love," point
> > here.
>
> If someone asks how tall I am, frown
> and measure with your fingers the space
> between the creases on your forehead.
> > This tall.
>
> The soul sometimes leaves the body, then returns.
> When someone doesn't believe that,
> walk back into my house.
> > Like this.
>
> When lovers moan,
> they're telling our story.
> > Like this.
>
> I am a sky where spirits live.
> Stare into this deepening blue,

while the breeze says a secret.
  Like this.

When someone asks what there is to do,
light the candle in his hand.
  Like this.

As Rumi reminds us, in this and countless other poems, the gateway to these other dimensions of reality – to the omnipresence of Spirit throughout the manifest world – is not elsewhere. It is here. Now. In these words. In your heart. In this moment of choice. In this gesture. In that sigh. In those clouds. Every moment on this planet is, in some sense, fated by conditions discussed at length in this chapter. Every moment on this planet is also a gateway to other dimensions in which the numinous potential of Ain Soph, the unconditioned state of all things, the Cosmic Zero of unlimited possibility yet untapped, reverberates. This is the great paradox that infuses our experience on this Earth with mystery and magic. If we fail to tap this magic with the gift of consciousness in the Realm of Nine, our Fate as we enter the Realm of Ten is the collapse of everything we understand to be real. If we enter the Realm of Nine with eyes wide open, then everything within this manifest creation becomes a portal through which it is possible to glimpse the omnipresence of Spirit, evolving and perpetually revolving through a kaleidoscope of forms.

Nine as the Ever-present Opening to the Magic Body

From this perspective, the Realm of Nine is like the bardo state between lives discussed in the *Tibetan Book of the Dead*, considered not just as a rite of passage at death, but as an ever-present opening to another way of seeing that changes everything. It is significant in this regard that Odin hung upon the world tree Yggdrasil for 9 days, and that Christ spent 40 days (1/9 of the year) in the desert undergoing his initiation, since both these experiences ushered these luminous beings into another dimension, in which everything was seen differently.

The ultimate goal of the spiritual quest is a return to the Light, alternately understood as the infusion of the Creation with Light. Still, at an intermediate stage – within the Realm of Nine – we have the opportunity to pierce the veils between the apparently solid parameters of our fate, to glimpse a freedom beyond ordinary comprehension. When the soul moves into position to be able to take advantage of this opportunity, this is referred to by Tibetan Buddhists as the cultivation of the magic body (Thurman 130, parenthetical phrases are mine):

This intense acceleration of evolution toward Buddhahood is possible because of
the great malleability and transformability of the soul and its subtle embodiment

◇◇◇

in the in between state (the Realm of Nine). The "mother reality" is the objective clear light, the actual transparency of ultimate reality directly experienced beyond the subject-object dichotomy, called "mother" because of her being the matrix of all possibility (Ain Sof, the Realm of Zero). The "child reality" is the semblant clear light, transparency still filtered through conceptuality, targeted by an accurate conceptual understanding, retaining still an instinctive sense of subjectivity and objectivity (Realms Two through Eight). It is the level of clear light experience that fits with the impure magic body presence. When it develops into the mother clear light, the magic body becomes purified, the supreme union between father magic body and mother clear light becoming the unexcelled Integration, the Great Seal, the Great Perfection (the Realm of Ten) – the evolutionary consummation that is Buddhahood.

A Personal Celebration of the Magic Body

While writing this chapter, a good friend of mine, Mau Blossom died, or as is the hope of every mortal, passed into another world free from pain and suffering. Mau was instrumental in helping me heal my lungs, and thus an important catalyst to my journey on the Odinic path. I remember Mau as one who took great delight in an ongoing process of self-discovery and the simple magic of the moment. In keeping with the spirit of Rumi's poem, she was not one to pontificate about the spiritual nature of the reality she lived. She just lived it, with as much consciousness as she could muster – whether making a drum or kayaking, playing her saxophone or sitting with a friend or client. *Like this. Like this.* As a practicing Buddhist, Mau approached her death bravely and with as much presence of mind as her pain medication would allow. Although not privileged to be at her bedside during her final days, I could easily imagine Mau moving in and out of the bardo state through the magic body into whatever dimensions of Light may exist beyond the portal of the Realm of Nine. I wrote about this in a poem I sent her before her death, which I am gratified to learn she read with relish:

> *You were there for me*
> *when breath came hard,*
> *you laughed with me*
> *when my drum took its sorry shape,*
> *you have been a true friend,*
> *more than that –*
> *you have been a light on my path*
> *and still are,*
> > *into the unknown and beyond*
> *the brave twinkle of your endless delight in life*

◇◇

and the joy of your perpetual self-discovery
everpresent in my heart,
a steady star of guidance and good news.

I wish I could hug you now,
tell you I love you and say goodbye,
wish you well upon your journey
    into the arms of angels.

In my dreams
    I shall watch you take flight
        free of pain,
      and join the flock of shining ones
        winging south to the place of unruffled peace.

I will miss you more than I can say,
    but I will see your face in the Full Moon,
      feel your touch in the first snow of winter,
and hear your song in the first cardinal of spring.
You are here with me,
    with us all
in this timeless place of love-sealed memory,
of treasured legacy,

      and always will be.

In this place of magic, beyond Atropos' final cutting of the thread, lies the promise of something no mind could possibly comprehend – the continuation of each within the all, the presence of Mau in the first snow of winter, the glimmer of recognition in you as you read these words, in the emergence of light from behind the clouds outside my window, the healing of ancient wounds in the awakening to the moment before they existed. We can only hope, in this place of transition between rounds of numbers, that when the collective dream becomes a nightmare, when death finally overtakes us, when Atropos cuts the final thread of Fate, beyond that is a Light into which everything dark reawakens to the innocence of untarnished rebirth.

Endnotes

1. Rulership is based on the affinity of a particular planet for a sign. Sun rules Leo; Moon rules Cancer; Mercury rules Gemini and Virgo; Venus rules Taurus and Libra; Mars

◇◇

rules Aries; Jupiter rules Sagittarius and is sometimes said to co-rule Pisces; Saturn rules Capricorn and co-rules Aquarius; Uranus rules Aquarius; Neptune rules Pisces; Pluto rules Scorpio. To the extent that any sign is implicated in a pattern of Fate, its ruler and the aspects formed to its ruler, is as well.

2. The personal planets are Sun, Moon, Mercury, Venus and Mars. Jupiter and Saturn are considered collective planets, meaning that they can reflect patterns of social adaptation and conditioning, as well as inherent personality dynamics. Transpersonal planets – Uranus, Neptune, and Pluto – are archetypal forces that move individuals and societies in complex ways.

3. The exact range of any astrological period is somewhat arbitrary, and depends upon the orb (or range of exactitude) allowed. For matters of personal growth, I tend to use an orb of 1–2°, while allowing a wider orb of 4-6° for historical trends. Richard Tarnas, whom I quote with reference to these larger collective cycles, uses a very wide orb of 15° for conjunctions and oppositions, and 10° for squares. For purposes of this example, I have chosen a much tighter orb of 4°.

The onset of this period coincides with the Gulf of Tonkin crisis, which led to then-president Lyndon Johnson's decision to bomb North Vietnam. Prior to this time, the US had been drifting ever more rapidly toward the waterfall of its collective fate in this war through a gradually accelerating military build-up. The decision to bomb was the point of no return, after which the gravity of the fall into the Abyss became impossible to reverse.

In the wake of this decision, public opinion polls showed 85% of Americans in support of Johnson's actions, and in early August, 1964, Congress passed the Gulf of Tonkin resolution, which gave Johnson a free hand in pursuing the escalating conflict for which he is known by history. By October 1967, one month after the 4° orb allowance had run its course, public support for the war had fallen to 46%. Beyond this point, presidents Johnson and Nixon pursued the war against mounting protest at home (The History Place).

Chapter Ten: The Principles of Arithmology

I n the preceding chapters, I have discussed each of the numbers from Zero to Nine, taking some liberty in using arithmological principles as a springboard for free association to ideas relevant to each number. This exercise was not intended to be a definitive or thorough investigation of these principles, but rather to be suggestive of the ways in which modern, more poetically-minded arithmologists might work. My goal has been to outline an astropoetic approach to number, which might serve as a basis for astrological inquiry. When combined with knowledge of the *chakras*, as discussed in *Tracking the Soul*, analysis by number will show the ways in which collective forces impact the evolution of the individual soul, and conversely how individual souls may potentially impact our collective evolution.

Before moving on to an application of astropoetic knowledge of number to the birthchart in Part Two, I want to present a comprehensive discussion of basic arithmology in this chapter, so that the reader can thoroughly appreciate the internal logic of the system that informs our astropoetic associations. Arithmologists themselves approached number astropoetically through metaphorical associations that at times waxed beyond the range of reason. There is also much overlap in application of principles, so that many of the same attributes are associated with more than one number. Underneath these free-form meanderings, however, is a set of principles upon which cogent arithmological associations can be made, and it is to these principles that I devote this chapter.

Types of Number

Arithmologists distinguished between different categories of number. Of greatest importance in the arithmological scheme were the primes, which are numbers divisible only by themselves and 1. Within the range of single-digit numbers from 1 to 9, these include 1, 2, 3, 5, and 7^1. Primes also include double-digit numbers such as 11, 13, 17, 19, and 23. Primes were generally considered to bring some new quality or force to the evolution of manifest Creation that could not be accounted for through a combination of more fundamental factors. Thus 1 introduces divinity, 2 polarity, 3 perspective, and 5 consciousness, etc.

Prime numbers exist in contrast to what the arithmologists called composite numbers, such as 4, 6, 8, and 9, which were factorable in terms of other numbers: $4 = 2 \times 2$; $6 = 2 \times 3$; $8 = 2 \times 2 \times 2$; $9 = 3 \times 3$, etc.

All numbers were considered to be either even or odd. Even numbers were those that could be divided into both equal and unequal parts, while odd numbers could be divided

◇◇

only into unequal parts. Thus the numbers 4 (2 + 2 or 1 + 3), 6 (3 + 3 or 2 + 4), and 8 (4 + 4 or 3 + 5) were considered even, while 3 (1 + 2), 5 (2 + 3 or 1 + 4), 7 (3 + 4, 2 + 5, or 1 + 6), and 9 (4 + 5, 3 + 6, 2 + 7, or 1 + 8) were odd. Odd numbers were generally considered to express themselves through Masculine (or *yang*) mechanisms, assuming a more initiatory, active, and extroverted role, while even numbers were generally considered Feminine (or *yin*), expressing themselves in a more responsive, passive, and introverted fashion. In this way, the function of the number 3 was a matter of actively mediating opposites, while the function of the number 4 was a matter of more passively embodying polarities in manifest form.

Portrayal of Numbers as Arrays of Dots

In the arithmological scheme, numbers were further classified according to how they arranged themselves as a series of dots. The *Tetraktys*, discussed in Chapter Four, is perhaps the most famous example of this logical strategy, representing four rows of dots, forming a triangular shape to depict the number Ten.

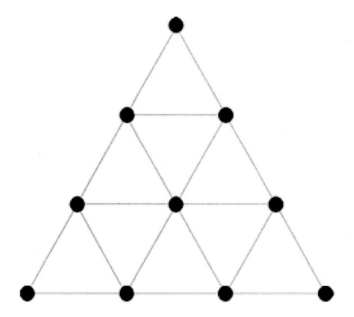

Figure 10: The Tetraktys (A Triangular Number)

The *Tetraktys* is the pre-eminent example of a general category of triangular numbers, which includes the series 3, 6, 10, 15, 21, etc. Each of these numbers lends itself to a triangular array when arranged as rows of dots. Triangular numbers generally seem to initiate an experience of resolution. The number 3, for example, mediates between the polar opposites, while the number 6 facilitates healing of core issues, and the number 10 represents completion, the full embodiment of Spirit in manifest Creation.

Square numbers are those that can be portrayed as a square of evenly spaced dots. Included in this series are the numbers 1, 4, 9, 16, 25, etc.

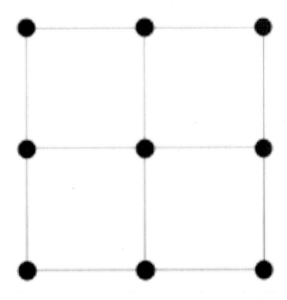

Figure 11: Nine-Dot Array (A Square)

Squares in general represent a solidification and further entrenchment of the principle factor being squared. This principle was discussed in Chapter Nine in relation to Fate, with the observation that the number 4 (2 squared) made more tangible the polarities arising in the Realm of Two, while the number 9 (3 squared) sealed the choices made in the Realm of Three as a set of cumulative consequences.

Cubes are those numbers that can be portrayed as a three-dimensional array of evenly spaced dots. These would include 1, 8, 27, 64, 125, etc.

Like triangular numbers, cubes bring resolution to the principles in the number being cubed, but there is also a further sense of redemption or even a reversal of original principles. We saw this process at work, for example, with the number 8, representing the embodiment of the wisdom contained with the number 2, which itself was fraught with ambivalence.

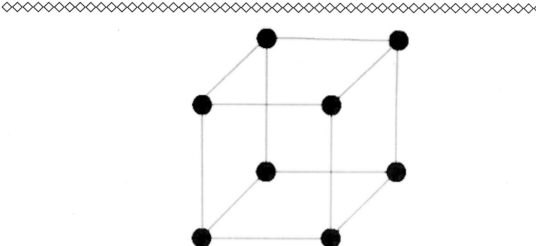

Figure 12: Eight-Dot Array (A Cube)

According to this strategy of arranging dots to determine the essence of a number, prime numbers were also called linear, given that they could only be arranged as a straight line of dots.

Figure 13: Seven-Dot Array (A Linear Number)

Oblong numbers (also called *gnomons*) were those that could be portrayed as a rectangular array, in which one side of the rectangle is one dot longer than the other. This series includes 6 (2 x 3), 12 (3 x 4), 20 (4 x 5), 30 (5 x 6), 42 (6 x 7), etc.

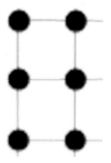

Figure 14: Six-Dot Array (A Gnomon)

Lastly, polygonal numbers are those that can be arranged in the shape of a polygon. For example, a series of polygonal numbers based on the number 5 would include 5, 12, 21, etc.

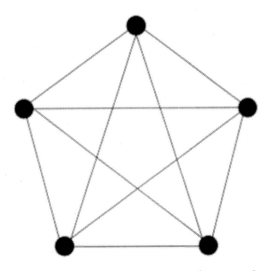

Figure 15: Five-Dot Array (A Polygonal Number)

Engenderment

All composite numbers were analyzed by arithmologists according to their factors – i.e. the prime numbers, which when multiplied, produced them: 2 x 3 = 6, 2 x 5 = 10, 3 x 5 = 15, etc. Primes were said to engender composite numbers; composite numbers were said to be engendered by primes. The numbers 1 and 2 were sometimes considered sources of numbers, rather than actual numbers.

Given that primes were associated with some unique fundamental force contributing to the process by which Spirit manifested in form, composite numbers were understood as derivative or secondary processes, in which primary forces combined in some synergistic way. The number 6, for example, could be understood as the mediation (3) of some polarity (2) capable of healing some core issue (6). The number 10 could be understood as the application of an awareness of natural law (5) to the realm of duality (2) that resulted in an integrated state of balance (10).

The number 1 was considered to have special properties in this regard, since 1 multiplied with any other number did not change that number, a fact that arithmologists took to be evidence of the omnipresence of the Creator throughout Creation.

CHAPTER TEN

◇◇◇

Generation By Combination

In addition to the process of engenderment by multiplication of factors, arithmologists also considered the possibilities inherent in adding numbers together: e.g., 1 + 2 = 3; 1 + 3 or 2 + 2 = 4; 1 + 4 or 2 + 3 = 5, etc. – a process generally referred as generation by combination. Generation by combination lent a certain symbolic repertoire to a number's range, since each distinct combination told the arithmologist something new. In considering the number 5, for example, arithmologists understood it to be a product of the Creator (1) entering creation (4), as well as a natural force mediating (3) the ambivalence of the Dyad (2).

Just as the number 1 was considered to have special properties because it did not engender other numbers (i.e. did not change them by multiplication), so too did the number 9 acquire special properties by virtue of the fact that the number 9 added to any other number resulted in that same number, once the digits composing that number were added again to produce a single-digit number: e.g. 9 + 6 = 15; 1 + 5 = 6. Although the arithmologists did not discuss the number 0, it should be noted that 0 also has this property. As noted in Chapter Nine, this implies that both the Abyss and the concept of Fate, which is the measure of a soul's relationship to the Abyss are ever-present throughout all the Realms in which number has significance.

Means

A third way the arithmologists understood number was through the concept of means. Arithmetical means were the simplest of three primary types of means. They were calculated according to the formula: $(a + b)/2$. The number 4, for example, was understood to be the arithmetical mean of 1 and 7, since $(1 + 7)/2 = 4$. This meant that the number 4 was the mediating point of balance between 1 and 7, or more specifically in symbolic terms, that the manifest creation (4) was an embodiment of the process of complex creativity (7) that the Creator (1) employed to produce it.

A second type of mean, called harmonic, was produced through the formula $2ab/(a + b)$. In this way 8 is the harmonic mean of 6 and 12 since $2 \times 6 \times 12/(6 + 12) = 144/18 = 8$.

A third type of mean, called geometric, was formed from the square root of ab. The geometric mean of 2 and 8, for example, would be 4, since the square root of $2 \times 8 = 4$.

Means in general were considered to be the ways in which the Creator mediated various forces set in motion (represented by the numbers) throughout Creation, in the same way that the soul itself mediated between the realms of Spirit and matter (Waterfield 17). Plato considered the geometrical means the most spiritual, meaning esoteric, relatively inscrutable by ordinary mortals and a matter of divine intervention outside the reach of human intent[2]. Harmonic means – considered by Plato to be psychic and anthropological – were those employed more consciously and deliberately by humans. Arithmetical means were those that were inherent within the physical nature of the manifest Creation itself, i.e. those that were most amenable to observation as an expression of natural law.

Arithmologists identified seven additional means, three subcontrary to the harmonic, two subcontrary to the geometric, and three others left unspecified (Waterfield 50, footnote 3), but these are all beyond the scope of this simple outline of basic principles. Of these additional means, harmonic means were instrumental in studies of music by which arithmological principles were embodied in sound. I will speak more about this later in this chapter.

Sequences of Numbers

Arithmologists were often enamored of the way in which sequences of numbers contributed to the generation of other numbers by combination. It was noted, for example, that squares were formed by the addition of successive odd numbers in sequence: e.g., $1 + 3 = 4$; $1 + 3 + 5 = 9$; $1 + 3 + 5 + 7 = 16$ – a generation by combination giving squares a Masculine or *yang* overtone, regardless of whether the number itself was even or odd. The arithmologists likewise noted that cubes were generated by a slightly different sequence of odd numbers: $3 + 5 = 8$; $7 + 9 + 11 = 27$; $13 + 15 + 17 + 19 = 64$; etc. The fact that both squares and cubes were generated through these sequences of Masculine numbers endeared them to the arithmologists, since Masculine numbers were considered to be of the nature of the Monad, which was the object of their reverence.

Other sequences were also noted to be of significance. Taking the sequence of doubles: 1, 2, 4, 8, 16, 32, 64, 128, 256, etc., arithmologists were impressed by the fact that every other double was a square: 1, 4, 16, 64, 256. In a similar way, they noted that a series of triples – 1, 3, 9, 27, 81, 243, 729, 2187, 6551, etc. – produced a cube every third term: 1, 27, 729 (9 cubed). These regularities were reassuring to the arithmologists, as they demonstrated that the Realms in which number operated were governed by demonstrable order.

The Lambda and the Tetraktys

As mentioned earlier, the arithmologists were fond of displaying numbers as a series of dots arranged in geometric patterns. They gained even more insight by assigning various numbers to each dot. One of the earliest patterns the arithmologists pondered was a triangular arrangement of seven dots, called the *Lambda*, a simplified version of the *Tetraktys* first introduced by Plato in his *Timaeus*.

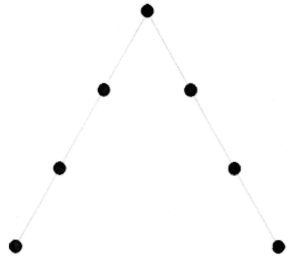

Figure 16: The Lambda

In order to demonstrate the way numbers were generated from the Monad, Plato assigned a series of doubles to the left-hand side of the *Lambda* and a series of triples to the right-hand side.

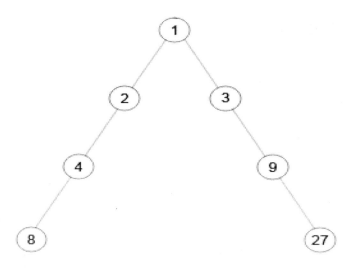

Figure 17: Plato's Numbered Lambda

This numbered *Lambda* immediately triggered the imagination of the arithmologists, who filled in the missing three dots to complete their sacred *Tetraktys*. In this way, the *Lambda* became a kind of early arithmological equation to be solved in the same way that later mathematicians solved more complex statements of numerical relationship. The difference is that, unlike mathematicians, arithmologists derived symbolic understanding of the interaction of primal forces impacting the ongoing process of creation from their solutions to the metaphysical problem posed by the *Lambda*. To solve the riddle posed by the *Lambda*, the arithmologists would have to discover the principles of generation by which other numbers entered into the creative process in order to facilitate the work of the Monad within Creation.

In practice, the problem was solved by assuming that the same logic that dictated the left-hand generation of doubles and the right-hand generation of triples would prevail through the generation of the missing dots. Since the process of left-hand generation involved multiplication by 2, the line of dots proceeding downward to the left from the number 3 would also have to proceed through multiplication by 2. According to this scheme, the middle dot in the third row would have to be 6, since 3 x 2 = 6, and the missing dot farthest to the left in the fourth row would have to be 12 since 6 x 2 = 12. By the same logic, the unnumbered dot farthest to the right in the fourth row would have to be 18, since 9 x 2 = 18, 9 being the number in the row above it from which a left-hand generation through multiplication by 2 proceeds.

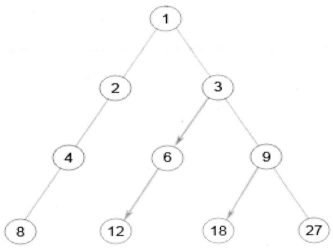

Figure 18: Plato's Lambda Solved
by Left-Hand Generation

So far, so good. But in order for this arithmological solution to the *Lambda* to work, it must also consistently fit the logic of right-hand generation through multiplication by 3 that proceeds down the other arm of the Lambda. Moving downward to the right, in other words, each of the numbers used to replace the unnumbered dots would have to be a multiple of 3 times the number to the left above it. To the great delight of the arithmologists, this scheme seemed to work, since the middle dot now assigned the number 6 represented 3 x 2, the number to the left above it, while the 12 assigned to the unnumbered dot farthest to the left in the fourth row was 3 x 4, and the 18 assigned to the unnumbered dot farthest to the right was 3 x 6.

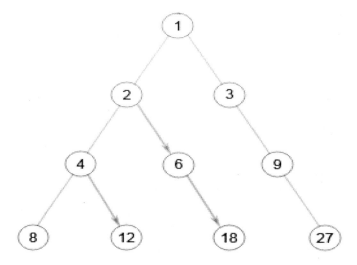

Figure 19: Plato's Lambda Solved by Right-Hand Generation

Using the Tetraktys to Read the Harmonies of Soul

While such exercises might seem a bit academic to the uninitiated, they revealed a great deal to the arithmologists. As Keith Critchlow suggests in his foreword to *The Theology of Arithmetic* (Waterfield 15-16, parenthetical phrases are mine):

> In summary, we might say that the point positions, one by one, represent the Decad (referring here to the Tetraktys); the sets of points represent the oneness, twoness, threeness and fourness of the constituent horizontal lines which we can call the first symmetry; the allocated number progressions represent 'states of being' of the

refraction of one, we might say, and particularly of two and three respectively as they move from plane to cubic form and thereby provide a pattern that has within it the essential numerical harmonies. In particular, this Decad contains those numerical harmonies proposed by Plato for such a vital story as the generation of the 'world soul' in the Timaeus...

We learn from Plato that 'soul' embraces both the mean proportionals (mechanisms within the Realm of Three) as well as the uniting principle of opposites (worked out in relation to the Realm of Two). We now can find out how the particular means, which Timaeus proposes the Divine Artificer uses to fill in the intervals between primary numbers, can be found 'in the symmetries,' or in other words how the 'harmonies' can be read.

Aside from the clues offered by Iamblichus in Waterfield's translation of *The Theology of Arithmetic*, much of the actual reading of the harmonies has been lost, if in fact it ever existed in the detail we would like. Nonetheless, from an astropoetic perspective, the *Tetraktys* built upon Plato's *Lambda* gives us a potent *dakini* syllable that we can mine for as much meaning as our intuitive imagination is capable of extracting, and in that sense is even more valuable than having the answers spelled out for us by previous speculators. The reader will gain a deeper sense of this when I show in Part Two how the numbers weave through the birthchart to provide layers of information not readily available through standard techniques, much less through cookbook interpretations of the symbols.

An Arithmological Analysis of the Tetraktys Extending From Plato's Lambda

Meanwhile, perhaps an illustration of how the *Lamda* serves as the point of departure for arithmological speculation will be useful. The number 6 takes its place at the center of our *Tetraktys* to fill in the first unnumbered dot in the exercise above. It completes the left-hand series moving from 3 to 12, where multiplication by 2 is the rule, and the right-hand series moving from 2 to 18, where multiplication by 3 is the rule. It also reveals a new function in its horizontal relationship on the third row of the *Tetraktys* as a mediating principle between 4 and 9.

In Chapter Nine, I suggested that these two numbers represent successive layers of Fate – Four in relation to the solidification of the material universe, and Nine in relation to the cumulative weight of choices made in all the other Realms. In Chapter Six, I associated the number Six with soul-work, that is to say, the conscious application of awareness (in the Realm of Three) to the seemingly intractable imbalances that arise in our relation to one or

◇◇

more polarities (in the Realm of Two). Pondering this new relationship between 4, 6 and 9, we can see that this soul-work is the mitigating factor that alters our Fate and makes it more malleable. From an arithmological perspective, it is now possible to see this assertion as a "reading of the harmonies" encoded in the way "number progressions represent 'states of being' of the refraction of one."

The same type of analysis can be applied to the numbers assigned to the unnumbered dots in the fourth row of this *Tetraktys*, although here the procedure is more complex. Three of the numbers in this row are double-digit composite numbers that must be reduced to some combination of single-digit principles. There are two ways to do this: the first, by considering which primes engender the number in question; and the second, by adding the digits composing the number together to arrive at a single-digit correlate.

If we consider the number 27, for example, which is the outermost bound of this Tetraktys in the right-hand series emanating from 1, we see that 27 is 3 cubed according to the first reduction strategy, and an octave (a higher recurrence of) the number 9 (i.e. 2 + 7 = 9). Given that the cubing of a number provides a sense of redemption or even a reversal of original principles, we might speculate that the cubing of 3 involves a perspective about perspectives that renders them more malleable or allows a degree of detachment from them that is not possible or desirable within the Realm of Three.

I discuss this kind of detachment in *Tracking the Soul* as an awakening of the sixth *chakra* (Landwehr 84) that liberates us to see the world in a number of ways, without necessarily being attached to our belief systems about the true nature of reality. Given this possibility, we might further speculate that the Realm of Twenty-Seven is where this awakening of the sixth *chakra* takes place, and where the mediation of the opposites that is necessary in the Realm of Three becomes less of an issue. Perhaps this is where, for example, the leaders of two nation states at war could come to a peace negotiation, learn to appreciate the perspective of their "enemy" and find the common ground that allowed them to begin to move toward peace. Whether or not these intuitive "truths" make sense to another arithmologist, arithmological principles give us a mechanism for contemplating the meaning of more complex harmonies posed by double-digit numbers.

Using the second strategy for doing this adds another level of information that we can now synthesize with our first impressions. Noting that the number 27 is an octave of 9, we know that 27 partakes in some way of the nature of 9. It occurs to me as I contemplate this, that the number 27 is related to the way in which the number 9 opens a window into other dimensions of reality parallel to this one, as discussed in Chapter Nine. If so, then perhaps the number 27 refers to the possibility of opening to other dimensions of reality that occurs when we suspend our beliefs and consider other points of view, along with the relativity of all points of view.

⬦⬦

This was a central practice, for example, in the work of Toltec shamans, as reported by Carlos Castaneda, when they moved what he referred to as the assemblage point, a kind of etheric correlate to the fixation of perspective that gets locked into place by some rigid belief system about "the way things are" (*The Fire From Within* 118):

> "*Every living being has an assemblage point,*" [Don Juan went on], "*which selects emanations for emphasis. Seers can see whether sentient beings share the same view of the world, by seeing if the emanations their assemblage points have selected are the same.*"
>
> "*The nagual's blow is of great importance . . . because it makes that point move. It alters its location . . . The assemblage point is totally dislodged, and awareness changes dramatically. But what is a matter of even greater importance is the proper understanding of the truths about awareness in order to realize that that point can be moved from within. The unfortunate truth is that human beings always lose by default. They simply don't know about their possibilities.*"
>
> "*How can one accomplish that change from within?*" I asked.
>
> "*The new seers say that realization is the technique,*" he said. "*They say that, first of all, one must become aware that the world we perceive is the result of our assemblage points' being located on a specific spot on the cocoon. Once that is understood, the assemblage point can move almost at will, as a consequence of new habits.*"

In *The Seven Gates of Soul* I reference philosopher David Hume's assertion that our accustomed interpretation of reality is not necessarily absolute truth, but rather a set of habits of mind, characteristic ways of seeing things that function unconsciously, and are consequently considered to be truth (Landwehr 179-180). If we can develop new habits of mind, then reality will rearrange itself to reflect those habits. At the mildest extreme of possibility, we gain the capacity to "create our own reality," as touted by New Age enthusiasts. At the most potent extreme, we gain the shamanic ability to shape-shift, and to bend various seemingly immutable laws of nature such as gravity, entropy, and limitations imposed by physical space and linear time. The bird shamans I speculated were behind the unexplained mysteries of Chaco Canyon, for example, undoubtedly pushed the envelope of these capacities. Here, to this discussion, I would add only the possibility that this capacity to shape-shift and to alter our perception of reality through changing our habits of mind is potentially the province of the Realm of Twenty-Seven, as determined by the preceding arithmological analysis of that number.

If so, then what we see, as we contemplate Plato's *Lambda*, tracking the emanations of 2 and 3 from the 1, is that at the fourth tier of the *Tetraktys* – which, as we saw in Chapter Four,

◇◇◇

has to do with bringing consciousness as deeply into the material realms as possible – we have a spectrum of possibilities ranging from the wisdom of experience along the left-hand path and the ability to enter other dimensions of reality along the right-hand path. The progression that stems from the Monad through the Dyad would thus appear to be about working out the difficulties inherent in a world conditioned by duality. Yet simultaneously, there is also a parallel process at work – progressing through the Triad – in which the object seems to be opening to a more fluid sense of reality that transcends duality altogether, revealing a world of mystical interconnection, magical possibility, and synergistic cooperation across barriers of space, time, and belief. All this is encoded in an intuitive reading of Plato's *Lambda*, from an arithmological perspective.

Lastly, if we note the intermediary terms, we derive additional information about the relationship between these two kinds of special knowledge: the hard-won wisdom of experience that comes through grappling with core issue; and the expanded awareness that comes through opening to altered states of reality. These intermediary terms – 12 and 18 – are both multiples of 6, which is the synthesis of the left-hand path and the right-hand path. The number 6, in turn, can be factored into 2 x 3, repeating the theme at a more fundamental level. The suggestion here is that the connecting link between the struggle to mediate duality and the capacity to gain liberating perspective that transcends duality is the work of soul. This work – which is about bringing a more focused and intentional awareness to bear upon our issues, problems and difficulties – yields a deepening reservoir of personal wisdom on the left-hand path and a certain receptivity to divine grace (more readily perceptible in altered states) on the right-hand path.

If we combine the single digits of our intermediary terms on this fourth level of the extended *Tetraktys* that derives from Plato's *Lambda*, we see that the deepening reservoir of personal wisdom that represents the culmination of the left-hand path is really a matter of gaining perspective about how our personal issues, problems and difficulties relate to the outworking of collective evolutionary challenges (1 + 2 = 3; 12 is an octave of 3), and ultimately how we might personally contribute to our collective resolution of common problems.

Meanwhile, we also see that the receptivity to divine grace that represents the culmination of the right-hand path requires a certain willingness to surrender to our fate, not out of resignation, but so that our lives can contribute to the outworking of collective evolutionary challenges. It is this willingness to transcend our sense of ourselves as separate beings, struggling to carve out separate lives, that allows Spirit to show us the parallel dimensions in which a more liberating sense of flexibility is possible in relation to Fate (1 + 8 = 9; 18 is an octave of 9).

While these are not the only conclusions that might be drawn from an arithmological reading of the numbered *Tetraktys*, the above discussion hopefully demonstrates how such an exercise might be used by an arithmologist, ancient or contemporary, to speculate intelligently about how the harmonies encoded by number within this manifest Creation might be read.

It is not necessary in reading these harmonies, to stop at Plato's extended *Lambda*. Other *Lambdas* will yield additional information. If, for example, we explore the numbered patterns of squares and cubes, or multiples of 5 and 7, we will gain additional information about the extended meaning of these factors and how they play their part in the revelation of the whole. Additional exercises can be imagined placing other numbers, particularly the primes at the apex of the *Lambda*, and filling in the missing numbers according to any number of possible schemes.

I will leave it to the reader to explore these possibilities on her own, noting only that entire volumes can be written through a full explication of the *terma* teachings encoded in the open-ended *Tetraktys*.

My Contemporary Reformulation of the Lambda

In this book – my contemporary reformulation of Pythagorean number theory – I have essentially re-ordered the *Lambda* by replacing the 1 historically positioned at the pinnacle by 0. This is a Lambda generated not from the Monad, but from the Abyss, which can also be understood as Ain Soph, the Source of Being as well as No-thingness, out of which the entire manifest Creation evolves, and to which it ultimately returns.

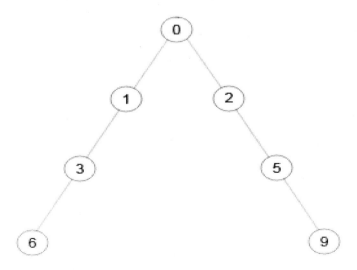

Figure 20: My Reformulated Lambda

In tier 1 of this *Lambda* is the term 0; in tier 2 are the terms 1 and 2; in tier 3 are the terms 3 and 5; in tier 4 are the terms 6 and 9. Left-hand generation down this Lambda proceeds by adding the tier number of the term to the term. Adding 1 (the tier number of 0) to 0 (the sole term in tier 1) gives 1; adding 2 (the tier number of 1) to 1 (the left-hand term in tier 2) gives 3; and adding 3 (the tier number of 3) to 3 (the left-hand term in tier 3) gives 6 (the left-hand term in tier 4).

Right-hand generation down this *Lambda* proceeds by adding the term to the tier number of the next tier down. Adding 0 (the sole term in tier 1) to 2 (the tier number of the next tier down) gives 2 (the right-hand term of tier 2); adding 2 (the right-hand term of tier 2) to 3 (the tier number of the next tier down) gives 5 (the right-hand term of tier 3); and adding 5 (the right-hand term of tier 3) to 4 (the tier number of the next tier down) gives 9 (the right-hand term of tier 4).

Moving down the *Lambda* by left-hand generation, we arrive at the following middle terms: 4 (the middle term of tier 3) = 2 (the tier number of 2) + 2 (the term); 7 (the left-hand middle term of tier 4) = 3 (the tier number of 4) + 4 (the term); and 8 (the right-hand middle term of tier 4) = 3 (the tier number of 5) + 5 (the term).

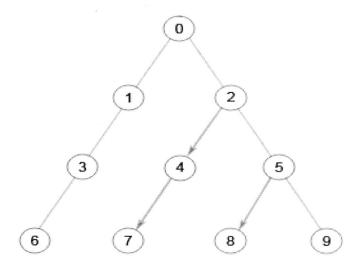

Figure 21: My Lambda Solved
by Left-Hand Generation

Moving down the *Lamda* by right-hand generation, we arrive at the same middle terms: 4 (the middle term of tier 3) = 1 (the left-hand term of tier 2) + 3 (the tier number of the next tier down); 7 (the left-hand middle term of tier 4) = 3 (the left-hand term of tier 3) + 4 (the tier number of the next term down); and 8 (the right-hand middle term of tier 4) = 4 (the middle term of tier 3) + 4 (the tier number of the next tier down).

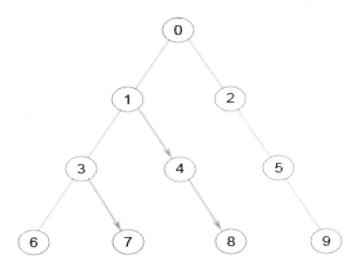

Figure 22: My Lambda Solved by Right-Hand Generation

Left-hand generation down this *Lambda* describes the process of intentionally following in the footsteps of one's chosen Creator god: by projection of the god on tier 2 in the Realm of One to mediation of the opposites on tier 3 in the Realm of Three; to the capacity for soul work and healing of the split between the One and the Two on tier 4 in the Realm of Six. This is essentially the pathway of Ascent on which we strive to become more godlike.

Right-hand generation down this *Lambda* describes the necessity for integrating the opposites inherent in Creation along whatever path we chose to follow: from our first encounter with the shadow, the Feminine, and Matter on tier 2 in the Realm of Two; to the discovery of divine intelligence within the embodied Creation on tier 3 in the Realm of Five; to the deeper discovery of the interdimensional nature of Fate on tier 4 in the Realm of Nine. This is essentially the Descent through which we bring Spirit more deeply into the manifest Creation through our increasingly skillful and astute exercise of consciousness.

At every step of this process and on every tier, the potential also exists for a refusal of divine intelligence, in which case this *Tetraktys* (the *Lambda* solved) yields a sure return to the Abyss – which is essentially a Creation devoid of Spirit. I will leave it to the reader to work

◇◇◇

out additional details of the *terma* teaching contained in this reformulated *Tetraktys*, using the abundant clues I have provided in this book. We will return to discuss the astrological implications of this *Tetraktys* in Part Two.

The Musical Correlates of Number

Aside from the purely arithmological analysis of numbers and arrays of numbers, the Pythagoreans also considered the musical correlates of number. The thrust of Part One of this book – creating a visceral sense of numbers that transcends their purely mathematical function to engage not only the mind, but also the senses and the emotions – is not just a whim, but rather a logical extension of the arithmological vision. One area in which it is possible to see the sensory logic of this vision is in the translation of number into sound that preoccupied the original arithmologists.

The concept of musical intervals – now widely employed by composers and musicians of all persuasions as a matter of course – is attributed to Pythagoras, who used a one-stringed instrument with a movable bridge, called a monochord, to investigate the ways in which number related to sound. Pythagoras considered the fact that numerical relationships could be used to produce the notes of a musical scale to be a clear demonstration of the way in which the unlimited Monad permeated creation through the vehicle of a series of limited Forms. If the string of the monochord represented a continuum of tonal flux that was unlimited in its potentiality, then dividing (or limiting) the string according to various numerical schemes allowed for the possibility of a musical scale amenable to the senses. With the monochord, numbers could actually be heard. In this way, numbers provided the limiting power (*peras*) that illuminated and contained the Unlimited (*apeiron*).

In theory, the string of the monochord can be divided at any point, yielding a spectrum of sound limited only by the range of human hearing. Yet when we divide the string according to numerical proportions, we arrive at the familiar tones that compose our musical scale. Dividing the string by 2 (or creating what is referred to as a 1:2 ratio), for example, produces an octave of the original tone produced when the string is undivided (allowed to remain as 1). This octave, produced by halving the string, vibrates at a frequency double the original tone. Dividing the string by 3 (referred to as a 2:3 ratio), creates the perfect fifth and by 4 (a 3:4 ratio), creates the perfect fourth. Similarly, division by 5 (4:5) produces the major third, and by 6 (5:6) the minor third. These six basic tones taken together account for what is commonly understood as the harmony of modern music.

To complete the musical scale, beyond the number 6, we experience increasing disharmony. The number 7 (6:7) produces a tone called the septenary third, which is more minor and discordant than the minor third. The number 8 (7:8) produces the septenary whole tone, which is more major than the major third, yet somehow also discordant in its own way. The number 9 (8:9) produces the whole tone, one step up the musical scale – melodious in its own right, but disharmonious when played simultaneously with the root tone of the scale. Lastly, the number 10 (9:10) produces the half-tone, which generates a similar disharmony with the root tone.

Harmony and Disharmony

The distinction between harmony and disharmony that derives from a consideration of the relationship of numbers to the scale is a useful metaphor that can shed additional light on the interrelationships between numbers. The number 1, for example, shares an octave relationship with 2, which is why the number 2 becomes an essential passageway through which the Monad enters Creation, despite the fact that 2 is in many ways antithetical to 1. This simple octave relationship is a profound statement of the great paradox that makes it possible for an unlimited omnipresent Spirit to dwell within a limited, fragmented reality.

The number 2 bears an octave relationship to 1, 4, and 8 (2 octaves above 2), with which we might say it is in harmonic resonance. It also bears the relationship of a perfect fifth to the number 3, which mediates the polarities and brings them into balance, that is to say, a state of harmony. It is this state of natural harmony between 2 and 3, for example, that makes Plato's *Lambda* such a pleasing template for understanding the intent of the Creator in entering Creation, since it extends this relationship into levels of ever-increasing subtlety and sophistication.

The number 2 forms no other whole-number ratios with any other single-digit number, and thus triggers no other relationship formed by numerical harmony. Conversely, any two numbers in an octave relationship will refer back to the number 2. It is this harmonic reference, for example, that allowed arithmologists to intuit the power of 5 to heal the ambivalence of 2, given that 5 shared an octave relationship with the number of perfection: 10.

Aside from its perfect fifth relationship to 2, the number 3 shares an octave relationship to 6 – which is why mediation of opposites is the modus operandi of choice in addressing core issues; and a perfect fourth relationship with the number 4 – the Realm of material manifestation, which is where the principle of mediation of opposites must be applied in order to be effective. One cannot attain harmony through perspective alone (also possible within the Realm of Three). One must enter into a ground-level relationship with real-world circumstances and mediate the opposites on an elemental level before harmony of soul can be achieved. In turn, this relationship between 3 and 4, understood harmonically, becomes the arithmological rationale for the embodiment of the soul, and in a more general way, for the entry of Spirit into matter.

CHAPTER TEN

◇◇◇

The number 4 shares an octave relationship with 1, 2 and with 8 – which is where the density of the former Realm becomes illuminated by the light of true understanding; and a perfect fifth relationship with the number 6 – which is where the material world comes into harmony through the application of consciousness to the concerns of the embodied life. As already discussed, 4 also shares the relationship of a perfect fourth with 3.   With 5, 4 shares the relationship of a major third, which is why consciousness is generally considered the fifth element, thus completing the Realm of Four through the entry of Spirit into it.

The number 5 shares an octave relationship with 10, where Spirit becomes fully embodied in Creation, and where Creation in turn becomes fully conscious, and a major third relationship with the number 4, as already discussed. Numbers 5 and 6 enjoy a minor third relationship, which is why the basis of soul work is coming into right relationship with natural law, and conversely (given the minor theme of this harmony), why the failure to live in accordance with natural law will produce the core issues in need of attention within the Realm of Six. In addition to its role within contemporary Western music, the number 5 also forms the basis for an alternate harmonic system, the pentatonic scale employed by Chinese music.

The number 6 is the octave of 3, in perfect fifth harmony with 4 and 9 – which is where Fate becomes a reflection of the degree of soul work that has been done, or not done; in major third relationship to 8 – which is where soul work accomplished leads to hard-won personal wisdom; and as discussed above, in minor third relationship to 5. In addition to these harmonious relationships, 6 also produces a disharmonious septenary third relationship to 7. We saw such disharmony reflected in Chapter Seven, for example, in the concept of complex creativity, which can be the expression of extreme polarization of the opposites, a state of being antithetical to their balancing and mediation within the Realm of Six, at least until the soul work of healing in the Realm of Six is done.

The number 7 is the first within the single-digit series for which no harmonious relationships exist with other single-digit numbers, thus setting the tone for subsequent primes with which it shares this quality. It does enjoy a septenary third relationship to 6, and a septenary whole tone relationship to 8. There are ways in which 8 seems to follow naturally from 7 (as is true of all the numbers) – say as immortality naturally accrues through the exercise of complex creativity. However, there are also ways in which what happens in the Realm of Seven is at odds with what happens in the Realm of Eight. The concept of spiritual initiation, for example, is at odds with the notion that wisdom gained is wisdom applied within the manifest creation, since the former generally implies a transcendence of the embodied life, while the latter requires a more conscious descent into it.

The fact that the number 7 is the first disharmonious prime is considered significant by some arithmologists, since beyond the number 6, which the arithmologists considered to be perfect, the harmony within the manifest universe begins breaking down. This occurs even as the process of creation proceeds into more complex domains and involves more sophisticated differentiations of the One. This sentiment is echoed by modern arithmologist Richard Heath, who asserts, "the prime numbers 3 and 5 are responsible for the principle of harmony; this means that all the higher primes are fated to occupy the voids left by the field of that harmony" (36). In those voids, what we perceive as disharmony must somehow be resolved. Or put another way, we must become conscious enough to see the true harmony within the apparent disharmony. This more challenging occupation prevails in Balinese gamelan and other ethnic forms of music as well as the more experimental forms employed by postmodern composers such as Walter Stockhausen, Morton Subotnik and Phillip Glass – where discordant harmonic systems based on the number 7 are more prevalent.

This deeper awareness of harmony within disharmony dawns within the Realm of Eight, which is why the arithmologists associated the number 8 with wisdom. This is not a wisdom of platitude, but one that has begun to incorporate the discordant anomalies of the Realm of Seven into a more genuine and complex truth, capable of embracing contradiction.

As already discussed, the number 8 is the octave of 4, 2, and 1, and forms a perfect fifth with 6 and a septenary whole tone with 7. It also forms a major third with the number 10, where the wisdom that is the pinnacle of human development meets with a dispensation of divine grace that transcends human effort. Lastly, the number 8 forms a whole tone with 9, where human effort no longer has the capacity to mitigate Fate.

The number 9 forms a perfect fifth with 6, a whole tone with 8, and a half-tone (or minor whole tone) with 10, which is where the concept of Fate ceases to have meaning in a manifest world become fully conscious and aligned with Spirit.

As with Plato's *Lambda* exercise, arithmologists also considered more intricate and detailed harmonies and disharmonies generated by double-digit numbers beyond 9. Pointing out that the number 12 was generally considered to be the "limit of the divine world," Richard Heath suggests that "if 2, 3 and 5 are 'the gods' of the numerical world, and the larger primes are 'the demons,' then traditionally there is a tug of war between them in the essential process of creation" (39). Within the range between 12 and 24, for example, musical articulation becomes more restricted, due to the prevalence of four discordant primes – 13, 17, 19, and 23 – as well as the new musical interval 15:16, a half-tone that makes possible the trance-like modal forms of music, which abandon the scale for more subtle meandering around a tonal center. The musical ratios that become possible within the range between 24 and 48 provide the foundation of the Greek modes, which return music to the scale, albeit in a manner that requires some adjustment of classically trained expectations.

All of these musical forms, born of complex numbers, illustrate the intricacies of creation, which begins with identifiable principles, but then extends into an infinitely intricate

◇◇◇

web of subtlety, where we are challenged to reconcile disharmonies that have no name. The path followed by any soul through this endless vale of numbers leads through the familiar landscape of collective culture into a timeless sacred grove, where intuitive shifts in perception resonate in sync with a silent rearrangement of molecules beyond the range of perception. Although the arithmologists largely kept to the familiar territory marked by principles, their practice potentially takes us into ever-increasing nuance, where one who is practiced in the art of listening for subtle shifts in tone outside the obvious domain of harmony, can use a knowledge of number to articulate the ineffable. Pure mathematicians and theoretical physicists such as Einstein and Bohm flirt with this possibility, although because the qualitative dimension of number studied by earlier arithmologists has been lost, they must often keep their more metaphysical speculations within a box too rigid to adequately allow them.

Platonic Solids

Arithmologists extended the concept of numerical harmony to the measurement of space. This practice is worthy of a book unto itself, and as with our discussion of music, we will necessarily limit ourselves here to a brief survey. The spatial understanding of number begins again with Plato's *Timaeus*, where, in addition to the *Lambda*, Plato identifies five regular solids extending number into the dimension of geometrical shape. Three of these Platonic solids – the tetrahedron (a 4-sided figure), the octahedron (an 8-sided figure), and the icosahedron (a 3-dimensional polygon with 20 sides or facets) – are based on the number 3; one – the cube (a 6-sided box) – on the number 4; and one – the dodecahedron (a 12-sided shape) – on the number 5. It was believed by Plato and other arithmologists that carried on Pythagoras' work that these five solids – which are the only polyhedra that can be constructed of the same regular polygons – showed how the Monad extended itself into space.

Each of these polyhedra was associated with an element – the tetrahedron with Fire, the octahedron with Air, the icosahedron with Water, the cube with Earth, and the dodecahedron with Aether, or consciousness. In this way, as space extended from the Monad through the numbers 3, 4 and 5, the manifest world – of which the elements were the primary building blocks – came into being. The solidity of the manifest world itself comes through the extension of the number 4 into space as Earth. Fire, Air and Water – associated with the number 3 – are mediating principles, operating within space through the Platonic solids with which they were associated. And Aether – associated with the number 5 – enters into three-dimensional space through the vehicle of the dodecahedron.

Exactly how this occurs is an occult mystery beyond the scope of this overview. Suffice it to say that what was important to the arithmologists was that the numbers associated with harmony (3 and 5) and Fate in its most elemental guise (4) made possible a three-dimensional physical universe in which the Fate of the embodied soul could be mitigated to conform with divine intention. The mechanisms for this were not only conceptual, but built into the very warp and woof of the material loom on which the embodied world was constructed.

In *Tracking the Soul*, I outline the relationship between the elements and the five *koshas* – or levels of embodiment between pure matter at one end and pure consciousness at the other. According to modern-day psychoacoustic pioneer, Michael Heleus – who has developed a way to render the Platonic solids as sound – these regular polyhedra hint at a more fundamental relationship between matter and consciousness than is recognized by science. As with everything studied by the arithmologists, number is the key to this relationship, manifesting simultaneously as "modes of relating in (outer physical) space" (Heleus qtd by Landwehr, *Reconnecting With the Cosmic Bearings of Life Through the Right Use of Sound*) and levels of receptivity to Spirit in inner psychological space. Through number, as it is employed by arithmologists investigating its extension into space, we begin to glimpse the possibility that what we call the manifest physical universe is actually a nexus of forms that embody and reflect the qualitative attributes and psycho-spiritual dynamics of the soul. In turn, both might be understood as parallel emanations of the One.

Sacred Geometry

This assertion that individual soul and the soul of the world (*anima mundi*) are one has implications that were well known to indigenous cultures steeped in participation mystique. The understanding that physical space and psychological space can not only reflect but also influence each other through numerical relationships, was the basis of construction for many early temples, sacred sites and buildings of ceremonial importance. The Great Pyramid, the lost Temple of Solomon, Stonehenge and other sacred sites throughout the British isles, Greek temples, the ancient Pueblo ruins of the American southwest, Gothic cathedrals, and even the construction of Washington, DC with the White House at its center (Heath 214) have been plumbed for secret numerical relationships at the core of their design.

The ratio of base to height of the Great Pyramid, for example, has been noted to be 11:7. An early numerical approximation of *pi* – the constant used to equate the circumference of a circle to its radius – was 22:7, a value that is amazingly 99.96% accurate. The ratio of base to height is thus half *pi*, or the symbolic equivalent of the arc distance between the poles of the Earth and its equator. This reference to *pi* further symbolizes the relationship between Heaven or the Realm of the Monad, which is circular or cyclical in nature, and Earth, where everything is measured in linear terms (Heath 57). Encoded within the Great Pyramid is a simple numerical relationship that symbolizes the arithmological notion that Heaven and Earth are connected by number.

◇◇

There is much more subtle numerical symbolism encoded within the Great Pyramid, as well as other sacred sites around the world, and an exploration of these relationships is far beyond the scope of this overview of arithmological principles. Suffice it to say that students of sacred geometry building upon arithmological principles and various measurements of astronomical cycles have demonstrated the possibility of intelligent design based on number, incorporated into many of the world's most sacred architectural wonders, and intended to mirror cosmological truths.

Arithmological Sequences in Nature

Simple numerical relationships have also been found to operate in nature in a wide range of settings that suggest an inherent design by number built into the construction of the manifest world. One such set of observations, for example, centers on the Fibonacci series – a sequence of numbers in which successive numbers are added to each other to produce the next term in the sequence: 0, 1, 2, 3, 5, 8, 13, 21, 34, 55, 89, etc. This simple sequence of numbers has been noted as the basis for the spiral, and is witnessed operating in sea shell shapes, flower petal arrangements, pineapples, pine cones and honeycombs, to name a few of its manifestations (Knott).

The number of petals on a flower, for example, tends to match some number in the Fibonacci series. The lily and the iris produce 3 petals (or sets of petals in 3); buttercup, wild rose, larkspur, columbine and vinca produce 5 petals; delphinium and coreopsis produce 8 petals; ragwort, marigold and cineraria produce 13 petals, and so on. Much more rare are flowers with 4, 6, 7, 9, 10, 11, 12 or some other number of petals not in the Fibonacci series (Thinkquest). The reason this is so is that the packing arrangement by objects in space afforded by numbers in the Fibonacci series is more efficient than numbers outside the series (Knott). While this can be understood scientifically as a simple law of nature, it can also be understood as evidence of the way in which design by nature is a numerical process, implying intelligence – possibly with arithmological overtones.

Although largely a lost art, the later Neoplatonists, including Iamblichus practiced a form of resonant magic, called theurgy, in which various plants, stones, animal spirits, and other elements of nature carried properties associated with various gods. By entering into a conscious relationship with these elements of nature, a practicing theurgist could evoke various deities for specific creative purposes. Given that deities were associated with numbers by the arithmologists, and the possibility suggested by modern arithmologist Richard Heath that the gods themselves were symbolic depictions of the archetypal powers of numbers, it is possible that within the lost art of theurgy, elements of nature were employed in ritual and ceremony to evoke these powers of number, as I have described them in this book.

The Numbers and the Gods

Although the full delineation of these associations between number, deity, and natural phenomena goes beyond the scope of this book, it is worth noting the correlations between number and deity that were specifically mentioned in *The Theology of Arithmetic* as a point of departure for further speculation about the nature of each number. These notations will be supplemented by additional references from Manly Palmer Hall, who claims to synthesize the musings of Nicomachus, Theon of Smyrna, Proclus, Porphyry, Plutarch, Clement of Alexandria, Aristotle and "other early authorities" (LXXI). Some of the references noted are not to deity per se, but to important figures in Greek mythology, all of whom play their metaphorical part in the outworking of cosmic processes. Occasionally there are references to deities or mythological figures from cultures other than Greek, primarily Egyptian.

One

The number One was of course associated with deity in general, through its more Platonic guise as the Monad, or source of all Creation. The arithmologists also associated the number One with a number of specific deities: Chaos, first principle of Hesiod's *Theogony*, the "yawning void" out of which emerge all the original Titans through which Creation proceeds (Morford & Lenardon 38); Zeus, the chief deity of the Olympian pantheon; Prometheus, the "artificer of life" (Waterfield 38), whose theft of fire for humans symbolizes the desire of an immanent god to infuse his creation with the power of the divine; Proteus, "the demigod in Egypt who could assume any form and contained the properties of everything, as the monad is the factor of each number" (Waterfield 40); and lastly, the Fate Atropos, encountered in Chapter Nine as the cutter of the thread of fate, suggesting that the ultimate cutting of this thread is the reunion of the individual soul with the One, into which it is subsumed, and through which the soul's ultimate Fate is consummated.

In addition to these references, Manly Palmer Hall claims the Pythagoreans associated One with Atlas, the brother of Prometheus, who sided with Cronus in the 10-year battle between the gods that led to the defeat of the Titans, and was subsequently punished by Zeus with the endless task of holding up the sky; and with Apollo, the Greek and Roman god of reason and intelligence, music, prophecy and medicine, often associated with the Sun (LXXI).

There is a long tradition in Greek and European philosophical history that associates One with the Sun, not surprising given the astronomical fact that the Sun is the center of our solar system – thus, like the Monad, making life possible upon this Earth. The association with Atlas is interesting, given that Prometheus – a key player on the other side of the 10-year

◇◇◇

battle – is also represented, as is Chaos, out of which the Titans who were defeated in this battle initially emerge. This constellation of symbols suggests that the Monad is not a static unity, but rather a dynamic and perpetually evolving state of equilibrium between opposing forces through which increasingly more inclusive levels of integration are achieved.

Two

The number Two was associated with Egyptian fertility goddess Isis, seen as the bringer of hope and new life through the myth of her quest for her fallen and dismembered husband-god, Osiris (Morford & Lenardon 309). Through this association, we might assume that the number Two could be related to the spiritual quest in general, and more specifically to reintegration of all the lost parts of self (relegated to the shadow) that must be reunited before a soul becomes whole. Arithmologists also associated Two with Rhea, the mother of all the gods, and specifically of Zeus. Here we encounter a paradox, only fully understood in the Realm of Eight, in which the Dyad becomes more fundamental to the life of the soul than the Monad, to which it gives birth – presumably through the fertile power of creative imagination. Lastly, the number Two is associated with the Greek muse Erato, muse of love poetry, hymns to the gods and lyre playing (music), suggesting an association of Two with a more Feminine path of devotion and worship through artistic expression.

Manly Palmer Hall associated Two with Phanes, an epithet of Eros, the Greek god of romantic love (Cupid in the Roman pantheon) – obviously a crucial factor in the law of attraction between the opposites; also understood as a primal force present at the beginning of Creation, the offspring of Chaos (Morford & Lenardon 38-39). The number Two is also associated with Cybele, the Phrygian mother-goddess, sprung from the Earth, who with her consort Attis, acted out yet another version of the Greek fertility cycle (parallel to the story of Isis and Osiris, or Aphrodite and Adonis). Like Rhea, worship of Cybele involved an orgiastic frenzy of devotion, accompanied by loud, discordant music and spontaneous acts of self-mutilation. In general, Two seems to be associated with a number of goddesses – aside from Isis, Rhea and Cybele, already mentioned; also fertility goddess Demeter; chaste goddess of the hunt Artemis; Aphrodite Pandemos[3]; Dione, the mother of Aphrodite Pandemos; Hera, Zeus' official consort and queen of the gods; and Maia, mother of Hermes.

Three

The number Three was not assigned a specific god by the arithmologists, although they did associate it with the triune nature of God, depicted in Christian mythology, for example, as Father, Son, and Holy Spirit; and in Hinduism as Brahma, Vishnu, and Shiva.

According to Manly Palmer Hall, the Pythagoreans associated the number Three with Cronus, primarily in his role as the god of time; Ophion, the great Serpent who ruled the world with his consort Eurynome, before being cast down by Cronus and Rhea; Thetis, the mother of Achilles; Hades (or Pluto), god of the Underworld; Hecate, an Underworld fertility goddess greatly feared for her skill in black magic; Athena, the virgin goddess of wisdom and war; Polyhymnia, muse of sacred music and dancing; Triton, son of Poseidon, a shape-shifting merman known as the trumpeter of the sea for his skill with the conch shell; and the Fates, Furies and Graces (each of which occurs in groups of three) (LXXII).

These associations suggest there is more to the number Three than meets the eye. In particular, its Underworld and underwater connections suggest that before the soul can aspire to the level of perspective at which mediation of the opposites is possible, a journey to the Underworld and an embrace of darkness is necessary. One cannot mediate the opposites until the shadows of Hades have been illuminated; fear of the power hidden there, symbolized by Hecate, has subsided; the rhythm of light and dark has been observed under the tutelage of Cronus and danced through a life inspired by Polyhymnia. Until then, Three exists as a dormant potential only, a redemption of the promise of the One that can only be achieved by unflinching passage through the Realm of Two.

Four

According to *The Theology of Arithmetic*, Four is associated with Hermes, whose square-cut statue marked many crossroads throughout Greece; with Aeolus, the keeper of the winds encountered by Odysseus; and with Heracles, the most popular of all the Greek heroes, whose twelve labors made him immortal. Like Mercury, Hercules (the Roman version of Heracles) was later considered a god of worldly commerce. According to Manly Palmer Hall, Four was also associated with Hephaestus, the lame master craftsman and oft-cuckolded husband of Aphrodite; Dionysus, god of the grape and wine, intoxication, sexual ecstasy, and other altered states of consciousness; and Urania, muse of astronomy.

Again, as with the number Three, these associations are intriguing, because they allude to a more esoteric function of the number Four. The reference to Aeolus harkens back to our discussion of the winds associated with the four directions of the Medicine Wheel in Chapter Four, through which our sense of direction becomes a gift of the gods, as it was to Odysseus in his arduous journey home. The reference to Hephaestus evokes The Magician card of the Tarot, in which all four elements – sword, wand, cup and pentacle – become sacred tools of creative magic. In Dionysus, we see reference to the world of apparently solid form as gateway beyond form, a theme that becomes more clearly articulated in the Realm of Nine, but which nonetheless is implied in the Realm of Four. Lastly, with the reference to

◇◇

Urania, muse of astronomy, comes the Hermetic suggestion – central to astrology – that "as above, so below," what happens in the Realm of Four, the most earthly of all the Realms, is but a reflection of the movements of the heavens.

Five

The number Five was assigned by Iamblichus to Aphrodite, "because it binds to each other a male and female number;" Pallas Athena, "because it reveals the fifth essence" (Waterfield 73-74); and Nemesis, the goddess of vengeance. To this, Manly Palmer Hall has nothing substantial to add. Surprising here is only the reference to Nemesis, which perhaps hints at the consequences for misuse of the fifth element, consciousness.

Six

The number Six was given to Amphitrite, wife of Poseidon, whose name means "separate triads," referring to the engendering factors of 6: 2 x 3; and Thaleia, the muse of comedy. Hall suggests that the number Six was particularly sacred to Orpheus, the son of Apollo, archetypal musician and poet who followed his wife Eurydice to Hades, and charmed the Underworld deities into letting her go (see Chapter Zero). Hall further associates Six with the Fate Lachesis, the measurer of fate, whom I suggested in Chapter Nine could be understood to work through the movement of various astrological cycles. As we will see in Part Two, the number Six is critical to an understanding of astrology, particularly from the perspective of the soul using the birthchart as a map to its journey.

Seven

According to *The Theology of Arithmetic* (Waterfield 99):

> *They called the heptad 'Athena' . . . because it is a virgin and unwed, just like Athena in myth, and it is born neither of mother (i.e. of even number) nor of father (i.e. odd number), but from the head of the father of all (i.e. from the monad, the head of number); and like Athena it is not womanish . . .*

It is worth noting here that Athena was also previously assigned to the numbers Three and Five, albeit for different reasons. With Three, the association is with Athena's capacity as a strategist in war; with Five, her wisdom as the embodiment of consciousness used in harmony with natural law; and with Seven, in her unnatural birth springing fully formed from the head of Zeus, a reference to 7's status as a prime. Both the cross-references between numbers and the multidimensionality of the Greek gods and goddesses attest to the complex web of interconnection between numbers and deity.

In addition to Athena, Hall associates Seven with the Egyptian god Osiris; Ares, the Greek god of war; and Clio, the muse of history and lyre playing. Osiris can be understood as an embodiment of complex creativity, having brought many of the accoutrements of civilization to Egyptian culture (along with his wife Isis), before being dismembered by his arch rival Seth. Ares "is generally depicted as a kind of divine swashbuckler ... not highly thought of, ... at times, little more than a butcher" (Morford & Lenardon 89). He is essentially a Greek cautionary tale, perhaps a harsh catalyst toward reconciliation between warring factions, if the disaster of war is taken seriously. As muse of history, Clio provides the same opportunity, as the cumulative litany of wars, genocides and other brutal acts of violence perpetrated by humans against each other, as well as our collective willful disregard of natural law, is contemplated in its overwhelming totality.

Eight

As discussed in Chapter Eight, arithmologists called the number Eight Cadmean, referring to Cadmus, the founder of Thebes proceeding by divine admonition. Since 8 is 2 cubed, they also assigned Eight to Rhea, who was previously invoked as a divine embodiment of Two. "In the case of Rhea, the myths tell us that Kronos (as the stories go) disposed of her children; and in the case of the ogdoad, labor in the eighth month is fruitless and hence is called 'untimely' " (Waterfield 103). Lastly, Eight is associated with Euterpe, muse of lyric poetry, tragedy and flute playing.

To this list, Hall adds Cybele, like Rhea also included in the list of deities resonant with Two, and likely for the same reason; Poseidon, the ferocious god of the sea, earthquakes, and sheep, bulls and horses; and Themis, a consort of Zeus with oracular powers and the mother of The Fates, discussed in Chapter Nine.

The inclusion of Poseidon in this grouping is particularly revealing, since he represents a patriarchal usurpation of a domain that previously belonged to a succession of goddesses. The earliest of these was Eurynome, in early Greek myth, the mother of Creation, arising out of Chaos, who divided the sea from the sky and stood upon the waves. To assuage her loneliness, Eurynome created the Great Serpent Ophion (associated earlier with the number Three), then laid the Universal Egg out of which Creation sprang (Greene 28). Within the origin of Poseidon in a primal fertility goddess (Realm of Two) married to a harbinger of the wisdom of the serpent ($8 = 2^3$), we have a symbolic depiction of the redemptive power that Eight had in the minds of the arithmologists.

◇◇

Nine

To Nine, the arithmologists assign another deity of the ocean, a Titan named Oceanus, who with his mate Tethys produced the Oceanids, 3,000 daughters and the same number of sons, each associated with a river, lake, or spring. Apparently the primary reference here was to the seemingly endless horizon of the ocean, which was once believed to pose the natural limit of the Earth in the same way that 9 is the natural limit of the numbers. It may also be possible to understand the 6,000 tributaries spawned by Oceanus and Tethys as parallel dimensions, water being the symbolic medium of choice for entry into them.

The arithmologists also called Nine Prometheus (previously associated with One), "because it prevents any number from producing further than itself[4];" Hephaestus (previously associated with Four), "because the way up to it is as it were, by smelting and evaporation;" Hera (previously associated with Two), "because the sphere of air falls under it[5];" the Curetes, eunuchs who originally hid the cries of Zeus, and later became associated with the frenzied worship of Rhea and Cybele; Persephone, for reasons not explained; Hyperion, Titan god of the Sun, "because it has gone beyond all the other numbers as regards magnitude[6];" and Terpsichore, the muse of choral dancing and flute playing, "because the recurrence of the principles and their convergence on it as if from an end to a mid-point and to the beginning is like the turning and revolution of a dance" (Waterfield 106-107). To this long list, Hall adds the already overworked Athena.

Ten

To the number Ten, the arithmologists assign Eros (also assigned to Two) – a curious designation given their mistrust of the Dyad; Urania, the muse of astronomy; Pan, the goat-like nature god, whose name means "all;" and Atlas (also associated with One). Hall adds Mnemosyne, a Titaness whose name means "memory," with whom Zeus mates to produce the Muses. The inclusion of Pan in this list suggests that a fully conscious creation is one that is perfectly aligned with natural law; and the presence of Mnemosyne on the list suggests that inspired creativity (facilitated by the Muses) partakes of the number Ten.

Many of these references to deity are rather obscure, and as with all arithmological principles, are meant to be suggestive of possibilities encoded by the numbers rather than dogmatic statements of fact. Nonetheless, within the full range of arithmological associations – from types of numbers and their mathematical combination to consideration of numbers in various arrays to music and spatial references to theurgical associations of numbers, minerals, plants and other natural phenomena to the overlapping correlation of number with deity – we see a coherent symbological system that can be employed to discern the qualitative dimension of numbers wherever they arise. In Part Two, we will look specifically at the arithmological implications of the numbers as they manifest within the language of astrology.

Endnotes

1. The number 2 was regarded by early arithmologists as prime, just as we do today. Later arithmologists considered 2 to be odd-even – an even number that becomes odd when divided by some power of 2, or even-odd – an even number whose half is odd. 2/2 = 1. Other odd-even (even-odd) numbers include 6, 10, 14, 18, etc. Some arithmologists considered any number that could not be divided into both equal and unequal parts to be a source of number, rather than a number itself. Such sources of number would include 1, 2 and 3 and all subsequent primes.

2. The geometric means of most numbers would be irrational. Irrational numbers are those that cannot be expressed as a simple fraction: a/b. The geometric mean of 1 and 2, for example, is the square root of 2 or 1.4142135... on to an infinite number of non-repeating decimal places. Legend has it that the Pythagorean scholar, Hippasus, who discovered the existence of irrational numbers was drowned at sea as punishment for contradicting Pythagoras' theory that all numbers could be expressed as the ratio of integers (Wikipedia, Irrational Numbers). Although the story is shrouded in mystery and contradiction, we might speculate that the shocking discovery of irrationality contributed to the ambivalence that Pythagoreans felt toward the Dyad, since it could not be reconciled with the Monad by geometric means without producing an irrational number, called by Greek mathematicians *alogos*, or inexpressible.

3. The Greek goddess Aphrodite was considered by the Greeks to exist in two distinct forms: Aphrodite Urania or Celestial Aphrodite, who sprang into existence out of the sea foam that bubbled up around Uranus' severed genitals; and Aphrodite Pandemos, Aphrodite of All the People or Common Aphrodite, believed to be born from a union between Zeus and Dione. In the *Symposium*, Plato elaborates on the difference between the two, considering Aphrodite Urania to be stronger, more intelligent and spiritual in nature, while Aphrodite Pandemos was thought to be a baser manifestation of the goddess, devoted primarily to sensual and sexual gratification.

4. The etymology of the name Prometheus is derived from the Greek word for "not outrunning."

5. The name Hera is etymologically linked with the Greek word *aer* (air).

6. According to Waterfield (107), "the word Hyperion could mean 'going beyond'."

I must confess that I incline to the view that numbers were as much found as invented, and that in consequence, they possess a relative autonomy analogous to that of the archetypes. They would then have, in common with the latter, the quality of being pre-existent to consciousness, and hence, on occasion, of conditioning it rather than being conditioned by it. . . . Accordingly, it would seem that natural numbers have an archetypal character. If so, then not only would certain numbers have a relation to and an effect on certain archetypes, but the reverse would also be true. The first case is equivalent to number magic, but the second is equivalent to inquiring whether numbers, in conjunction with the combination of archetypes found in astrology, would show a tendency to behave in a special way.

from *Synchronicity* by Carl Jung, Princeton, NJ: Princeton University Press, 1960.

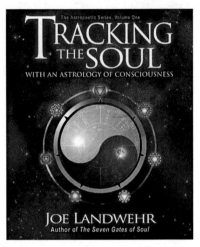

The Astropoetic School of Soul-Discovery

Where Your Life is the Classroom

Learn astrology through intensive study of your own chart
and the correlation of symbolism with your own life experiences.

Forget keywords.
Learn to think and feel astrologically.

Memorize less.
Observe and self-reflect.

Rediscover the poetry of everyday life
and the everchanging wonder of the cosmic dance.

**24 lessons with homework assignments.
90-minute phone consultation with each lesson.
3-day annual workshop with other students.
Advanced placement possible.**

Joe has the uncanny ability to see through the mask of an individual and gently prod them onto a deeper, more relevant and profound exploration of their chart symbolism. I personally know him to be a genuine spiritual seeker, as well as a gifted writer and astrologer of the highest moral caliber. He is a master of the art of astropoetic interpretation with the ability to skillfully interrelate mind and body, heart and soul into his teachings, which he pursues with passion. On this uncertain journey towards a deeper understanding of the meaning of life, I think Joe is as close as one can get! - M.A, Columbia Station, OH

See www.astropoetics.com for more information.

I am an astrologer of nearly 40 years experience, seeking an eclectic integration of astrology, spiritual psychology and ancient wisdom teachings. Author of two award-winning books and Director of **The Astropoetic School of Soul-Discovery**, I bring a wide-ranging spiritual background to my work, which in addition to astrology draws from yogic philosophy, Jungian psychology, Native American teachings, and the runic lore of Teutonic mythology. I thrive on meaningful contact with readers, students, peers and colleagues, and invite you to connect with me in whatever way suits you best.

Email me at:

jlandwehr@ancient-tower-press.com or
jlandwehr@astropoetics.com

Follow me on my blogs:

The Astropoetic School of Soul-Discovery: www.astropoetics.com - for information about the school, my correspondence courses, and workshops, as well as blog series on topics of interest to astrologers and astrology students.

The Sky Is My Mirror: www.theskyismymirror.com - for my personal blog on which I share my experiences using astropoetics (an intuitive, soul-based, self-reflective approach to astrology that I teach at The Astropoetic School) for my own personal growth.

Ancient Tower Press: www.ancient-tower-press.com - for information about my publishing company, my books (including introductions, reviews, and ordering), my vision of Fair Trade Publishing, as well as occasional articles about topics of interest to self-published authors and independent presses.

Connect with me on:

Facebook   Linked In   Plaxo   Twitter

Figure 23: The Blank Rune

Odin, No-Thingness, the Abyss

Appendices

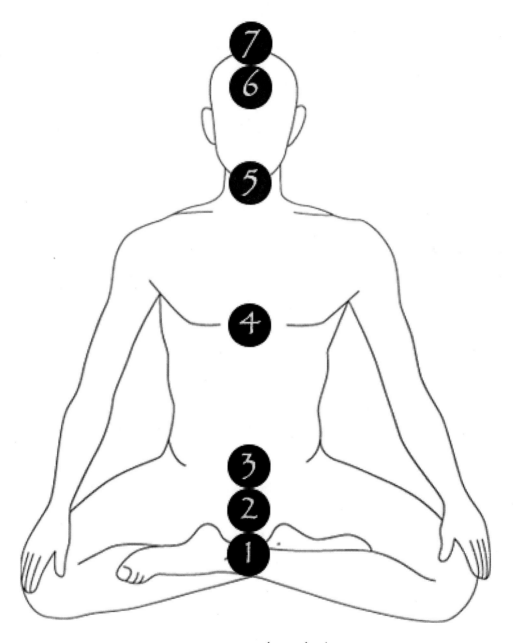

Figure 24: The Chakras

The Chakras

1. The first *chakra* is concerned primarily with matters of **survival** – both on the physical plane, and on every other level on which survival appears to be an issue. When the first *chakra* is emphasized, we secure our survival when it appears to be threatened, and establish a safe and secure perimeter to our existence in which it is possible to construct a life of stability and continuity

2. The second *chakra* is concerned primarily with the creation of a pleasurable and fulfilling existence. The second *chakra* is associated with sexuality, our natural predispositions, innate desires, and natural rhythms. When the second *chakra* is emphasized, we maximize **pleasure**, minimize pain, and organize our lives in ways that are enjoyable, emotionally rewarding, rich and abundant.

3. The third *chakra* is concerned with the cultivation of **ego,** personal competence, talent and/or expertise, learning, acquiring life skills, establishing our place within the world, and making a creative contribution. When the third *chakra* is emphasized, we pursue opportunities for advancement within the world and face challenges and obstacles to our advancement.

4. In the fourth *chakra*, we learn how to negotiate our **relationships,** seek a deeper connection to our **calling,** and cultivate awareness of being on a spiritual path. When the fourth *chakra* is emphasized, we begin to understand how we are connected to everyone and everything around us.

5. In the fifth *chakra*, we harvest and utilize whatever **wisdom** we have earned from life's experiences, learn to walk our talk and to share what we have learned. When the fifth *chakra* is emphasized, we recognize everyone we meet and everything that happens to us as an opportunity to learn and to grow in consciousness.

6. In the sixth *chakra*, we see the relative nature of our most cherished **belief systems,** develop flexibility in our understanding of reality and our capacity to respond to it, and shed our egos to identify with the *anima mundi*, or soul of the world. When the sixth *chakra* is emphasized, we see through the illusion that we are separate from the world, from each other, and from Spirit.

7. In the seventh *chakra*, we identify completely with **Spirit,** take the vow of the *bodhisattva* (of service to others) and work for the enlightenment of all sentient beings. When the seventh *chakra* is emphasized, we see the world through a set of perceptual filters in which this conundrum resolves itself beyond the reach of any words we might use to articulate it.

The Number Realms
as summarized by Sara Firman, Editor

In choosing a deity to follow in the Realm of One,

In choosing equality over inequality in the Realm of Two,

In choosing to mediate the opposites in the Realm of Three,

In choosing a more conscious relationship with the elements
in the Realm of Four,

In choosing to live in harmony with natural law in the Realm of Five,

In choosing to address core issues through the soul work required of us
in the Realm of Six,

In choosing to exercise complex creativity in the Realm of Seven,

In choosing to use hard-earned wisdom to create a life of harmony
in the Realm of Eight,

We are basically exercising our power to modify our Fate.

The Realm of Nine is the crystallization of Fate.

In the Realm of Ten Spirit and Matter are indistinguishable.

The Chakras and the Number Realms

In the Realm of Zero, our task is to make a shift
from the 1st chakra to the 4th and 6th

In the Realm of One, our task is to make a shift
from the 4th and 6th chakras to the 5th

In the Realm of Two, our task is to make a shift
from the 3rd chakra to the 2nd and 4th

In the Realm of Three, our task is
to integrate the 3rd and 6th chakras

In the Realm of Four, our task is
to integrate the 1st and 7th chakras

In the Realm of Five, our task is to shift
the Realm of Four integration (1st and 7th chakras) into the 4th

In the Realm of Six, our task is to cultivate
a more conscious relationship
to our personal core issues, centered in one or more chakras

In the Realm of Seven, our task is
to integrate the 2nd and 5th chakras

In the Realm of Eight, our task is
to integrate the 5th, 6th, and 7th chakras

In the Realm of Nine, our task is to shift
the Realm of Eight integration back down to the 1st chakra.

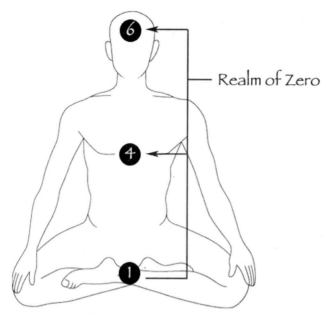

Figure 25: The Chakras and the Realm of Zero

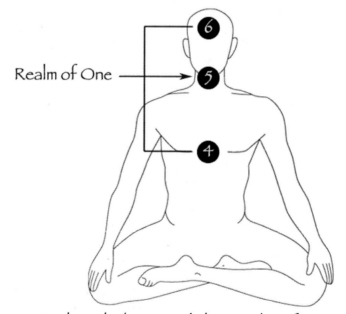

Figure 26: The Chakras and the Realm of One

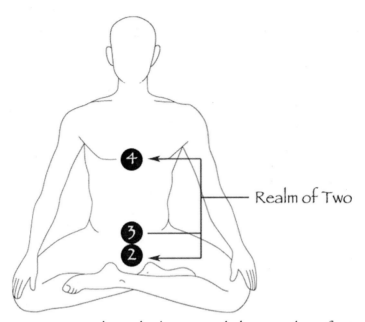

Figure 27: The Chakras and the Realm of Two

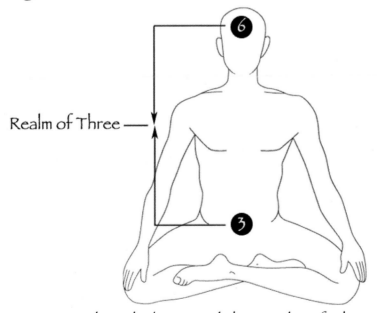

Figure 28: The Chakras and the Realm of Three

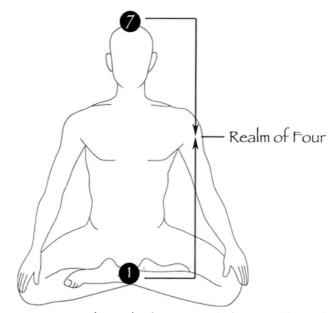

Figure 29: The Chakras and the Realm of Four

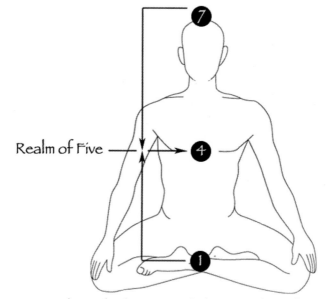

Figure 30: The Chakras and the Realm of Five

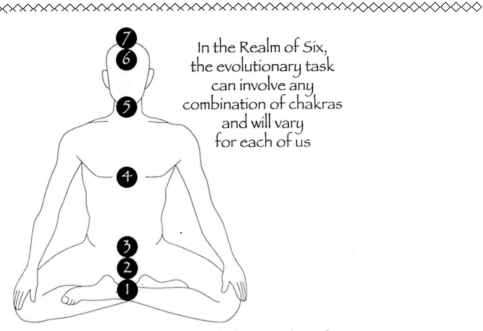

In the Realm of Six,
the evolutionary task
can involve any
combination of chakras
and will vary
for each of us

Figure 31: The Chakras and the Realm of Six

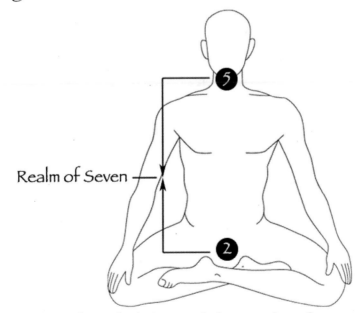

Realm of Seven

Figure 32: The Chakras and the Realm of Seven

283

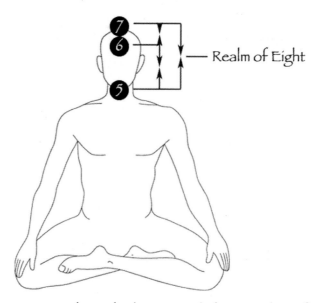

Figure 33: The Chakras and the Realm of Eight

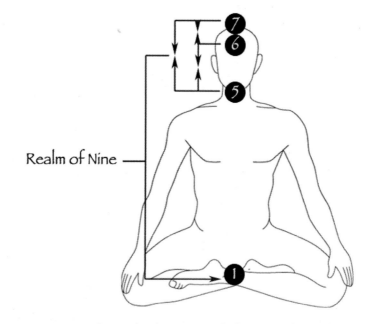

Figure 34: The Chakras and the Realm of Nine

Glossary

Because this book spans a number of discipines, the glossary is divided into four sections: 1) Metaphysical terms from various spiritual traditions, psychology and cutting edge science; 2) terms relevant to a study of arithmology; 3) astrological terms; and 4) astropoetic terms.

Metaphysical Terms from Various Spiritual Traditions, Psychology and Cutting Edge Science

anandamaya kosha – state of bliss in which a full embodiment of Spirit is being realized in each moment; the deepest possible penetration of Spirit into matter, according to yogic philosophy.

anima mundi – literally the "soul of the world;" an experience of the world as a soulful place through the projection of one's imagination into it.

annamaya kosha – the physical dimension of experience; the shallowest level of penetration of Spirit into matter according to yogic philosophy.

Anthropic Principle – a theory developed by astrophysicist Brandon Carter, in which the universe we know is the only one accessible to human consciousness.

archetype – a Jungian term referring to universal psychic principles at work within the evolution of consciousness.

baktun – a 5,120-year cycle used in the Mayan calendar.

bardo state – in Tibetan Buddhism, the liminal passage of the soul between death and rebirth.

biomimicry – a design principle in which nature is taken as the model.

bodhichitta – a Buddhist term for the innate vulnerability at the heart of the human condition.

chakra – one of seven subtle nerve plexuses in yogic philosophy, the activation of which results in a predisposition to see the world through a particular set of psychospiritual filters.

consciousness – the innate intelligence of the soul; a set of perceptual filters that condition our sense of who we are, what we are doing with our lives, and what our soul's journey is about.

GLOSSARY

◇◇

controlled folly – in the teachings of Carlos Castenada, the will to follow a path of heart, knowing it will ultimately make no difference to the state of the world.

Creator god – the chief deity in any tradition, responsible for the creation of the manifest world, including human beings, and the animation of Creation through the infusion of divine intelligence into it.

dakini language – a concentrated code through which a few seed syllables might contain within them volumes of information.

dual unity – a term used by consciousness researcher Stanislav Grof to describe a merging of subject and object without loss of self-awareness, such as is often experienced in prenatal regressions, tantric sexual practices, and the use of psychotropic substances.

emergentism – the recognition that complex phenomena such as consciousness emerge at higher levels of system organization and cannot be discerned or analyzed at the level of the system's essential building blocks.

Gaia Hypothesis – the idea, developed by James Lovelock and Lynn Margulis, that the Earth is a living organism, infused with consciousness and intelligence.

hieros gamos – a Jungian term used to describe the reconciliation of opposites; a state of functional balance and psychological integration.

holism – the understanding that wholes are more than the sum of their parts, and have their own identity that transcends that of the parts.

implicate order – a term used by theoretical physicist David Bohm to describe the underlying self-generating structure of reality that gives rise to any number of outer phenomena.

Intelligent Design – a theory normally ascribed to fundamentalist Christians, offered as an alternative to Darwinism, in which the manifest universe was literally created in seven days by a Creator god; in its broader implications, any theory that accepts intelligence as an innate quality of the manifest universe.

iti iti – a Hindu term to denote the realization that the immanent Creator God is omnipresent.

Kabala – a Jewish mystical tradition built around the esoteric function of numbers.

karma – an Eastern term referring to the consequences of actions, sometimes carrying over from one lifetime to the next.

kosha – a yogic term indicating one of five levels of penetration by Spirit into matter.

kundalini – the life force, conceptualized by yogic practitioners as a coiled sleeping serpent at the base of the spine, which becomes activated through spiritual practice.

kundalini yoga – a form of yoga, involving physical posture, meditation, chanting and other spiritual practices, specifically designed to awaken kundalini.

magic – the ability to align one's intentions with natural law and the divine intelligence at work beneath the surface of things.

magic body – an enlightened state of being in Tibetan Buddhism, in which it is possible to glimpse the omnipresence of Light within ordinary reality.

manomaya kosha – the emotional, psychological realm, understood as an interaction between Spirit (as consciousness) and material reality within the context of yogic philosophy.

maya – a Hindu term evoked to describe the illusory nature of the world of duality.

Medicine Wheel – a Native American psycho-spiritual navigational scheme, oriented around the four cardinal directions and their metaphorical implications.

mythopoetic worldview – a pre-scientific perceptual filter, informed by participation mystique, in which the interaction between gods and humans is a common everyday experience.

nadis – subtle nerve currents activated through yogic practice.

natural law – the observable ways in which divine intelligence is built into the natural world.

neti neti – a Hindu term, referring to the spiritual practice of negation, questing for a deeper understanding of Spirit through the process of elimination, i.e. Spirit is "not this, not that."

nirvana - enlightenment in the Buddhist tradition, after which rebirth is no longer necessary.

Odin – the Teutonic god of wind, a perpetual wanderer in search of hidden knowledge.

participation mystique – a term coined by anthropologist Lucien Levy Bruhl to describe the mystical identity between subject and object, that characterizes many indigenous cultures.

path of heart – a term used by Carlos Castenada to describe a life that revolves around one's dharma or calling; in the shamanic tradition, a primary source of empowerment.

permaculture – a system of ecological design developed by Bill Mollison, based on observation and application of natural principles.

power-from-within – a term used by Pagan eco-activist Starhawk to denote the power that comes from an internal alignment with Spirit.

power over – a term used by Pagan eco-activist Starhawk to denote the power of domination.

pranamaya kosha – according to yogic philosophy, the dimension of experience encompass-

GLOSSARY

◇◇◇

ing the movement of energy within the body, where awareness is directed to various physiological changes in one's state of being and their symbolic correlates.

samsara – a Buddhist term denoting the endless cycle of birth, death and rebirth, necessitated by the unresolved karma of previous existences.

science – an empirical approach to objective knowledge.

Second Attention – a term used in Toltec shamanic practice to describe a state of mind in which the spiritual significance of the events of ordinary life is revealed.

Selfish Biocosm Principle – a theory developed by James N. Garner and Seth Shostak, in which highly evolved intelligences with a superior knowledge of physics are able to create new "baby universes" hospitable to intelligent life.

shadow – a Jungian term to describe the repressed contents of consciousness, generally in relation to the rejected parts of the psyche, and often the unconscious cause of unresolved psychological issues.

sipapu – in Pueblo cultures, a great hole out of which the first ancestors emerged to begin their quest for the navel, or center of the world.

soul – the inhabitation of Spirit (Conscious Intelligence, both immanent and transcendent) within a mortal body.

Spirit – an omnipresent Divine Intelligence immanent within both the natural world and the human psyche.

sunyata – a Buddhist term roughly translated as "emptiness," but also simultaneously meaning "pregnant with unlimited possibilities."

tapas – a manifestation of kundalini, usually expereinced as a burning sensation in the spine, activated when a chakra is being cleansed.

terma – profound and extensive teachings encoded symbolically in a few pithy symbols.

terton – a translator of dakini language; one who studies and learns from terma.

vijnanamaya kosha – the symbolic realm, as understood in yogic philosophy as the attempt of Spirit to communicate with the mind.

World Tree – in shamanic tradition, the center of the manifest universe; often symbolized by a particular tree used for rites of shamanic initiation.

worldview – a mindset about the nature of reality that in turn colors one's experience of reality.

wu-wei – a Taoist word used to describe the art of doing by not-doing; going with the flow;

getting out of the way, so that Spirit can operate through one, unimpeded by the self-important ego.

xibalba – a dark bifurcation of the Milky Way by interstellar clouds, considered to be a shamanic opening by the Mayans; also known as the Black Road.

Yggdrasil – the World Tree on which Odin hung upside down for 9 days and nights, while channeling the runes.

yuga – one of four ages in Hindu cosmology, each encompassing a period of 4,320,000 years.

Terms Relevant to a Study of Arithmology

Abyss – the Void, Underworld, No-thingness; the Realm of Zero; a state of abject despair in which destruction, chaos, and psychic confusion prevail; Hell on Earth.

Ain Soph – the name for God in the Kabala, understood as the pre-existence Void out of which Creation arises, literally meaning "without ceasing to exist."

arithmetical means – the simplest of three primary types of means, calculated according to the formula: $(a + b)/2$.

arithmology – an application of Pythagorean ideas to number, arithmetic and geometry distinct from, but related to, the essence of mathematics as a cosmic art.

Ascent – a process of spiritual purification by which the Pythagoreans attempted to return to the One.

Awakening Phase – the contemplation of the Divine in the objects of sensory perception; the beginning of the Pythagorean Ascent.

composite numbers – numbers that can be factored into two or more primes.

cubes – numbers that can be portrayed as a three-dimensional array of evenly spaced dots, thought by Pythagoreans to bring resolution to the principles in the number being cubed, along with a sense of redemption or even a reversal of original principles.

Descent – a counter-flow to the Pythagorean Ascent by which Spirit enters more deeply into matter and becomes more conscious within it.

dodecahedron – a 12-sided polyhedron.

engenderment – an arithmological term to describe the ways in which numbers combine by multiplication to form other numbers, e.g. $2 \times 3 = 6$.

even numbers – numbers that can be divided into both equal and unequal parts.

generation by combination – an arithmological term to describe the ways in which numbers

GLOSSARY

◇◇

combine arithmetically to form other numbers, e.g. 3 + 4 = 7.

geometric means – an intermediary term between two numbers, formed from the square root of ab.

geometry – the projection of numbers into three-dimensional space.

gnomons – oblong numbers.

half-tone – in music, a tone produced along a string divided by a 9:10 ratio.

harmonic means – an intermediary term between two numbers, produced through the formula 2ab/(a + b).

harmony – the relationship between numbers, often expressed as ratios, capable of being perceived musically.

icosahedron – a three-dimensional polygon with 20 sides or facets.

Inebriation of Love – the emotional goal of the Pythagorean Ascent, in which there is no separation between the soul and the One.

Kathartic Excellences – virtues associated with control and suppression of the body, aspired to by the Pythagoreans as part of their rites of purification.

Lambda – an open, simplified Tetraktys with seven dots instead of ten; an early arithmological equation to be solved by calculation of the middle terms.

left-hand generation – the calculation of terms along the left arm of a Lambda, or parallel to it, using a common formula.

linear numbers – prime numbers, which can only be portrayed by dots in a one-line array.

major third – in music, a tone produced along a string divided by a 4:5 ratio.

means – various arithmological methods for calculating intermediary terms between two numbers, considered to represent the ways in which the Creator mediated various forces set in motion (represented by the numbers) throughout Creation

oblong numbers – numbers that can be portrayed as a rectangular array, in which one side of the rectangle is one dot longer than the other, also called gnomons.

octahedron – an 8-sided polyhedra.

octave – the relationship between a double-digit number and the sum of its digits; in music, a tone produced along a string divided by a 1:2 ratio.

odd numbers – numbers that can be divided only into unequal parts.

perfect fifth – in music, a tone produced along a string divided by a 2:3 ratio.

perfect fourth – in music, a tone produced along a string divided by a 3:4 ratio.

prime number – any number divisible only by itself and 1.

Purification Phase – after the Awakening Phase, a stage of the Pythagorean Ascent, in which the ascending soul sought to "calm the non-reasoning mind," the sensory self, filled with desire, moved by emotion, and driven by instinct.

right-hand generation – the calculation of terms along the right arm of a Lambda, or parallel to it, using a common formula.

septenary third – in music, a tone produced along a string divided by a 6:7 ratio.

septenary whole tone – in music, a tone produced along a string divided by a 7:8 ratio.

squares – numbers that can be portrayed as a square array of dots, thought by the Pythagoreans to result in a solidification and further entrenchment of the principle factor being squared.

tetrahedron – a 4-sided polyhedra, a pyramid.

Tetraktys – an arrangement of ten dots in a pyramid composed of four rows, understood by the Pythagoreans to illustrate the unity of the first ten numbers.

triangular numbers – numbers, like the Tetraktys, that can be portrayed as a triangular array of dots, thought by the Pythagoreans to initiate an experience of resolution.

whole tone – in music, a tone produced along a string divided by a 8:9 ratio.

Astrological Terms

eclipse – an alignment of Sun, Earth and Moon along the nodal axis in such a way that the light of Sun or Moon is blocked by shadow.

ecliptic – the path of the Earth's orbit around the Sun, containing the constellations of the astrological zodiac.

elements – the four primary "subtances" out of which all of manifest creation is composed, each with a rich metaphorical lexicon of pscyho-spiritual correlates: Earth, Air, Fire, and Water.

equinox – a time of year in spring and fall when the day is equal to the night.

lunar nodes – the intersection of the Moon's orbit around the Earth with the Earth's orbit around the Sun.

lunar standstill – the point of maximum and minimum declination in the Moon's orbit around the Earth, at which it appears to stand still in its position relative to the Earth's

GLOSSARY

orbit, before shifting back in the opposite direction.

Saros cycle – a series of eclipses 18.6 years apart in which the alignment of Sun, Earth and Moon with the nodal axis of the Moon, and the distance between the Earth and the Moon are similar.

solstice – a time of year in summer and winter when either day or night is at its maximum length.

syzygy – union of the opposites, depicted astrologically by a New or Full Moon.

Astropoetic Terms

astro-chakra system – a fusion of astrology and the chakra system of yogic philosophy, creating a model for the assessment of the consciousness expressed by an individual soul in relation to specific astrological dynamics within his/her birthchart.

astropoetics – a hybrid word meant to describe a poetic approach to astrology, applied to the human quest for meaning, purpose, and a deeper sense of connection to all of life and to the larger universe of which it is part.

chakra signature – the astrological pattern(s) in a birthchart that refer to the function of a particular chakra.

core wound – a complex of psychological issues, set in motion by some seminal experience in childhood, that subsequently serves as the pivot point for a lifetime of learning and spiritual growth.

counter-flow – a conscious adjustment in the Realm of Three, along a path of equality, in which a given polarity is balanced through resonance by contrast.

creative chakras – the 2nd and 5th chakras working in combination with each other.

creative synthesis – an opportunity for synergistic blending of opposites in the Realm of Three, along a path of equality.

cyclical history – an astropoetic technique involving remembering events and processes related to a specific planetary cycle, useful in determining real life correlations to a given astrological dynamic.

deficiency – a condition in which too little of any side of a given polarity, accumulated through resonance by affinity with its opposite, leads to imbalance and wounding.

excess – a condition in which too much of any side of a given polarity, accumulated through resonance by affinity, leads to imbalance and wounding.

mediation of opposites – an antidote in the Realm of Three, to conflict on the path of inequality, possible through conscious observation of a given polarity accompanied by intentional adjustment of excesses and deficiencies.

navigation – the art of self-direction, or orientation in soul space, using the four directions of the Medicine Wheel, astrology, dreams, the runes, or some other combination of metaphorical tools as an intuitive guide.

number Realm – an archetypal dimension of number conditioning our experience of the anima mundi, or soul of the world.

path of equality – an intentional choice to view the opposites as equal in value and both necessary to the wellbeing of the whole, an attitude that contributes to caring, cooperation, peace, tolerance of differences, and harmony.

path of inequality – a choice, often unconscious, to view the opposites as unequal in value, with one superior to the other, one good and one bad, or one preferable and the other rejected; an attitude that leads to conflict, war, genocide, destruction, and damage to the environment.

power chakras – the 3rd and 6th chakras working in combination with each other.

resonance – the capacity of each soul to attract to itself that which is most conducive to its growth

resonance by affinity – the power of the soul to attract that which exists in harmony with its nature.

resonance by contrast – the power of the soul to attract that which exists in contrast to its nature, i.e. opposites attracting.

resonance by wounding – the power of the soul to attract people, experiences and circumstances that force a reckoning with core issues.

secondary elements – an astropoetic term used to describe an alchemical mix of primary elements, e.g. earth and water create mud, paste or slurry; fire and water create steam.

soul space – the environment, internal and external, in which a soul seeks to learn, grow, and more consciously embody Spirit.

soul work – a conscious and intentional pursuit of healing in relation to core wounds by a wide variety of means.

Figure 35: The Rune Raido

Communication, a journey, the reunion of the soul & the divine

Works Cited

Abram, David. <u>The Spell of the Sensuous</u>. New York: Vintage Books, 1996.

Advancement Project Los Angeles. "Peace on the Streets." 17 July 2007. http://www.advanceproj.org/016.html.

Anti-War. "Casualties in Iraq." 25 August 2010. http://antiwar.com/casualties.

Barclay, Olivia. <u>Horary Astrology Rediscovered</u>. Atglen, PA: Whitford Press, 1997.

Barks, Coleman. <u>The Essential Rumi</u>. New York: Quality Paperback Book Club, 1995.

Barringer, Mark. "The Antiwar Movement in the United States." Department of English of University of Illinois – Urbana-Champaign. 27 September 2010. http://www.english.illinois.edu/maps/vietnam/antiwar.html.

Bauer, Shane. "The Ecology of Genocide." <u>E: The Environmental Magazine</u>. Volume XVIII, Number 2: March/April 2007, pp 14-16.

Benyus, Janine. "What is Biomimicry?" Biomimicry Institute. 21 September 2010. http://www.biomimicryinstitute.org/about-us/what-is-biomimicry.html.

Blum, Ralph. <u>The Book of Runes: A Handbook for the Use of an Ancient Oracle: The Viking Runes</u>. New York: St. Martin's Press, 1982.

Bohm, David. <u>On Dialogue</u>. London & New York: Routledge, 1996.

___. <u>Wholeness and the Implicate Order</u>. London & New York: Routledge & Kegan Paul, 1980.

Bright Future: Solutions for a Better Tomorrow. "Canadian Environmentalists and Logging Companies Create Historic Partnership." 17 July 2007. http://www.brightfuture.us/content/view/103/30.

Butler, Judith. "Uncritical Exuberance." Angry White Kid. 27 September 2010. http://angrywhitekid.blogs.com/weblog/2008/11/uncritical-exuberance-judith-butlers-take-on-obama.html.

Cameron, Alister. <u>The Pythagorean Background of the Theory of Recollection</u>. Menasha, WI: George Banta, 1938.

Campbell, Joseph. <u>The Hero With a Thousand Faces</u>. 2nd Ed. Boligen Series XVII. New York: Princeton University Press, 1968.

___. <u>The Masks of God: Creative Mythology</u>. New York: Penguin, 1968.

___. <u>The Masks of God: Occidental Mythology</u>. New York: Viking, 1964.

◇◇

___. The Masks of God: Oriental Mythology. New York: Penguin, 1962.

___. The Masks of God: Primitive Mythology. New York: Penguin, 1969.

___. The Mythic Image. Princeton, NJ: Princeton University Press, 1974.

Cancer Research UK News and Resources. "CancerStats." 12 October 2007. http//:info.cancerresearchuk.org/cancerstats.

Cardinale, Matthew. Fears Grow Over Oil Spill's Long-Term Effects on Food Chain." Guardian. 28 September 2010. http://www.guardian.co.uk/environment/2010/jun/01/bp-oil-spill-wildlife.

Carey, Ken. Return of the Bird Tribes. Kansas City, MO: Uni-Sun, 1988.

Castenada, Carlos. A Separate Reality: Further Conversations With Don Juan. New York, NY: Simon & Schuster, 1971.

___. The Fire From Within. New York: Simon and Schuster, 1984.

Chodron, Pema. Start Where You Are: A Guide to Compassionate Living. Boston, MA: Shambala, 1994.

Cleland, Corey L. "Pain," Biology Reference Encyclopedia. 25 August 2010. http://www.biologyreference.com/Oc-Ph/Pain.html.

Cockburn, Patrick. "Opium: Iraq's deadly new export," The Independent World. 25 August 2010. http://www.independent.co.uk/news/world/middle-east/opium-iraqs-deadly-new-export-449962.html.

Conze, E., Tr. "The Heart Sutra." Buddhism.org. 4 October 2010. http://kr.buddhism.org/zen/sutras/conze.htm.

Cornelius, Geoffrey. The Moment of Astrology: Origins in Divination. London, England: Wessex Astrologer, 2002.

Cowen, Tyler. Department of Economics, George Mason University. "Civilization Renewed: A Pluralistic Approach to a Free Society." 9 August 2007. http://64.233.169.104/search?q=cache:4xNh84MywIwJ:www.gmu.edu/jbc/Tyler/civ.doc+duration+of+civilizations&hl=en&ct=clnk&cd=106&gl=us.

Csikszentmihalyi, Mihaly. Creativity: Flow and the Psychology of Discovery and Invention. New York: HarperCollins, 1996.

Cumont, Franz. "Astrology and Religion among the Greeks and Romans." American Lectures on the History of Religions. New York and London: G. P. Putnam's Sons, 1912. 16 August 2010. http://www.valentino-salvato.com/Astrology/pdf/Astrology_and_Religion_Among_the_Greeks_and_Romans-Franz_Cumont.pdf.

Dignubia. "Bookshelf: Nubia/Egyptian Gods and Goddesses." 21 August 2008. http://www.dignubia.org/bookshelf/goddesses.php?god_id=00013.

Earthwatch Institute. Home Page. 19 December 2007. http://www.earthwatch.org/site/pp.asp?c=dsJSK6PFJnH&b=386443.

Eck, Michael. The Book of Threes: A Subject Reference Tricyclopedia. 11 July 2007. http://threes.com/cms/index.php.

Eliade, Mircea. A History of Religious Ideas: Vol 2 – From Gautama Buddha to the Triumph of Christianity. Tr. Willard R. Trask. Chicago: University of Chicago Press, 1982.

Fontaine, Ray. "The Triumph of Nature's God in My Life." Deism. 17 September 2010. http://www.deism.com/to-natures-god.net.

Frankl, Viktor E. Man's Search For Meaning. New York: Pocket Books, 1963.

Frawley, John. The Horary Textbook. London, England: Apprentice Books, 2003.

Freud, Sigmund. "Obsessive Acts and Religious Practices." The Collected Papers of Sigmund Freud. Vol. 9. Tr. J Riviere. Ed. J. Strachey. London: Hogarth Press, 1907.

Garner, James N. and Seth Shostak. Biocosm: The New Scientific Theory of Evolution: Intelligent Life Is the Architect of the Universe. Makawao, HI: Inner Ocean Publishing, 2003.

Global Ministries. "Globalization Timeline." 24 December 2008. http://new.gbgm-umc.org/media/missionstudies/pdf/globalizationtimeline.pdf.

Global Policy Forum. "Globalization." 24 December 2008. http://www.globalpolicy.org/globaliz/index.htm.

Goetz, Thomas. "The Bug Detector." Wired. August 2007, pp 92-97.

Goldenberg, Suzanne. "BP Oil Spill: Obama Administration's Scientists Admit Alarm Over Chemicals." Guardian. 28 September 2010. http://www.guardian.co.uk/environment/2010/aug/03/gulf-oil-spill-chemicals-epa.

Grof, Stanislav. The Adventure of Self-Discovery: Dimensions of Consciousness and New Perspectives in Psychotherapy and Inner Exploration. Albany, NY: State University of New York Press, 1988.

Grun, Bernard. The Timetables of History. 3rd Rev. Ed. New York: Simon & Schuster, 1991.

Gundarsson, Kveldulf. Teutonic Magic: The Magical & Spiritual Practices of the Germanic Peoples. St. Paul, MN: Llewellyn Publications, 1990.

Guthrie, Kenneth Sylvan. The Pythagorean Sourcebook and Library. Grand Rapids, MI:

◇◇◇

Phanes Press, 1988.

Halifax, Joan. The Fruitful Darkness: Reconnecting With the Body of the Earth. New York: Harper Collins, 1993.

Hall, Manly P. The Secret Teachings of All Ages: An Encyclopedic Outline of Masonic, Hermetic, Qabbalistic and Rosicrucian Symbolical Philosophy. Los Angeles, CA: Philosophical Research Society, 1988.

Hamourtziadou, Lily. "Week in Iraq." Iraq Body Count. 1 March 2007. http://www.iraqbodycount.org/editorial/weekiniraq/.

Harris, William. "The Golden Mean." Middlebury College Community. 28 April 2008. http://community.middlebury.edu/~harris/Humanities/TheGoldenMean.html.

Harrison, George. "All Things Must Pass." All Things Must Pass. Apple Records, 1970.

Heath, Richard. Sacred Number and the Origins of Civilization: The Unfolding of History Through the Mystery of Number. Rochester, Vermont: Inner Traditions, 2007.

Heathcote Community. "Permaculture Intro – Lesson Four: What Are Some Principles of Permaculture." 10 August 2007. http://www.heathcote.org/PCIntro/4Principles.htm.

Heinrich, Bernd. Mind of the Raven: Investigations and Adventures With Wolf-Birds. New York, NY: Harper Perennial, 1999.

Hender, Robert A. "Economic Collapse – Martial Law – 24 Experts Warn of 2010 Meltdown." Morning Liberty. 17 September 2010. http://www.morningliberty.com/2010/05/16/economic-collapse-martial-law-24-experts-warn-of-2010-meltdown.

Henderson, Joseph L. Thresholds of Initiation. Middletown, CT: Weslyan University Press, 1967.

Hillman, James. The Soul's Code: In Search of Character and Calling. New York: Warner Books, 1996.

History Place, The. "The Vietnam War." 10 January 2008 & 27 September 2010. http://www.historyplace.com/unitedstates/vietnam.

Hobbes, Thomas. Leviathan. Ed. A. P. Martinich. Peterborough, Ontario: Broadview Press, 2002.

Hobson, J. Allan. Consciousness. New York, NY: Scientific American Library, 1999.

Huffman, Carl. "Pythagoreanism." Stanford Encyclopedia of Philosophy. 13 March 2007. http://www.plato.stanford.edu/entries/pythagoreanism.

Iraq Body Count. 1 March 2007, 25 August 2010. http://www.iraqbodycount.org/.

◇◇

Jaffe, Greg. "Gates Sees Momentum in Afghanistan But Plays Down Prospects for Reconciliation." The Washington Post. 27 September 2010. http://www.washingtonpost.com/wp-dyn/content/article/2010/03/08/AR2010030801693.html.

Jenkins, John Major. "The How and the Why of the Mayan End Date in 2012." The Mountain Astrologer, Issue #52, Dec-Jan 1994-95, pp 52-57+.

Jerusalem Bible, Reader's Edition. Garden City, NY: Doubleday, 1968.

Jung, Carl. Archetypes of the Collective Unconscious. Tr. R.F.C. Hull. Bolligen Series XX, Vol. 9, Part 1. Princeton, NJ: Princeton University Press, 1969.

___. Memories, Dreams and Reflections. Tr. Richard and Clara Winston, Ed. Aniela Jaffé, New York: Vintage Books, 1965.

___. Psychological Types. Tr. R.F.C. Hull. Bolligen Series XX, Vol. 6. Princeton, NJ: Princeton University Press, 1971.

___. Symbols of Transformation. Tr. R.F.C. Hull. Bolligen Series XX, Vol. 5. Princeton, NJ: Princeton University Press, 1956.

___. Synchronicity. Tr. R.F.C Hull. Bolligen Series XX, Vol. 8. Princeton, NJ: Princeton University Press, 1960.

Kaplan, Aryeh. Sefer Yetzirah: The Book of Creation in Theory and Practice. York Beach, ME: Samuel Weiser, 1997.

Kaufmann, William J. and Roger A. Freedman. Universe. 5th Ed. New York: W.H. Freeman and Co., 1999.

Killoran, Elyse Hope. "The Secret Controversy Heats Up – Vol. 1." Prosperity From the Inside Out EZine. Choosing Prosperity. 17 July 2007. http://www.choosingprosperity.com/archive/pftio030107.htm.

Kim, Alan. "Wilhelm Maximilian Wundt". The Stanford Encyclopedia of Philosophy (Summer 2006 Ed.). Ed. Edward N. Zalta. 27 February 2007. http://plato.stanford.edu/archives/sum2006/entries/wilhelm-wundt.

Knott, Ron. "Fibonacci Numbers and Golden sections in Nature." 19 January 2008. http://www.mcs.surrey.ac.uk/Personal/R.Knott/Fibonacci/fib.html.

Kochunas, Brad. "Peaks and Vales." Sermon delivered to Miami Valley Unitarian Universalist Fellowship. Miamisburg, Ohio, March 11, 2007.

Lao Tsu. Tao Te Ching. Tr. Gia-Fu Feng and Jane English. New York: Vintage Books, 1972.

Landwehr, Joe. Birth of the Shining One: Moving Beyond Apocalypse Into Godbeing. Kansas City, MO: Uni-Sun, 1988.

◇◇

___. "Hiding From the Wind." Full Moon Meditations: Musings on the Journey Toward Godbeing. Mountain View, MO: Light of the Forest Primeval, 1993.

___. "The Lightning of Opheukos." Full Moon Meditations: Musings on the Journey Toward Godbeing. Mountain View, MO: Light of the Forest Primeval, 1993.

___. "Nicaraguan Dialogue Evaluates Peace Process." Dialogue. Vol. 1, Issue 3: Spring, 1989, pp 1-2.

___. "Reconnecting With the Cosmic Bearings of Life Through the Right Use of Sound." The Whole Network Journal, Issue 2: Summer, 1988.

___. Security 2000: A Dialogue for America's Future – Briefing Booklet. Santa Fe, NM: The Trinity Forum, 1989.

___. The Seven Gates of Soul: Reclaiming the Poetry of Everyday Life. Abilene, TX: Ancient Tower Press, 2004.

___. Tracking the Soul With An Astrology of Consciousness. Mountain View, MO: Ancient Tower Press, 2007.

Learner, Michele. Bread for the World. "Budgeting for Justice." 14 August 2007. http://www.bread.org/learn/background-papers/2006/budgeting-for-justice.html.

Lehmann, Arthur C. and James E. Myers. Magic, Witchcraft, and Religion: An Anthropological Study of the Supernatural. 3rd Ed. Mountain View, CA: Mayfield, 1993.

Leonard, Scott and Michael McClure. Myth and Knowing: An Introduction to World Mythology. New York: McGraw Hill, 2004.

Lilly, William. Christian Astrology. Abingdon, MD: Astrology Center of America, 2004.

Lüthi, Lorenz M. "Mao Zedong." Enotes. 17 September 2010. http://www.enotes.com/genocide-encyclopedia/mao-zedong.

Malville, J. McKim and Claudia Putnam. Prehistoric Astronomy in the Southwest. Rev. Ed. Boulder, CO: Johnson Books, 1993.

Marx, Karl. Marxists Internet Archive. "The English Middle Classes." 7 May 2008. http://marxists.org/archive/marx/works/1854/08/01.htm.

Matt, Daniel C. The Essential Kabbalah. New York: Quality Paperback Book Club, 1995.

McDonald, G. Jeffrey. USA Today. "Does Maya calendar predict 2012 apocalypse?" 10 October 2008. http://www.usatoday.com/tech/science/2007-03-27-maya-2012_n.htm.

Metzner, Ralph. The Well of Remembrance: Rediscovering the Earth Wisdom Myths of Northern Europe. Boston: Shamabala Pubns, 1994.

Mideast Web. "The Iraq Crisis – Timeline: Modern Chronology of Iraqi History." 10 January 2008. http://www.mideastweb.org/iraqtimeline.htm.

Mollison, Bill and David Holmgren. Permaculture One: A Perennial Agriculture for Human Settlements. Maryborough, Australia: Dominion Press-Hedges & Bell, 1984.

Moore, Thomas. Care of the Soul: A Guide for Cultivating Depth and Sacredness in Everyday Life. New York: Harper Perennial, 1992.

Morford, Mark P.O. and Robert J. Lenardon. Classical Mythology. 5th Ed. White Plains, NY: Longman, 1995.

Muktananda, Swami. Play of Consciousness. South Fallsburg, NY: SYDA Foundation, 1978.

National Center for Health Statistics, Center For Disease Control. "Health Insurance Coverage." 14 August 2007. http://www.cdc.gov/nchs/fastats/hinsure.htm.

National Poverty Center, University of Michigan, Gerald R, Ford School of Public Policy, "Poverty in the United States: Frequently Asked Questions." 14 August 2007. http://www.npc.umich.edu/poverty/#2.

National Priorities Project. "Turning Data Into Action: The Calculator." 17 July 2007. http://costofwar.com/numbers.html.

___. "Cost of War." 13 September 2010. http://www.nationalpriorities.org/costofwar_home.

Nelson, James B. "Homosexuality and the Church." Religion Online. 17 September 2010. http://www.religion-online.org/showarticle.asp?title=430.

Neumann, Erich. The Origins and History of Consciousness. Bolligen Series XLII. Princeton, NJ: Princeton University Press, 1954.

NewsOK.com. "Heavy rains cause flooding, evacuations, two deaths and possibly three more." 9 October 2007. http://www.newsok.com/article/3106692.

Norbu, Namkhai. The Crystal and the Way of Light: Sutra, Tantra, and Dzogchen: The Teachings of Namkhai Norbu, Ithaca, NY: Snow Lion Pubns, 2000.

Oil-Price.Net. "Crude Oil and Commodity Prices." 13 September 2010. http://oil-price.net.

Online Etymology Dictionary. 16 October 2007. http://www.etymonline.com/index.php?l=h&p=2.

On This Day. "1989: Berliners celebrate the fall of the Wall." 19 May 2008. http://news.bbc.co.uk/onthisday/hi/dates/stories/november/9/newsid_2515000/2515869.stm.

Opsopaus, John. "A Summary of Pythagorean Theology." Bibliotecha Arcana. 13 March

2007. https://www.cs.utk.edu/~mclennan/BA/EMP/I-V.html.

Otto, Rudolf. The Idea of the Holy: An Inquiry into the Non-rational Factor in the Idea of the Divine and Its Relation to the Rational. London: Oxford University Press, 1923.

Oxtoby, Willard G., Ed. World Religions: Eastern Traditions. New York: Oxford University Press, 2002.

___. World Religions: Western Traditions. New York: Oxford University Press, 2002.

Paul, James A. "The Iraq Oil Bonanza: Estimating Future Profits." Global Policy Forum. 17 July 2007. http://www.globalpolicy.org/security/oil/2004/0128oilprofit.htm.

___. "Oil in Iraq: The Heart of the Crisis." Global Policy Forum. 17 July 2007. http://www.globalpolicy.org/security/oil/2002/12heart.htm.

Pelikan, Jaroslav, Ed. Sacred Writings, Vol. 1: Judaism – The Tanakh. New York: Quality Paperback Book Club, 1992.

Powell, Barry B. Classical Myth. Englewood Cliffs, NJ: Prentice Hall, 1995.

Radin, Paul. The Trickster: A Study in American Indian Mythology. New York, NY: Shocken Books, 1956.

Ram Dass and Paul Gorman. How Can I Help? Stories and Reflections on Service. New York: Knopf, 1985.

Ray, Paul H. and Sherry Ruth Anderson. The Cultural Creatives. New York: Harmony Books, 2000.

___. "FAQ." Cultural Creatives. 15 October 2007. http://www.culturalcreatives.org/faq.html.

Rifkin, Jeremy. The Empathic Civilization: The Race to Consciousness in a World in Crisis. New York, NY: Tarcher/Penguin, 2009.

Rodden, Lois. Astrodatabank. "Jimmy Carter." 30 December 2008. http://www.astrodatabank.com/NM/CarterJimmy.htm.

Rukeyser, Muriel. "The Speed of Darkness," The Speed of Darkness. New York, NY: Vintage Press, 1971.

Ryan, Robert. Rense.com. "Cancer Research – A Super Fraud?" 12 October 2007. http://www.rense.com/general9/cre.htm.

Sams, Jamie. Sacred Path Cards: The Discovery of Self Through Native Teachings. San Francisco, CA: Harper, 1990.

___. & David Carson. Medicine Cards: The Discovery of Power Through the Ways of

Animals. Santa Fe, NM: Bear & Company, 1988.

Scaruffi, Piero. "The Worst Natural Disasters Ever; 1900: A Century of Genocides; Disasters Related to the Energy Industry. Politics. 1 March 2007. http://www.scaruffi.com/politics/disaster.html.

Schulman, Daniel. "Don't Whistle While You Work." Mother Jones. Vol. 32, No. 3: May/June 2007, pp 52-57+.

Sharer, Robert J. The Ancient Maya. 5th Ed. Stanford, CA: Stanford University Press, 1994.

Simon, Dennis M. "The War in Vietnam: 1965-1968." Department of Political Science, Southern Methodist University. 27 September 2010. http://faculty.smu.edu/dsimon/Change-Viet2.html.

Smith, John. "America's Foreclosure Statistics are Muddled Still." Real Estate Pro Articles. 27 September 2010. http://www.realestateproarticles.com/Art/16607/265/America-s-foreclosure-statistics-are-muddled-still.html.

Snell, Marilyn Berlin. "Climate Change: Cool Heads Tackle Our Hottest Issue." Sierra. Vol. 92, No. 3, May/June, 2007, pp 44-53+.

Stanford Encyclopedia of Philosophy. "Relational Quantum Mechanics." 17 July 2007. http://plato.stanford.edu/entries/qm-relational.

Starhawk. Dreaming the Dark: Magic, Sex and Politics. New Ed. Boston: Beacon Press, 1988.

Sturgis, Sue. "Gulf Coast Reconstruction Watch." Institute for Southern Studies. 1 March 2007. http://www.southernstudies.org/gulfwatch/.

Substance Abuse & Mental Health Services Administration, US Department of Health and Human Services. "Homelessness Statistics and Data." 14 August 2007. http://www.samhsa.gov/Matrix/statistics_homeless.aspx.

Sutton, Mark Q. An Introduction to Native North America. 2nd Ed. Boston, MA: Pearson Education, 2004.

Talbot, Michael. The Holographic Universe. New York, NY: HarperCollins, 1991.

Tarnas, Richard. Cosmos and Psyche: Intimations of a New World View. New York: Viking, 2006.

Tedlock, Dennis. The Popol Vuh: The Definitive Edition of the Mayan Book of the Dawn of Life and the Glories of Gods and Kings. New York: Simon & Schuster, 1985.

Tester, Jim. A History of Western Astrology. New York: Ballentine Books, 1987.

Thinkquest. "The Fibonacci Series." 19 January 2008. http://library.thinkquest.org/27890/

◇◇

applications5.html.

Thurman, Robert A. F., Tr. <u>The Tibetan Book of the Dead</u>. New York: Quality Paperback Book Club, 1994.

Tutu, Desmond. "Let South Africa Show the World How to Forgive." <u>Knowledge of Reality Magazine</u>, Issue 19. 12 March 2007. http://www.sol.com.au/kor/19_03.htm.

UNAIDS. "Global Facts and Figures 06." 12 October 2007. http://data.unaids.org/pub/EpiReport/2006/20061121_EPI_FS_GlobalFacts_en.pdf.

USA History. "American Chronology: Timeline for the Period of the Revolution." 19 May 2008. http://www.usahistory.info/timeline/revolution.html.

Vargo, Andrew. "Response to Islamic Awareness: The Queen of Sheba and Sun Worship." 21 August 2008. http://www.answering-islam.org/Responses/Saifullah/sheba_moon.htm.

Von Franz, Marie-Louise. <u>Creation Myths</u>. Dallas, TX: Spring Publications, 1972.

Waterfield, Robin, Tr. <u>The Theology of Arithmetic</u>. Grand Rapids, MI: Phanes Press, 1988.

Whitty, Julia. "Gone." <u>Mother Jones</u>. May – June, 2007, pp. 36-45+.

Wikipedia Online Encyclopedia. "Anthropic principle." 3 August 2007. http://en.wikipedia.org/wiki/Anthropic_principle.

___. "Biocosm Hypothesis." 2 August 2007. http://en.wikipedia.org/wiki/Biocosm.

___. Cro-Magnon. 11 January 2008. http://en.wikipedia.org/wiki/Cro-Magnon.

___. "Gaia Hypothesis." 2 August 2007. http://en.wikipedia.org/wiki/Gaia_hypothesis#_note-3.

___. "List of Natural Disasters." 25 August 2010. http://en.wikipedia.org/wiki/List_of_natural_disasters#Ten_deadliest_natural_disasters_of_the_past_century.

___. "Intelligent Design." 3 August 2007. http://en.wikipedia.org/wiki/Intelligent_design.

___. "Irrational Numbers." 30 September 2010. http://en.wikipedia.org/wiki/Irrational_numbers.

___. "Iriquois." 9 August 2007. http://en.wikipedia.org/wiki/Iroquois.

___. "Joe Darby." 12 September 2010. http://en.wikipedia.org/wiki/Joe_Darby.

___. "Kent State Shootings." 27 September 2010. http://en.wikipedia.org/wiki/Kent_State_shootings.

___. "List of Wars and Disasters by Death Toll." 1 March 2007. http://en.wikipedia.

◇◇

org/wiki/Death_toll.

___. "A Study of History." 9 August 2007. http://en.wikipedia.org/wiki/A_Study_of_ History.

___. "Thomas Hobbes: Leviathan." 17 September 2010. http://en.wikipedia.org/wiki/ Thomas_Hobbes#Leviathan.

___. "World War I: Economics and Manpower Issues". 24 December 2008. http:// en.wikipedia.org/wiki/World_War_I#Economics_and_manpower_issues.

Williamson, Ray A. Living the Sky: The Cosmos of the American Indian. Norman, OK: University of OK Press, 1984.

Wilson Center. "Guenter Schabowski's Press Conference in the GDR International Press Center." 8 June 2008. http://www.wilsoncenter.org/cwihp/documentreaders/ eotcw/891109f.pdf.

Wintle, Prier. "Five Jugglers." Considerations. Vol XVIII: 4 – November, 2003 - February, 2004.

Wolff, Glenn. The Bird in the Waterfall: A Natural History of Oceans, Rivers, and Lakes. New York: Harper Collins, 1996.

World Health Organization. "HIV and AIDS estimate and data, 2007 and 2001." 23 September 2010. http://data.unaids.org/pub/GlobalReport/2008/jc1510_2008_global_ report_pp211_234_en.pdf.

Wray, Richard. "Abandoned Oil Wells Make Gulf of Mexico 'Environmental Minefield'." Guardian. 28 September 2010. http://www.guardian.co.uk/business/2010/jul/07/ abandoned-oil-wells-gulf-mexico.

Zussy, Nancy. "Chief Seattle Speech: Washington State Library." The Nomadic Spirit: Critical Thinkers Resources. 17 November 2010. http://www.synaptic.bc.ca/ejournal/seattle.htm.

Figure 36: The Rune Mannaz

The Self, the Divine human, the anima mundi.

Index

A

INDEX

◇◇

INDEX

◇◇

◇◇

◇◇

◇◇◇

N

◇◇

◇◇

Q

R

◇◇

S

INDEX